# A STUBBORNLY
# PERSISTENT
# ILLUSION

# A STUBBORNLY
# PERSISTENT
# ILLUSION

The Essential Scientific Writings of Albert Einstein

RUNNING PRESS
PHILADELPHIA • LONDON

9  8  7  6  5  4  3  2  1
Digit on the right indicates the number of this printing

Library of Congress   Control Number: 2007935658
ISBN-13   978-0-7624-3003-1
ISBN-10   0-7624-3003-6

Cover design by Bill Jones
Interior design by Bill Jones and Aptara, Inc.
Edited by Jennifer Kasius
with David Goldberg
Typography: Adobe Garamond

This book may be ordered by mail from the publisher.
Please include $2.50 for postage and handling.
**But try your bookstore first!**

Running Press Book Publishers
2300 Chestnut Street
Philadelphia, PA 19103-4317
Visit us on the web!
www.runningpress.com

# TEXT CREDITS

Selections from THE EVOLUTION OF PHYSICS: Reprinted with the permission of Simon & Schuster Adult Publishing Group, from *The Evolution of Physics* by Albert Einstein and Leopold Infeld. Copyright © 1938 by Albert Einstein and Leopold Infeld. Copyright © renewed 1966 by Albert Einstein and Leopold Infeld. All Rights Reserved.

Selection from THE MEANING OF RELATIVITY: Courtesy of Princeton University Press.

Autobiographical notes reprinted by permission of Open Court Publishing Company, a division of Carus Publishing Company, Peru, IL, from *A. Einstein: Autobiographical Notes* translated and edited by Paul Arthur Schilpp, first published in *Albert Einstein: Philosopher-Scientist* in The Library of Living Philosophers Series Volume VII, (c) 1949, 1951, 1970, 1979 by The Library of Living Philosophers, Inc., and the Estate of Albert Einstein.

Selections from OUT OF MY LATER YEARS: Reprinted with permission from The Philosophical Library, New York.

# CONTENTS

INTRODUCTION BY STEPHEN HAWKING    ix

*THE PRINCIPLE OF RELATIVITY*    1

*RELATIVITY, THE SPECIAL AND GENERAL THEORY*    125

*SIDELIGHTS ON RELATIVITY*    235

*SELECTION FROM THE MEANING OF RELATIVITY,* "SPACE AND TIME IN PRE-RELATIVITY PHYSICS"    263

*SELECTIONS FROM THE EVOLUTION OF PHYSICS,* "RELATIVITY, FIELD" AND "QUANTA"    283

*AUTOBIOGRAPHICAL NOTES*    337

*SELECTIONS FROM OUT OF MY LATER YEARS*    383

INDEX    457

# INTRODUCTION

## BY STEPHEN HAWKING

A few years ago the world celebrated the 100th anniversary of Einstein's miracle year, the year in which he revolutionized physics in multiple ways with a series of astonishing new ideas that brought about profound changes in the way physicists view the universe. Human intuition tells us that space is a stage upon which the events of our lives play out, that time is governed by a universal clock. But in 1905 and the decade that followed, Einstein showed that space and time do not have identical meanings for observers sitting in a chair and those flying on a plane, those orbiting with us on earth, those having tea somewhere in the Virgo cluster, or those being sucked into a black hole.

Einstein's ideas once stunned physicists. Today they are automatically incorporated into the equations and formalism learned by every undergraduate physics major. As long as those ideas stand up, Einstein wrote in one of the articles in this collection, the Germans will call him a "German savant," and the English will call him a "Swiss Jew." But if his ideas were ever discredited, he wrote, he would be a "Swiss Jew" for the Germans and a "German savant" for the English. Today there are few physicists left who remember Einstein as a living, breathing, and witty human being. Today his ideas of space and time intertwined are ingrained in popular culture, and described by writers several generations down. But the most lucid, not to mention entertaining, proponent of Einstein's ideas has always been Einstein himself.

As he describes in this volume, Einstein's 1905 special theory of relativity grew out of a simple observation. The theory of electromagnetism discovered by James Clerk Maxwell in the 1860s showed that whether you are moving toward or away from a beam of light, the light will always approach you at the same speed. This is not true of our experience in the everyday world. If you race away from an onrushing train you will survive for a few more seconds than if you race toward it (assuming that you don't get the idea to jump sideways).

In the former case the train's speed of approach will be the difference between its speed and your speed (relative to the track). In the latter case its speed of approach will be the sum of your speeds. The same, according to Maxwell's theory, does not apply to the light emitted from the train's headlamps. How could the speed of light not appear slower in the former case and faster in the latter?

By speed we mean distance traveled divided by the time of the trip. And so, Einstein realized, if we are to take Maxwell's theory at face value, we must alter our ideas of space and time. They are not fixed and unchanging, but adjust according to the observer, bending or stretching in just the way necessary to keep the speed of light constant. The same bending and stretching means of course that the speed at which the train itself approaches is also not the simple sum or difference I described above. But at speeds far less than the speed of light the difference in adding and subtracting derived by Einstein has only negligible effect. The same chain of logic, when taken further, requires also the equivalence of mass and energy, the reason that we can have atomic energy, and, unfortunately, atomic weapons. The details of Einstein's reasoning, and the simple algebra behind it, are explained nowhere better than as found here, in Einstein's own words.

Einstein's theory of general relativity also grew from a simple observation. In Newton's laws of motion there appears a quantity called the mass, which determines how easily an object accelerates when a force is applied. A massive truck is far more difficult to bring to speed than a far less massive Volkswagen. In Newton's day three forces were known: electricity, magnetism, and gravity. The resistance to changing velocity in Newton's laws of motion does not depend on which of those forces is applied. But Newton also discovered a law governing one of those forces, the force of gravity. In that law there appears another quantity which determines the amount of gravitational pull an object exerts, and the amount of gravitational pull it feels when in the presence of another object. That quantity is also called the mass. The two definitions of mass play quite different roles, but they are both called mass for good reason: it turns out they are one and the same. Why should they be equivalent? That question, plus Einstein's brilliant logic, led to his realization that the scaffolding of space and time reacts to the presence of matter and energy.

"At a time like the present," Einstein wrote, "when experience forces us to seek a newer and more solid foundation, the physicist cannot simply surrender to the philosopher the critical contemplation of the theoretical foundations; for, he himself knows best, and feels more surely where the shoe pinches." Einstein was not narrowly interested in science, but also in the philosophy and language of science, and even its ethical implications. Some of his writings on those subjects, too, are included here. And. though Einstein wrote the above words in 1936, today is also a time in which physicists seek a new foundation, and a time in which such metaphysical issues have as much relevance as they did then. Today, since Einstein described space and time as dynamic variables, we see the universe as having not just one, but every possible history. We contemplate not only warped space and time, but whether the universe has additional dimensions. And we speculate about the very meaning of those concepts, and whether they are well defined or only approximate. We seek today a unified theory of all forces, as well as the framework of space and time in which we experience that the universe unfolds. It is a quest of which Einstein would have approved, and for which the remarkable work in this volume provides the foundation.

# The Principle
# of Relativity

We can sometimes be fooled into thinking that great scientific breakthroughs, such as Einstein's theory of relativity, were made from whole cloth and were completely independent of the work that came before. In "The Principle of Relativity," we see the context out of which Einstein developed his theory, including some of the fundamental papers on which it was based.

In order to put this work in context, it is best to consider the state of physics at the turn of the twentieth century. In 1864, James Clerk Maxwell developed a complete theory of electricity and magnetism, and demonstrated that an electrical field is generated by a stationary charge, and a magnetic field is generated by a moving charge. These were seen as fundamentally different forces.

Hendrick A. Lorentz, in a series of papers published in 1895 and 1904, asked a seemingly simple question. What happens if a charge is sitting still, and we are running past it? Lorentz showed that to a moving observer, a stationary charge will "look like" a moving charge, and thus, an electric field will look like a magnetic one. Lorentz showed that to a moving observer, an electromagnetic wave will propagate at the same speed as to a stationary observer: the speed of light.

In 1905, Einstein reached a similar conclusion, that electric and magnetic forces are fundamentally related to one another and can appear in different proportions to observers moving at different speeds. But Einstein showed much more. He postulated that all physical laws must be equally valid in any "inertia reference frame" (traveling at fixed speed and direction) and that for any such observer the speed of light will be a constant.

These assumptions were well supported by both Maxwell's theory and in the experimental work of Michelson and Morley, who showed that light travels at a constant speed regardless of the motion of the earth. Einstein posited that two observers with identical clocks and meter sticks who are moving relative to one another will each measure the meter stick of the other as foreshortened, and will measure the clock of the other as running slow. This seeming paradox lies at the heart of relativity.

The transformations between moving frames, conventionally known as the Lorentz transforms, lead to another important correction to Sir Isaac Newton's laws of motion. According to Newton, applying a constant force to a body will accelerate it, and doing so indefinitely will increase the speed of the body without limit. However, Einstein's theory of relativity showed that nothing can exceed the speed of light—Newton was wrong, but only in the limit where speeds approached that of light.

Einstein recognized that relativity was incomplete. It only accounted for systems where bodies moved at constant speeds, whereas in gravitational fields, bodies are constantly being accelerated. He thus developed his "general theory of relativity" in several landmark papers from 1911 to 1916, the principal results of which are described in chapters VII and VIII of "The Principle of Relativity."

In one of his "thought experiments," Einstein postulated that there should be no difference between any experiment conducted in an elevator sitting still on the surface of the earth, and one being accelerated from below in deep space. Since an accelerated frame will cause all projectiles, including light beams, to be bent, Einstein showed that light will be bent by gravitational fields. In fact, the general theory states that it is *space and time* that are curved, and light or any other object simply follows a "straight line" through space and time.

As John Archibald Wheeler has put it, "Matter tells space-time how to curve, and space-time tells matter how to move." Einstein recognized that his equations could govern not only light beams and stars, but also the universe as a whole. He realized that the universe could not be static and should either expand or collapse, and thus general relativity forms the basis for the field now known as cosmology as described in chapter X.

In order to force the universe into an eternally static state, Einstein introduced an *ad hoc* term into his field equations, known as the "cosmological constant." When Edwin Hubble discovered the expanding universe in 1929, Einstein realized his error, and referred to the cosmological constant as "the greatest blunder of my life." In recent years, the cosmological constant has been reintroduced into cosmology in a new form—a "dark energy" that pervades the universe. Recent observations of distant supernovae suggest that dark energy is fueling an acceleration of the universe.

The model that Einstein came up with is still very current, and has not yet failed any observational tests on large scales. When we read through his thoughts on the matter, what remains so remarkable is how much he, and subsequent thinkers, were able to infer from such simple starting assumptions.

# ON THE ELECTRODYNAMICS OF MOVING BODIES

BY

A. EINSTEIN

*Translated from "Zur Elektrodynamik bewegter Körper,"*
*Annalen der Physik,* 17, 1905.

IT is known that Maxwell's electrodynamics—as usually understood at the present time—when applied to moving bodies, leads to asymmetries which do not appear to be inherent in the phenomena. Take, for example, the reciprocal electrodynamic action of a magnet and a conductor. The observable phenomenon here depends only on the relative motion of the conductor and the magnet, whereas the customary view draws a sharp distinction between the two cases in which either the one or the other of these bodies is in motion. For if the magnet is in motion and the conductor at rest, there arises in the neighbourhood of the magnet an electric field with a certain definite energy, producing a current at the places where parts of the conductor are situated. But if the magnet is stationary and the conductor in motion, no electric field arises in the neighbourhood of the magnet. In the conductor, however, we find an electromotive force, to which in itself there is no corresponding energy, but which gives rise—assuming equality of relative motion in the two cases discussed—to electric currents of the same path and intensity as those produced by the electric forces in the former case.

Examples of this sort, together with the unsuccessful attempts to discover any motion of the earth relatively to the "light medium," suggest that the phenomena of electrodynamics as well as of mechanics possess no properties corresponding to the idea of absolute rest. They suggest rather that, as has already been shown to the first order of small quantities, the same laws of electrodynamics and optics will be valid for all frames of reference for which the equations of mechanics hold

4

good.* We will raise this conjecture (the purport of which will hereafter be called the "Principle of Relativity") to the status of a postulate, and also introduce another postulate, which is only apparently irreconcilable with the former, namely, that light is always propagated in empty space with a definite velocity $c$ which is independent of the state of motion of the emitting body. These two postulates suffice for the attainment of a simple and consistent theory of the electrodynamics of moving bodies based on Maxwell's theory for stationary bodies. The introduction of a "luminiferous ether" will prove to be superfluous inasmuch as the view here to be developed will not require an "absolutely stationary space" provided with special properties, nor assign a velocity-vector to a point of the empty space in which electromagnetic processes take place.

The theory to be developed is based—like all electrodynamics—on the kinematics of the rigid body, since the assertions of any such theory have to do with the relationships between rigid bodies (systems of co-ordinates), clocks, and electromagnetic processes. Insufficient consideration of this circumstance lies at the root of the difficulties which the electrodynamics of moving bodies at present encounters.

# I. KINEMATICAL PART

## § I. DEFINITION OF SIMULTANEITY

Let us take a system of co-ordinates in which the equations of Newtonian mechanics hold good.† In order to render our presentation more precise and to distinguish this system of co-ordinates verbally from others which will be introduced hereafter, we call it the "stationary system."

If a material point is at rest relatively to this system of co-ordinates, its position can be defined relatively thereto by the employment of rigid standards of measurement and the methods of Euclidean geometry, and can be expressed in Cartesian co-ordinates.

If we wish to describe the *motion* of a material point, we give the values of its co-ordinates as functions of the time. Now we must bear carefully in mind that a mathematical description of this kind has no

---

*The preceding memoir by Lorentz was not at this time known to the author.
†i.e. to the first approximation.

physical meaning unless we are quite clear as to what we understand by "time." We have to take into account that all our judgments in which time plays a part are always judgments of *simultaneous events*. If, for instance, I say, "That train arrives here at 7 o'clock," I mean something like this: "The pointing of the small hand of my watch to 7 and the arrival of the train are simultaneous events."*

It might appear possible to overcome all the difficulties attending the definition of "time" by substituting "the position of the small hand of my watch" for "time." And in fact such a definition is satisfactory when we are concerned with defining a time exclusively for the place where the watch is located; but it is no longer satisfactory when we have to connect in time series of events occurring at different places, or—what comes to the same thing—to evaluate the times of events occurring at places remote from the watch.

We might, of course, content ourselves with time values determined by an observer stationed together with the watch at the origin of the co-ordinates, and co-ordinating the corresponding positions of the hands with light signals, given out by every event to be timed, and reaching him through empty space. But this co-ordination has the disadvantage that it is not independent of the standpoint of the observer with the watch or clock, as we know from experience. We arrive at a much more practical determination along the following line of thought.

If at the point A of space there is a clock, an observer at A can determine the time values of events in the immediate proximity of A by finding the positions of the hands which are simultaneous with these events. If there is at the point B of space another clock in all respects resembling the one at A, it is possible for an observer at B to determine the time values of events in the immediate neighbourhood of B. But it is not possible without further assumption to compare, in respect of time, an event at A with an event at B. We have so far defined only an "A time" and a "B time." We have not defined a common "time" for A and B, for the latter cannot be defined at all unless we establish *by*

---

*We shall not here discuss the inexactitude which lurks in the concept of simultaneity of two events at approximately the same place, which can only be removed by an abstraction.

*definition* that the "time" required by light to travel from A to B equals the "time" it requires to travel from B to A. Let a ray of light start at the "A time" $t_A$ from A toward B, let it at the "B time" $t_B$ be reflected at B in the direction of A, and arrive again at A at the "A time" $t'_A$.

In accordance with definition the two clocks synchronize if

$$t_B - t_A = t'_A - t_B.$$

We assume that this definition of synchronism is free from contradictions, and possible for any number of points; and that the following relations are universally valid:—

1. If the clock at B synchronizes with the clock at A, the clock at A synchronizes with the clock at B.

2. If the clock at A synchronizes with the clock at B and also with the clock at C, the clocks at B and C also synchronize with each other.

Thus with the help of certain imaginary physical experiments we have settled what is to be understood by synchronous stationary clocks located at different places, and have evidently obtained a definition of "simultaneous," or "synchronous," and of "time." The "time" of an event is that which is given simultaneously with the event by a stationary clock located at the place of the event, this clock being synchronous, and indeed synchronous for all time determinations, with a specified stationary clock.

In agreement with experience we further assume the quantity

$$\frac{2\,AB}{t'_A - t_A} = c,$$

to be a universal constant—the velocity of light in empty space.

It is essential to have time defined by means of stationary clocks in the stationary system, and the time now defined being appropriate to the stationary system we call it "the time of the stationary system."

## § 2. On the Relativity of Lengths and Times

The following reflexions are based on the principle of relativity and on the principle of the constancy of the velocity of light. These two principles we define as follows:—

1. The laws by which the states of physical systems undergo change are not affected, whether these changes of state be referred to

the one or the other of two systems of coordinates in uniform trans-latory motion.

2. Any ray of light moves in the "stationary" system of co-ordinates with the determined velocity $c$, whether the ray be emitted by a stationary or by a moving body. Hence

$$\text{velocity} = \frac{\text{light path}}{\text{time interval}}$$

where time interval is to be taken in the sense of the definition in § 1.

Let there be given a stationary rigid rod; and let its length be $l$ as measured by a measuring-rod which is also stationary. We now imagine the axis of the rod lying along the axis of $x$ of the stationary system of co-ordinates, and that a uniform motion of parallel translation with velocity $v$ along the axis of $x$ in the direction of increasing $x$ is then imparted to the rod. We now inquire as to the length of the moving rod, and imagine its length to be ascertained by the following two operations:—

(a) The observer moves together with the given measuring-rod and the rod to be measured, and measures the length of the rod directly by superposing the measuring-rod, in just the same way as if all three were at rest.

(b) By means of stationary clocks set up in the stationary system and synchronizing in accordance with § 1, the observer ascertains at what points of the stationary system the two ends of the rod to be measured are located at a definite time. The distance between these two points, measured by the measuring-rod already employed, which in this case is at rest, is also a length which may be designated "the length of the rod."

In accordance with the principle of relativity the length to be discovered by the operation (a)—we will call it "the length of the rod in the moving system"—must be equal to the length $l$ of the stationary rod.

The length to be discovered by the operation (b) we will call "the length of the (moving) rod in the stationary system." This we shall determine on the basis of our two principles, and we shall find that it differs from $l$.

Current kinematics tacitly assumes that the lengths determined by these two operations are precisely equal, or in other words, that a moving rigid body at the epoch $t$ may in geometrical respects be perfectly represented by *the same* body *at rest* in a definite position.

We imagine further that at the two ends A and B of the rod, clocks are placed which synchronize with the clocks of the stationary system, that is to say that their indications correspond at any instant to the "time of the stationary system" at the places where they happen to be. These clocks are therefore "synchronous in the stationary system."

We imagine further that with each clock there is a moving observer, and that these observers apply to both clocks the criterion established in § 1 for the synchronization of two clocks. Let a ray of light depart from A at the time* $t_A$, let it be reflected at B at the time $t_B$, and reach A again at the time $t'_A$. Taking into consideration the principle of the constancy of the velocity of light we find that

$$t_B - t_A = \frac{r_{AB}}{c - v} \text{ and } t'_A - t_B = \frac{r_{AB}}{c + v}$$

where $r_{AB}$ denotes the length of the moving rod—measured in the stationary system. Observers moving with the moving rod would thus find that the two clocks were not synchronous, while observers in the stationary system would declare the clocks to be synchronous.

So we see that we cannot attach any *absolute* signification to the concept of simultaneity, but that two events which, viewed from a system of co-ordinates, are simultaneous, can no longer be looked upon as simultaneous events when envisaged from a system which is in motion relatively to that system.

## § 3. THEORY OF THE TRANSFORMATION OF CO-ORDINATES AND TIMES FROM A STATIONARY SYSTEM TO ANOTHER SYSTEM IN UNIFORM MOTION OF TRANSLATION RELATIVELY TO THE FORMER

Let us in "stationary" space take two systems of coordinates, i.e. two systems, each of three rigid material lines, perpendicular to one

---

*"Time" here denotes "time of the stationary system" and also "position of hands of the moving clock situated at the place under discussion."

another, and issuing from a point. Let the axes of X of the two systems coincide, and their axes of Y and Z respectively be parallel. Let each system be provided with a rigid measuring-rod and a number of clocks, and let the two measuring-rods, and likewise all the clocks of the two systems, be in all respects alike.

Now to the origin of one of the two systems ($k$) let a constant velocity $v$ be imparted in the direction of the increasing $x$ of the other stationary system (K), and let this velocity be communicated to the axes of the co-ordinates, the relevant measuring-rod, and the clocks. To any time of the stationary system K there then will correspond a definite position of the axes of the moving system, and from reasons of symmetry we are entitled to assume that the motion of $k$ may be such that the axes of the moving system are at the time $t$ (this "$t$" always denotes a time of the stationary system) parallel to the axes of the stationary system.

We now imagine space to be measured from the stationary system K by means of the stationary measuring-rod, and also from the moving system $k$ by means of the measuring-rod moving with it; and that we thus obtain the co-ordinates $x$, $y$, $z$, and $\xi$, $\eta$, $\zeta$ respectively. Further, let the time $t$ of the stationary system be determined for all points thereof at which there are clocks by means of light signals in the manner indicated in § 1; similarly let the time $\tau$ of the moving system be determined for all points of the moving system at which there are clocks at rest relatively to that system by applying the method, given in § 1, of light signals between the points at which the latter clocks are located.

To any system of values $x$, $y$, $z$, $t$, which completely defines the place and time of an event in the stationary system, there belongs a system of values $\xi$, $\eta$, $\zeta$, $\tau$, determining that event relatively to the system $k$, and our task is now to find the system of equations connecting these quantities.

In the first place it is clear that the equations must be *linear* on account of the properties of homogeneity which we attribute to space and time.

If we place $x' = x - vt$, it is clear that a point at rest in the system $k$ must have a system of values $x'$, $y$, $z$, independent of time. We first define $\tau$ as a function of $x'$, $y$, $z$, and $t$. To do this we have to express in equations that $\tau$ is nothing else than the summary of the data of clocks at rest in system $k$, which have been synchronized according to the rule given in § 1.

From the origin of system $k$ let a ray be emitted at the time $\tau_0$ along the X-axis to $x'$, and at the time $\tau_1$ be reflected thence to the origin of the co-ordinates, arriving there at the time $\tau_2$; we then must have $\frac{1}{2}(\tau_0 + \tau_2) = \tau_1$, or, by inserting the arguments of the function $\tau$ and applying the principle of the constancy of the velocity of light in the stationary system:—

$$\frac{1}{2}\left[\tau(0,0,0,t) + \tau\left(0,0,0,t + \frac{x'}{c-v} + \frac{x'}{c+v}\right)\right] = \tau\left(x',0,0,t + \frac{x'}{c-v}\right).$$

Hence, if $x'$ be chosen infinitesimally small,

$$\frac{1}{2}\left(\frac{1}{c-v} + \frac{1}{c+v}\right)\frac{\partial\tau}{\partial t} = \frac{\partial\tau}{\partial x'} + \frac{1}{c-v}\frac{\partial\tau}{\partial t},$$

or

$$\frac{\partial\tau}{\partial x'} + \frac{v}{c^2-v^2}\frac{\partial\tau}{\partial t} = 0.$$

It is to be noted that instead of the origin of the co-ordinates we might have chosen any other point for the point of origin of the ray, and the equation just obtained is therefore valid for all values of $x'$, $y$, $z$.

An analogous consideration—applied to the axes of Y and Z—it being borne in mind that light is always propagated along these axes, when viewed from the stationary system, with the velocity $\sqrt{(c^2 - v^2)}$, gives us

$$\frac{\partial\tau}{\partial y} = 0, \frac{\partial\tau}{\partial z} = 0.$$

Since $\tau$ is a *linear* function, it follows from these equations that

$$\tau = a\left(t - \frac{v}{c^2-v^2}x'\right)$$

where $a$ is a function $\phi(v)$ at present unknown, and where for brevity it is assumed that at the origin of $k$, $\tau = 0$, when $t = 0$.

With the help of this result we easily determine the quantities $\xi, \eta, \zeta$ by expressing in equations that light (as required by the principle of the constancy of the velocity of light, in combination with the principle of relativity) is also propagated with velocity $c$ when measured in the moving system. For a ray of light emitted at the time $\tau = 0$ in the direction of the increasing $\xi$

$$\xi = c\tau \text{ or } \xi = ac\left(t - \frac{v}{c^2 - v^2}x'\right).$$

But the ray moves relatively to the initial point of $k$, when measured in the stationary system, with the velocity $c - v$, so that

$$\frac{x'}{c - v} = t.$$

If we insert this value of $t$ in the equation for $\xi$, we obtain

$$\xi = a\frac{c^2}{c^2 - v^2}x'.$$

In an analogous manner we find, by considering rays moving along the two other axes, that

$$\eta = c\tau = ac\left(t - \frac{v}{c^2 - v^2}x'\right)$$

when

$$\frac{y}{\sqrt{(c^2 - v^2)}} = t, x' = 0.$$

Thus

$$\eta = a\frac{c}{\sqrt{(c^2 - v^2)}}y \text{ and } \zeta = a\frac{c}{\sqrt{(c^2 - v^2)}}z.$$

Substituting for $x'$ its value, we obtain

$$\tau = \phi(v)\beta(t - vx/c^2),$$
$$\xi = \phi(v)\beta(x - vt),$$
$$\eta = \phi(v)y,$$
$$\zeta = \phi(v)z,$$

where

$$\beta = \frac{1}{\sqrt{(1 - v^2/c^2)}},$$

and $\phi$ is an as yet unknown function of $v$. If no assumption whatever be made as to the initial position of the moving system and as to the zero point of $\tau$, an additive constant is to be placed on the right side of each of these equations.

We now have to prove that any ray of light, measured in the moving system, is propagated with the velocity $c$, if, as we have assumed, this is the case in the stationary system; for we have not as yet furnished the proof that the principle of the constancy of the velocity of light is compatible with the principle of relativity.

At the time $t = \tau = 0$, when the origin of the co-ordinates is common to the two systems, let a spherical wave be emitted therefrom, and be propagated with the velocity $c$ in system K. If $(x, y, z)$ be a point just attained by this wave, then

$$x^2 + y^2 + z^2 = c^2 t^2.$$

Transforming this equation with the aid of our equations of transformation we obtain after a simple calculation

$$\xi^2 + \eta^2 + \zeta^2 = c^2 \tau^2.$$

The wave under consideration is therefore no less a spherical wave with velocity of propagation $c$ when viewed in the moving system. This shows that our two fundamental principles are compatible.*

In the equations of transformation which have been developed there enters an unknown function $\phi$ of $v$, which we will now determine.

For this purpose we introduce a third system of co-ordinates K', which relatively to the system $k$ is in a state of parallel translatory motion parallel to the axis of X, such that the origin of co-ordinates of system $k$ moves with velocity $-v$ on the axis of X. At the time $t = 0$ let all three origins coincide, and when $t = x = y = z = 0$ let the time $t'$ of the system K' be zero. We call the co-ordinates, measured in the system K', $x'$, $y'$, $z'$, and by a twofold application of our equations of transformation we obtain

$$
\begin{aligned}
t' &= \phi(-v)\beta(-v)(\tau + v\xi/c^2) &&= \phi(v)\phi(-v)t, \\
x' &= \phi(-v)\beta(-v)(\xi + v\tau) &&= \phi(v)\phi(-v)x, \\
y' &= \phi(-v)\eta &&= \phi(v)\phi(-v)y, \\
z' &= \phi(-v)\zeta &&= \phi(v)\phi(-v)z.
\end{aligned}
$$

Since the relations between $x'$, $y'$, $z'$ and $x$, $y$, $z'$ do not contain the time $t$, the systems K and K' are at rest with respect to one another,

---

* The equations of the Lorentz transformation may be more simply deduced directly from the condition that in virtue of those equations the relation $x^2 + y^2 + z^2 = c^2 t^2$ shall have as its consequence the second relation $\xi^2 + \eta^2 + \zeta^2 = c^2 \tau^2$.

and it is clear that the transformation from K to K′ must be the identical transformation. Thus

$$\phi(v)\phi(-v) = 1.$$

We now inquire into the signification of $\phi(v)$. We give our attention to that part of the axis of Y of system $k$ which lies between $\xi = 0$, $\eta = 0$, $\zeta = 0$ and $\xi = 0$, $\eta = l$, $\zeta = 0$. This part of the axis of Y is a rod moving perpendicularly to its axis with velocity $v$ relatively to system K. Its ends possess in K the co-ordinates

$$x_1 = vt, \; y_1 = \frac{l}{\phi(v)}, \; z_1 = 0$$

and

$$x_2 = vt, \; y_2 = 0, \; z_2 = 0.$$

The length of the rod measured in K is therefore $l/\phi(v)$; and this gives us the meaning of the function $\phi(v)$. From reasons of symmetry it is now evident that the length of a given rod moving perpendicularly to its axis, measured in the stationary system, must depend only on the velocity and not on the direction and the sense of the motion. The length of the moving rod measured in the stationary system does not change, therefore, if $v$ and $-v$ are interchanged. Hence follows that $l/\phi(v) = l/\phi(-v)$, or

$$\phi(v) = \phi(-v).$$

It follows from this relation and the one previously found that $\phi(v) = 1$, so that the transformation equations which have been found become

$$\tau = \beta(t - vx/c^2),$$
$$\xi = \beta(x - vt),$$
$$\eta = y,$$
$$\zeta = z,$$

where

$$\beta = 1/\sqrt{(1 - v^2/c^2)}.$$

## § 4. Physical Meaning of the Equations Obtained in Respect to Moving Rigid Bodies and Moving Clocks

We envisage a rigid sphere* of radius R, at rest relatively to the moving system $k$, and with its centre at the origin of co-ordinates of $k$.

*That is, a body possessing spherical form when examined at rest.

The equation of the surface of this sphere moving relatively to the system K with velocity $v$ is

$$\xi^2 + \eta^2 + \zeta^2 = R^2.$$

The equation of this surface expressed in $x$, $y$, $z$ at the time $t = 0$ is

$$\frac{x^2}{(\sqrt{(1 - v^2/c^2)})^2} + y^2 + z^2 = R^2.$$

A rigid body which, measured in a state of rest, has the form of a sphere, therefore has in a state of motion—viewed from the stationary system—the form of an ellipsoid of revolution with the axes

$$R\sqrt{(1 - v^2/c^2)}, R, R.$$

Thus, whereas the Y and Z dimensions of the sphere (and therefore of every rigid body of no matter what form) do not appear modified by the motion, the X dimension appears shortened in the ratio $1 : \sqrt{(1 - v^2/c^2)}$, i.e. the greater the value of $v$, the greater the shortening. For $v = c$ all moving objects—viewed from the "stationary" system—shrivel up into plain figures. For velocities greater than that of light our deliberations become meaningless; we shall, however, find in what follows, that the velocity of light in our theory plays the part, physically, of an infinitely great velocity.

It is clear that the same results hold good of bodies at rest in the "stationary" system, viewed from a system in uniform motion.

Further, we imagine one of the clocks which are qualified to mark the time $t$ when at rest relatively to the stationary system, and the time $\tau$ when at rest relatively to the moving system, to be located at the origin of the co-ordinates of $k$, and so adjusted that it marks the time $\tau$. What is the rate of this clock, when viewed from the stationary system?

Between the quantities $x$, $t$, and $\tau$, which refer to the position of the clock, we have, evidently, $x = vt$ and

$$\tau = \frac{1}{\sqrt{(1 - v^2/c^2)}} (t - vx/c^2).$$

Therefore,

$$\tau = t\sqrt{(1 - v^2/c^2)} = t - (1 - \sqrt{(1 - v^2/c^2)})t$$

whence it follows that the time marked by the clock (viewed in the stationary system) is slow by $1 - \sqrt{(1 - v^2/c^2)}$ seconds per second, or—neglecting magnitudes of fourth and higher order—by $\frac{1}{2}v^2/c^2$.

From this there ensues the following peculiar consequence. If at the points A and B of K there are stationary clocks which, viewed in the stationary system, are synchronous; and if the clock at A is moved with the velocity $v$ along the line AB to B, then on its arrival at B the two clocks no longer synchronize, but the clock moved from A to B lags behind the other which has remained at B by $\frac{1}{2}tv^2/c^2$ (up to magnitudes of fourth and higher order), $t$ being the time occupied in the journey from A to B.

It is at once apparent that this result still holds good if the clock moves from A to B in any polygonal line, and also when the points A and B coincide.

If we assume that the result proved for a polygonal line is also valid for a continuously curved line, we arrive at this result: If one of two synchronous clocks at A is moved in a closed curve with constant velocity until it returns to A, the journey lasting $t$ seconds, then by the clock which has remained at rest the travelled clock on its arrival at A will be $\frac{1}{2}tv^2/c^2$ second slow. Thence we conclude that a balance-clock* at the equator must go more slowly, by a very small amount, than a precisely similar clock situated at one of the poles under otherwise identical conditions.

## § 5. THE COMPOSITION OF VELOCITIES

In the system $k$ moving along the axis of X of the system K with velocity $v$, let a point move in accordance with the equations

$$\xi = w_\xi\tau, \ \eta = \omega_\eta\tau, \ \zeta = 0,$$

where $w_\xi$ and $w_\eta$ denote constants.

Required: the motion of the point relatively to the system K. If with the help of the equations of transformation developed in § 3 we

---

*Not a pendulum-clock, which is physically a system to which the Earth belongs. This case had to be excluded.

introduce the quantities $x$, $y$, $z$, $t$ into the equations of motion of the point, we obtain

$$x = \frac{w_\xi + v}{1 + vw_\xi/c^2} t,$$

$$y = \frac{\sqrt{(1 - v^2/c^2)}}{1 + vw_\xi/c^2} w_\eta t,$$

$$z = 0.$$

Thus the law of the parallelogram of velocities is valid according to our theory only to a first approximation. We set

$$V^2 = \left(\frac{dx}{dt}\right)^2 + \left(\frac{dy}{dt}\right)^2,$$

$$w^2 = w_\xi^2 + w_\eta^2,$$

$$a = \tan^{-1} w_y/w_x,$$

$a$ is then to be looked upon as the angle between the velocities $v$ and $w$. After a simple calculation we obtain

$$V = \frac{\sqrt{\left[(v^2 + w^2 + 2vw \cos a) - (vw \sin a/c^2)^2\right]}}{1 + vw \cos a/c^2}.$$

It is worthy of remark that $v$ and $w$ enter into the expression for the resultant velocity in a symmetrical manner. If $w$ also has the direction of the axis of X, we get

$$V = \frac{v + w}{1 + vw/c^2}.$$

It follows from this equation that from a composition of two velocities which are less than $c$, there always results a velocity less than $c$. For if we set $v = c - \kappa$, $w = c - \lambda$, $\kappa$ and $\lambda$ being positive and less than $c$, then

$$V = c\frac{2c - \kappa - \lambda}{2c - \kappa - \lambda + \kappa\lambda/c} < c.$$

It follows, further, that the velocity of light $c$ cannot be altered by composition with a velocity less than that of light. For this case we obtain

$$V = \frac{c + w}{1 + w/c} = c.$$

We might also have obtained the formula for V, for the case when $v$ and $w$ have the same direction, by compounding two transformations in accordance with § 3. If in addition to the systems K and *k* figuring in § 3 we introduce still another system of co-ordinates *k'* moving parallel to *k*, its initial point moving on the axis of X with the velocity *w*, we obtain equations between the quantities *x, y, z, t* and the corresponding quantities of *k'*, which differ from the equations found in § 3 only in that the place of "*v*" is taken by the quantity

$$\frac{v + w}{1 + vw/c^2};$$

from which we see that such parallel transformations—necessarily—form a group.

We have now deduced the requisite laws of the theory of kinematics corresponding to our two principles, and we proceed to show their application to electrodynamics.

# II. ELECTRODYNAMICAL PART

### § 6. TRANSFORMATION OF THE MAXWELL-HERTZ EQUATIONS FOR EMPTY SPACE. ON THE NATURE OF THE ELECTROMOTIVE FORCES OCCURRING IN A MAGNETIC FIELD DURING MOTION

Let the Maxwell-Hertz equations for empty space hold good for the stationary system K, so that we have

$$\frac{1}{c}\frac{\partial X}{\partial t} = \frac{\partial N}{\partial y} - \frac{\partial M}{\partial z}, \quad \frac{1}{c}\frac{\partial L}{\partial t} = \frac{\partial Y}{\partial z} - \frac{\partial Z}{\partial y},$$

$$\frac{1}{c}\frac{\partial Y}{\partial t} = \frac{\partial L}{\partial z} - \frac{\partial N}{\partial x}, \quad \frac{1}{c}\frac{\partial M}{\partial t} = \frac{\partial Z}{\partial x} - \frac{\partial X}{\partial z},$$

$$\frac{1}{c}\frac{\partial Z}{\partial t} = \frac{\partial M}{\partial x} - \frac{\partial L}{\partial y}, \quad \frac{1}{c}\frac{\partial N}{\partial t} = \frac{\partial X}{\partial y} - \frac{\partial Y}{\partial x},$$

where (X, Y, Z) denotes the vector of the electric force, and (L, M, N) that of the magnetic force.

If we apply to these equations the transformation developed in § 3, by referring the electromagnetic processes to the system of co-ordinates there introduced, moving with the velocity $v$, we obtain the equations

$$\frac{1}{c}\frac{\partial X}{\partial \tau} = \frac{\partial}{\partial \eta}\left\{\beta\left(N - \frac{v}{c}Y\right)\right\} - \frac{\partial}{\partial \zeta}\left\{\beta\left(M + \frac{v}{c}Z\right)\right\},$$

$$\frac{1}{c}\frac{\partial}{\partial \tau}\left\{\beta\left(Y - \frac{v}{c}N\right)\right\} = \frac{\partial L}{\partial \xi} \qquad\qquad - \frac{\partial}{\partial \zeta}\left\{\beta\left(N - \frac{v}{c}Y\right)\right\}.$$

$$\frac{1}{c}\frac{\partial}{\partial \tau}\left\{\beta\left(Z + \frac{v}{c}M\right)\right\} = \frac{\partial}{\partial \xi}\left\{\beta\left(M + \frac{v}{c}Z\right)\right\} - \frac{\partial L}{\partial \eta},$$

$$\frac{1}{c}\frac{\partial L}{\partial \tau} = \frac{\partial}{\partial \zeta}\left\{\beta\left(Y - \frac{v}{c}N\right)\right\} - \frac{\partial}{\partial \eta}\left\{\beta\left(Z + \frac{v}{c}M\right)\right\},$$

$$\frac{1}{c}\frac{\partial}{\partial \tau}\left\{\beta\left(M + \frac{v}{c}Z\right)\right\} = \frac{\partial}{\partial \xi}\left\{\beta\left(Z + \frac{v}{c}M\right)\right\} - \frac{\partial X}{\partial \zeta},$$

$$\frac{1}{c}\frac{\partial}{\partial \tau}\left\{\beta\left(N - \frac{v}{c}Y\right)\right\} = \frac{\partial X}{\partial \eta} \qquad\qquad - \frac{\partial}{\partial \xi}\left\{\beta\left(Y - \frac{v}{c}N\right)\right\},$$

where

$$\beta = 1/\sqrt{(1 - v^2/c^2)}.$$

Now the principle of relativity requires that if the Maxwell-Hertz equations for empty space hold good in system K, they also hold good in system $k$; that is to say that the vectors of the electric and the magnetic force—(X', Y', Z') and (L', M', N')—of the moving system $k$, which are defined by their ponderomotive effects on electric or magnetic masses respectively, satisfy the following equations:—

$$\frac{1}{c}\frac{\partial X'}{\partial \tau} = \frac{\partial N'}{\partial \eta} - \frac{\partial M'}{\partial \zeta}, \qquad \frac{1}{c}\frac{\partial L'}{\partial \tau} = \frac{\partial Y'}{\partial \zeta} - \frac{\partial Z'}{\partial \eta},$$

$$\frac{1}{c}\frac{\partial Y'}{\partial \tau} = \frac{\partial L'}{\partial \zeta} - \frac{\partial N'}{\partial \xi}, \qquad \frac{1}{c}\frac{\partial M'}{\partial \tau} = \frac{\partial Z'}{\partial \xi} - \frac{\partial X'}{\partial \zeta},$$

$$\frac{1}{c}\frac{\partial Z'}{\partial \tau} = \frac{\partial M'}{\partial \xi} - \frac{\partial L'}{\partial \eta}, \qquad \frac{1}{c}\frac{\partial N'}{\partial \tau} = \frac{\partial X'}{\partial \eta} - \frac{\partial Y'}{\partial \xi}.$$

Evidently the two systems of equations found for system $k$ must express exactly the same thing, since both systems of equations are equivalent to the Maxwell-Hertz equations for system K. Since, further, the equations of the two systems agree, with the exception of the symbols for the vectors, it follows that the functions occurring in the systems of equations at corresponding places must agree, with the exception of a factor $\psi(v)$, which is common for all functions of the one system of equations, and is independent of $\xi, \eta, \zeta$ and $\tau$ but depends upon $v$. Thus we have the relations

$$X' = \psi(v)X, \qquad\qquad L' = \psi(v)L,$$

$$Y' = \psi(v)\beta\left(Y - \frac{v}{c}N\right), \qquad M' = \psi(v)\beta\left(M + \frac{v}{c}Z\right),$$

$$Z' = \psi(v)\beta\left(Z + \frac{v}{c}M\right), \qquad N' = \psi(v)\beta\left(N - \frac{v}{c}Y\right).$$

If we now form the reciprocal of this system of equations, firstly by solving the equations just obtained, and secondly by applying the equations to the inverse transformation (from $k$ to K), which is characterized by the velocity $-v$, it follows, when we consider that the two systems of equations thus obtained must be identical, that $\psi(v)\psi(-v) = 1$. Further, from reasons of symmetry* $\psi(v) = \psi(-v)$. and therefore

$$\psi(v) = 1,$$

and our equations assume the form

$$X' = X, \qquad\qquad L' = L,$$

$$Y' = \beta\left(Y - \frac{v}{c}N\right), M' = \beta\left(M + \frac{v}{c}Z\right),$$

$$Z' = \beta\left(Z + \frac{v}{c}M\right), N' = \beta\left(N - \frac{v}{c}Y\right).$$

As to the interpretation of these equations we make the following remarks: Let a point charge of electricity have the magnitude "one" when measured in the stationary system K, i.e. let it when at rest in the stationary system exert a force of one dyne upon an equal quantity of electricity at a distance of one cm. By the principle of relativity this electric charge is also of the magnitude "one" when measured in the moving system. If this quantity of electricity is at rest relatively to the stationary system, then by definition the vector (X, Y, Z) is equal to the force acting upon it. If the quantity of electricity is at

*If, for example, X = Y = Z = L = M = O, and N ≠ O, then from reasons of symmetry it is clear that when $v$ ohanges sign without changing its numerical value, Y' must also change sign without changing its numerical value.

rest relatively to the moving system (at least at the relevant instant), then the force acting upon it, measured in the moving system, is equal to the vector (X', Y', Z'). Consequently the first three equations above allow themselves to be clothed in words in the two following ways:—

1. If a unit electric point charge is in motion in an electromagnetic field, there acts upon it, in addition to the electric force, an "electromotive force" which, if we neglect the terms multiplied by the second and higher powers of $v/c$, is equal to the vector-product of the velocity of the charge and the magnetic force, divided by the velocity of light. (Old manner of expression.)

2. If a unit electric point charge is in motion in an electromagnetic field, the force acting upon it is equal to the electric force which is present at the locality of the charge, and which we ascertain by transformation of the field to a system of co-ordinates at rest relatively to the electrical charge. (New manner of expression.)

The analogy holds with "magnetomotive forces." We see that electromotive force plays in the developed theory merely the part of an auxiliary concept, which owes its introduction to the circumstance that electric and magnetic forces do not exist independently of the state of motion of the system of co-ordinates.

Furthermore it is clear that the asymmetry mentioned in the introduction as arising when we consider the currents produced by the relative motion of a magnet and a conductor, now disappears. Moreover, questions as to the "seat" of electrodynamic electromotive forces (unipolar machines) now have no point.

## § 7. THEORY OF DOPPLER'S PRINCIPLE AND OF ABERRATION

In the system K, very far from the origin of co-ordinates, let there be a source of electrodynamic waves, which in a part of space containing the origin of co-ordinates may be represented to a sufficient degree of approximation by the equations

$$X = X_0 \sin \Phi, \quad L = L_0 \sin \Phi,$$
$$Y = Y_0 \sin \Phi, \quad M = M_0 \sin \Phi,$$
$$Z = Z_0 \sin \Phi, \quad N = N_0 \sin \Phi,$$

where

$$\Phi = \omega\left\{t - \frac{1}{c}(lx + my + nz)\right\}.$$

Here $(X_0, Y_0, Z_0)$ and $(L_0, M_0, N_0)$ are the vectors defining the amplitude of the wave-train, and $l$, $m$, $n$ the direction-cosines of the wave-normals. We wish to know the constitution of these waves, when they are examined by an observer at rest in the moving system $k$.

Applying the equations of transformation found in § 6 for electric and magnetic forces, and those found in § 3 for the co-ordinates and the time, we obtain directly

$$X' = X_0 \sin \Phi', \qquad\qquad L' = L_0 \sin \Phi',$$
$$Y' = \beta(Y_0 - vN_0/c) \sin \Phi', \quad M' = \beta(M_0 + vZ_0/c) \sin \Phi',$$
$$Z' = \beta(Z_0 + vM_0/c) \sin \Phi', \quad N' = \beta(N_0 - vY_0/c) \sin \Phi',$$

$$\Phi' = \omega'\left\{\tau - \frac{1}{c}(l'\xi + m'\eta + n'\zeta)\right\}$$

where

$$\omega' = \omega\beta(1 - lv/c),$$
$$l' = \frac{l - v/c}{1 - lv/c},$$
$$m' = \frac{m}{\beta(1 - lv/c)},$$
$$n' = \frac{n}{\beta(1 - lv/c)}.$$

From the equation for $\omega'$ it follows that if an observer is moving with velocity $v$ relatively to an infinitely distant source of light of frequency $\nu$, in such a way that the connecting line "source—observer" makes the angle $\phi$ with the velocity of the observer referred to a system of co-ordinates which is at rest relatively to the source of light, the frequency $\nu'$ of the light perceived by the observer is given by the equation

$$\nu' = \nu\frac{1 - \cos\phi \cdot v/c}{\sqrt{(1 - v^2/c^2)}}.$$

This is Doppler's principle for any velocities whatever. When $\phi = 0$ the equation assumes the perspicuous form

$$\nu' = \nu\sqrt{\frac{1 - v/c}{1 + v/c}}.$$

We see that, in contrast with the customary view, when $v = -c, \nu' = \infty$.

If we call the angle between the wave-normal (direction of the ray) in the moving system and the connecting line "source—observer" $\phi'$, the equation for $l'$ assumes the form

$$\cos\phi' = \frac{\cos\phi - v/c}{1 - \cos\phi \cdot v/c}.$$

This equation expresses the law of aberration in its most general form. If $\phi = \frac{1}{2}\pi$, the equation becomes simply

$$\cos\phi' = -v/c.$$

We still have to find the amplitude of the waves, as it appears in the moving system. If we call the amplitude of the electric or magnetic force A or A' respectively, accordingly as it is measured in the stationary system or in the moving system, we obtain

$$A'^2 = A^2\frac{(1 - \cos\phi \cdot v/c)^2}{1 - v^2/c^2}$$

which equation, if $\phi = 0$, simplifies into

$$A'^2 = A^2\frac{1 - v/c}{1 + v/c}.$$

It follows from these results that to an observer approaching a source of light with the velocity $c$, this source of light must appear of infinite intensity.

## § 8. TRANSFORMATION OF THE ENERGY OF LIGHT RAYS. THEORY OF THE PRESSURE OF RADIATION EXERTED ON PERFECT REFLECTORS

Since $A^2/8\pi$ equals the energy of light per unit of volume, we have to regard $A'^2/8\pi$, by the principle of relativity, as the energy of light in the moving system. Thus $A'^2/A^2$ would be the ratio of the "measured in motion" to the "measured at rest" energy of a given light complex, if the volume of a light complex were the same, whether measured in

K or in $k$. But this is not the case. If $l$, $m$, $n$ are the direction-cosines of the wave-normals of the light in the stationary system, no energy passes through the surface elements of a spherical surface moving with the velocity of light:—

$$(x - lct)^2 + (y - mct)^2 + (z - nct)^2 = R^2.$$

We may therefore say that this surface permanently encloses the same light complex. We inquire as to the quantity of energy enclosed by this surface, viewed in system $k$, that is, as to the energy of the light complex relatively to the system $k$.

The spherical surface—viewed in the moving system—is an ellipsoidal surface, the equation for which, at the time $\tau = 0$, is

$$(\beta\xi - l\beta\xi v/c)^2 + (\eta - m\beta\xi v/c)^2 + (\zeta - n\beta\xi v/c)^2 = R^2.$$

If S is the volume of the sphere, and S' that of this ellipsoid, then by a simple calculation

$$\frac{S'}{S} = \frac{\sqrt{1 - v^2/c^2}}{1 - \cos\phi \cdot v/c}.$$

Thus, if we call the light energy enclosed by this surface E when it is measured in the stationary system, and E' when measured in the moving system, we obtain

$$\frac{E'}{E} = \frac{A'^2 S'}{A^2 S} = \frac{1 - \cos\phi \cdot v/c}{\sqrt{(1 - v^2/c^2)}},$$

and this formula, when $\phi = 0$, simplifies into

$$\frac{E'}{E} = \sqrt{\frac{1 - v/c}{1 + v/c}}.$$

It is remarkable that the energy and the frequency of a light complex vary with the state of motion of the observer in accordance with the same law.

Now let the co-ordinate plane $\xi = 0$ be a perfectly reflecting surface, at which the plane waves considered in § 7 are reflected. We seek for the pressure of light exerted on the reflecting surface, and for the direction, frequency, and intensity of the light after reflexion.

Let the incidental light be defined by the quantities A, $\cos\phi$, $\nu$ (referred to system K). Viewed from $k$ the corresponding quantities are

$$A' = A \frac{1 - \cos\phi \cdot v/c}{\sqrt{(1 - v^2/c^2)}},$$

$$\cos\phi' = \frac{\cos\phi - v/c}{1 - \cos\phi \cdot v/c},$$

$$v' = v \frac{1 - \cos\phi \cdot v/c}{\sqrt{(1 - v^2/c^2)}}.$$

For the reflected light, referring the process to system $k$, we obtain

$$A'' = A'$$
$$\cos\phi'' = -\cos\phi'$$
$$v'' = v'$$

Finally, by transforming back to the stationary system K, we obtain for the reflected light

$$A''' = A'' \frac{1 + \cos\phi'' \cdot v/c}{\sqrt{(1 - v^2/c^2)}} = A \frac{1 - 2\cos\phi \cdot v/c + v^2/c^2}{1 - v^2/c^2},$$

$$\cos\phi''' = \frac{\cos\phi'' + v/c}{1 + \cos\phi'' \cdot v/c} = -\frac{(1 + v^2/c^2)\cos\phi - 2v/c}{1 - 2\cos\phi \cdot v/c + v^2/c^2}$$

$$v''' = v'' \frac{1 + \cos\phi'' v/c}{\sqrt{(1 - v^2/c^2)}} = v \frac{1 - 2\cos\phi \cdot v/c + v^2/c^2}{1 - v^2/c^2}.$$

The energy (measured in the stationary system) which is incident upon unit area of the mirror in unit time is evidently $A^2(c\cos\phi - v)/8\pi$. The energy leaving the unit of surface of the mirror in the unit of time is $A'''^2(-c\cos\phi''' + v)/8\pi$. The difference of these two expressions is, by the principle of energy, the work done by the pressure of light in the unit of time. If we set down this work as equal to the product $Pv$, where P is the pressure of light, we obtain

$$P = 2 \cdot \frac{A^2}{8\pi} \frac{(\cos\phi - v/c)^2}{1 - v^2/c^2}.$$

In agreement with experiment and with other theories, we obtain to a first approximation

$$P = 2 \cdot \frac{A^2}{8\pi} \cos^2\phi.$$

All problems in the optics of moving bodies can be solved by the method here employed. What is essential is, that the electric and magnetic force of the light which is influenced by a moving body, be transformed into a system of co-ordinates at rest relatively to the body. By this means all problems in the optics of moving bodies will be reduced to a series of problems in the optics of stationary bodies.

## § 9. Transformation of the Maxwell-Hertz Equations when Convection-Currents are Taken into Account

We start from the equations

$$\frac{1}{c}\left\{\frac{\partial X}{\partial t} + u_x\rho\right\} = \frac{\partial N}{\partial y} - \frac{\partial M}{\partial z}, \quad \frac{1}{c}\frac{\partial L}{\partial t} = \frac{\partial Y}{\partial z} - \frac{\partial Z}{\partial y},$$

$$\frac{1}{c}\left\{\frac{\partial Y}{\partial t} + u_y\rho\right\} = \frac{\partial L}{\partial z} - \frac{\partial N}{\partial x}, \quad \frac{1}{c}\frac{\partial M}{\partial t} = \frac{\partial Z}{\partial x} - \frac{\partial X}{\partial z},$$

$$\frac{1}{c}\left\{\frac{\partial Z}{\partial t} + u_z\rho\right\} = \frac{\partial M}{\partial x} - \frac{\partial L}{\partial y}, \quad \frac{1}{c}\frac{\partial N}{\partial t} = \frac{\partial X}{\partial y} - \frac{\partial Y}{\partial x},$$

where

$$\rho = \frac{\partial X}{\partial x} + \frac{\partial Y}{\partial y} + \frac{\partial Z}{\partial z}$$

denotes $4\pi$ times the density of electricity, and $(u_x, u_y, u_z)$ the velocity-vector of the charge. If we imagine the electric charges to be invariably coupled to small rigid bodies (ions, electrons), these equations are the electromagnetic basis of the Lorentzian electrodynamics and optics of moving bodies.

Let these equations be valid in the system K, and transform them, with the assistance of the equations of transformation given in §§ 3 and 6, to the system $k$. We then obtain the equations

$$\frac{1}{c}\left\{\frac{\partial X'}{\partial \tau} + u_\xi\rho'\right\} = \frac{\partial N'}{\partial \eta} - \frac{\partial M'}{\partial \zeta}, \quad \frac{1}{c}\frac{\partial L'}{\partial \tau} = \frac{\partial Y'}{\partial \zeta} - \frac{\partial Z'}{\partial \eta},$$

$$\frac{1}{c}\left\{\frac{\partial Y'}{\partial \tau} + u_\eta\rho'\right\} = \frac{\partial L'}{\partial \zeta} - \frac{\partial N'}{\partial \xi}, \quad \frac{1}{c}\frac{\partial M'}{\partial \tau} = \frac{\partial Z'}{\partial \xi} - \frac{\partial X'}{\partial \zeta},$$

$$\frac{1}{c}\left\{\frac{\partial Z'}{\partial \tau} + u_\zeta\rho'\right\} = \frac{\partial M'}{\partial \xi} - \frac{\partial L'}{\partial \eta}, \quad \frac{1}{c}\frac{\partial N'}{\partial \tau} = \frac{\partial X'}{\partial \eta} - \frac{\partial Y'}{\partial \xi},$$

where

$$u_\xi = \frac{u_x - v}{1 - u_x v/c^2}$$

$$u_\eta = \frac{u_y}{\beta(1 - u_x v/c^2)}$$

$$u_\zeta = \frac{u_z}{\beta(1 - u_x v/c^2)},$$

and

$$\rho' = \frac{\partial X'}{\partial \xi} + \frac{\partial Y'}{\partial \eta} + \frac{\partial Z'}{\partial \zeta}$$

$$= \beta(1 - u_x v/c^2)\rho.$$

Since—as follows from the theorem of addition of velocities (§ 5)—the vector ($u_\xi$, $u_\eta$, $u_\zeta$) is nothing else than the velocity of the electric charge, measured in the system $k$, we have the proof that, on the basis of our kinematical principles, the electrodynamic foundation of Lorentz's theory of the electrodynamics of moving bodies is in agreement with the principle of relativity.

In addition I may briefly remark that the following important law may easily be deduced from the developed equations: If an electrically charged body is in motion anywhere in space without altering its charge when regarded from a system of co-ordinates moving with the body, its charge also remains—when regarded from the "stationary" system K—constant.

## § 10. Dynamics of the Slowly Accelerated Electron

Let there be in motion in an electromagnetic field an electrically charged particle (in the sequel called an "electron"), for the law of motion of which we assume as follows:—

If the electron is at rest at a given epoch, the motion of the electron ensues in the next instant of time according to the equations

$$m\frac{d^2x}{dt^2} = \varepsilon X$$

$$m\frac{d^2y}{dt^2} = \varepsilon Y$$

$$m\frac{d^2z}{dt^2} = \varepsilon Z$$

where $x$, $y$, $z$ denote the co-ordinates of the electron, and $m$ the mass of the electron, as long as its motion is slow.

Now, secondly, let the velocity of the electron at a given epoch be $v$. We seek the law of motion of the electron in the immediately ensuing instants of time.

Without affecting the general character of our considerations, we may and will assume that the electron, at the moment when we give it our attention, is at the origin of the co-ordinates, and moves with the velocity $v$ along the axis of X of the system K. It is then clear that at the given moment $(t = 0)$ the electron is at rest relatively to a system of co-ordinates which is in parallel motion with velocity $v$ along the axis of X.

From the above assumption, in combination with the principle of relativity, it is clear that in the immediately ensuing time (for small values of $t$) the electron, viewed from the system $k$, moves in accordance with the equations

$$m\frac{d^2\xi}{d\tau^2} = \varepsilon X',$$

$$m\frac{d^2\eta}{d\tau^2} = \varepsilon Y',$$

$$m\frac{d^2\zeta}{d\tau^2} = \varepsilon Z',$$

in which the symbols $\xi$, $\eta$, $\zeta$, $\tau$, X', Y', Z' refer to the system $k$. If, further, we decide that when $t = x = y = z = 0$ then $\tau = \xi = \eta = \zeta = 0$, the transformation equations of §§ 3 and 6 hold good, so that we have

$$\xi = \beta(x - vt), \eta = y, \zeta = z, \tau = \beta(t - vx/c^2)$$
$$X' = X, Y' = \beta(Y - vN/c), Z' = \beta(Z + vM/c).$$

With the help of these equations we transform the above equations of motion from system $k$ to system K, and obtain

$$\left.\begin{array}{l} \dfrac{d^2x}{dt^2} = \dfrac{\varepsilon}{m\beta^3}X \\[2ex] \dfrac{d^2y}{dt^2} = \dfrac{\varepsilon}{m\beta}\left(Y - \dfrac{v}{c}N\right) \\[2ex] \dfrac{d^2z}{dt^2} = \dfrac{\varepsilon}{m\beta}\left(Z + \dfrac{v}{c}M\right) \end{array}\right\} \quad . \quad . \quad . \text{ (A)}$$

Taking the ordinary point of view we now inquire as to the "longitudinal" and the "transverse" mass of the moving electron. We write the equations (A) in the form

$$m\beta^3 \frac{d^2x}{dt^2} = \varepsilon X = \varepsilon X',$$

$$m\beta^2 \frac{d^2y}{dt^2} = \varepsilon\beta\left(Y - \frac{v}{c}N\right) = \varepsilon Y',$$

$$m\beta^2 \frac{d^2z}{dt^2} = \varepsilon\beta\left(Z + \frac{v}{c}M\right) = \varepsilon Z',$$

and remark firstly that $\varepsilon X'$, $\varepsilon Y'$, $\varepsilon Z'$ are the components of the ponderomotive force acting upon the electron, and are so indeed as viewed in a system moving at the moment with the electron, with the same velocity as the electron. (This force might be measured, for example, by a spring balance at rest in the last-mentioned system.) Now if we call this force simply "the force acting upon the electron,"* and maintain the equation—mass × acceleration = force—and if we also decide that the accelerations are to be measured in the stationary System K, we derive from the above equations

$$\text{Longitudinal mass} = \frac{m}{\left(\sqrt{1 - v^2/c^2}\right)^3}.$$

$$\text{Transverse mass} = \frac{m}{1 - v^2/c^2}.$$

With a different definition of force and acceleration we should naturally obtain other values for the masses. This shows us that in comparing different theories of the motion of the electron we must proceed very cautiously.

We remark that these results as to the mass are also valid for ponderable material points, because a ponderable material point can be made into an electron (in our sense of the word) by the addition of an electric charge, *no matter how small.*

We will now determine the kinetic energy of the electron. If an electron moves from rest at the origin of co-ordinates of the system

---

*The definition of force here given is not advantageous, as was first shown by M. Planck. It is more to the point to define force in such a way that the laws of momentum and energy assume the simplest form.

K along the axis of X under the action of an electrostatic force X, it is clear that the energy withdrawn from the electrostatic field has the value $\int \varepsilon X\, dx$. As the electron is to be slowly accelerated, and consequently may not give off any energy in the form of radiation, the energy withdrawn from the electrostatic field must be put down as equal to the energy of motion W of the electron. Bearing in mind that during the whole process of motion which we are considering, the first of the equations (A) applies, we therefore obtain

$$W = \int \varepsilon X dx = m \int_0^v \beta^3 v dv$$

$$= mc^2 \left\{ \frac{1}{\sqrt{1 - v^2/c^2}} - 1 \right\}.$$

Thus, when $v = c$, W becomes infinite. Velocities greater than that of light have—as in our previous results—no possibility of existence.

This expression for the kinetic energy must also, by virtue of the argument stated above, apply to ponderable masses as well.

We will now enumerate the properties of the motion of the electron which result from the system of equations (A), and are accessible to experiment.

1. From the second equation of the system (A) it follows that an electric force Y and a magnetic force N have an equally strong deflective action on an electron moving with the velocity $v$, when $Y = Nv/c$. Thus we see that it is possible by our theory to determine the velocity of the electron from the ratio of the magnetic power of deflexion $A_m$ to the electric power of deflexion $A_e$, for any velocity, by applying the law

$$\frac{A_m}{A_e} = \frac{v}{c}.$$

This relationship may be tested experimentally, since the velocity of the electron can be directly measured, e.g. by means of rapidly oscillating electric and magnetic fields.

2. From the deduction for the kinetic energy of the electron it follows that between the potential difference, P, traversed and the acquired velocity $v$ of the electron there must be the relationship

$$P = \int X\,dx = \frac{m}{\varepsilon}\,c^2 \left\{ \frac{1}{\sqrt{1 - v^2/c^2}} - 1 \right\}$$

3. We calculate the radius of curvature of the path of the electron when a magnetic force N is present (as the only deflective force), acting perpendicularly to the velocity of the electron. From the second of the equations (A) we obtain

$$-\frac{d^2y}{dt^2} = \frac{v^2}{R} = \frac{\varepsilon}{m}\frac{v}{c}N\sqrt{1 - \frac{v^2}{c^2}}$$

or

$$R = \frac{mc^2}{\varepsilon} \cdot \frac{v/c}{\sqrt{(1 - v^2/c^2)}} \cdot \frac{1}{N}.$$

These three relationships are a complete expression for the laws according to which, by the theory here advanced, the electron must move.

In conclusion I wish to say that in working at the problem here dealt with I have had the loyal assistance of my friend and colleague M. Besso, and that I am indebted to him for several valuable suggestions.

# DOES THE INERTIA OF A BODY DEPEND UPON ITS ENERGY-CONTENT?

BY

A. EINSTEIN

*Translated from "Ist die Trägheit eines Körpers von seinem Energiegehalt abhängig?" Annalen der Physik, 17, 1905.*

The results of the previous investigation lead to a very interesting conclusion, which is here to be deduced.

I based that investigation on the Maxwell-Hertz equations for empty space, together with the Maxwellian expression for the electromagnetic energy of space, and in addition the principle that:—

*The laws by which the states of physical systems alter are independent of the alternative, to which of two systems of coordinates, in uniform motion of parallel translation relatively to each other, these alterations of state are referred (principle of relativity).*

With these principles* as my basis I deduced *inter alia* the following result (§ 8):—

Let a system of plane waves of light, referred to the system of coordinates $(x, y, z)$, possess the energy $l$; let the direction of the ray (the wave-normal) make an angle $\phi$ with the axis of $x$ of the system. If we introduce a new system of co-ordinates $(\xi, \eta, \zeta)$ moving in uniform parallel translation with respect to the system $(x, y, z)$, and having its origin of co-ordinates in motion along the axis of $x$ with the velocity $v$, then this quantity of light—measured in the system $(\xi, \eta, \zeta)$—possesses the energy

$$l^* = l\frac{1 - \frac{v}{c}\cos\phi}{\sqrt{1 - v^2/c^2}}$$

---

*The principle of the constancy of the velocity of light is of course contained in Maxwell's equations.

where $c$ denotes the velocity of light. We shall make use of this result in what follows.

Let there be a stationary body in the system $(x, y, z)$, and let its energy—referred to the system $(x, y, z)$—be $E_0$. Let the energy of the body relative to the system $(\xi, \eta, \zeta)$, moving as above with the velocity $v$, be $H_0$.

Let this body send out, in a direction making an angle $\phi$ with the axis of $x$, plane waves of light, of energy $\frac{1}{2}L$ measured relatively to $(x, y, z)$, and simultaneously an equal quantity of light in the opposite direction. Meanwhile the body remains at rest with respect to the system $(x, y, z)$. The principle of energy must apply to this process, and in fact (by the principle of relativity) with respect to both systems of co-ordinates. If we call the energy of the body after the emission of light $E_1$ or $H_1$ respectively, measured relatively to the system $(x, y, z)$ or $(\xi, \eta, \zeta)$ respectively, then by employing the relation given above we obtain

$$E_0 = E_1 + \tfrac{1}{2}L + \tfrac{1}{2}L,$$

$$H_0 = H_1 + \tfrac{1}{2}L\frac{1 - \dfrac{v}{c}\cos\phi}{\sqrt{1 - v^2/c^2}} + \tfrac{1}{2}L\frac{1 + \dfrac{v}{c}\cos\phi}{\sqrt{1 - v^2/c^2}}$$

$$= H_1 + \frac{L}{\sqrt{1 - v^2/c^2}}.$$

By subtraction we obtain from these equations

$$H_0 - E_0 - (H_1 - E_1) = L\left\{\frac{1}{\sqrt{1 - v^2/c^2}} - 1\right\}.$$

The two differences of the form $H - E$ occurring in this expression have simple physical significations. H and E are energy values of the same body referred to two systems of co-ordinates which are in motion relatively to each other, the body being at rest in one of the two systems (system $(x, y, z)$). Thus it is clear that the difference $H - E$ can differ from the kinetic energy K of the body, with respect to the other system $(\xi, \eta, \zeta)$, only by an additive constant C, which depends on the choice of the arbitrary additive constants of the energies H and E. Thus we may place

$$H_0 - E_0 = K_0 + C,$$
$$H_1 - E_1 = K_1 + C,$$

since C does not change during the emission of light. So we have

$$K_0 - K_1 = L\left\{\frac{1}{\sqrt{1 - v^2/c^2}} - 1\right\}.$$

The kinetic energy of the body with respect to $(\xi, \eta, \zeta)$ diminishes as a result of the emission of light, and the amount of diminution is independent of the properties of the body. Moreover, the difference $K_0 - K_1$, like the kinetic energy of the electron (§ 10), depends on the velocity.

Neglecting magnitudes of fourth and higher orders we may place

$$K_0 - K_1 = \frac{1}{2}\frac{L}{c^2}v^2.$$

From this equation it directly follows that:—

*If a body gives off the energy L in the form of radiation, its mass diminishes by L/c². *The fact that the energy withdrawn from the body becomes energy of radiation evidently makes no difference, so that we are led to the more general conclusion that

The mass of a body is a measure of its energy-content; if the energy changes by L, the mass changes in the same sense by $L/9 \times 10^{20}$, the energy being measured in ergs, and the mass in grammes.

It is not impossible that with bodies whose energy-content is variable to a high degree (e.g. with radium salts) the theory may be successfully put to the test.

If the theory corresponds to the facts, radiation conveys inertia between the emitting and absorbing bodies.

# On the Influence of Gravitation on the Propagation of Light

## By

### A. Einstein

*Translated from "Über den Einfluss der Schwerkraft auf die Ausbreitung des Lichtes," Annalen der Physik, 35, 1911.*

In a memoir published four years ago* I tried to answer the question whether the propagation of light is influenced by gravitation. I return to this theme, because my previous presentation of the subject does not satisfy me, and for a stronger reason, because I now see that one of the most important consequences of my former treatment is capable of being tested experimentally. For it follows from the theory here to be brought forward, that rays of light, passing close to the sun, are deflected by its gravitational field, so that the angular distance between the sun and a fixed star appearing near to it is apparently increased by nearly a second of arc.

In the course of these reflexions further results are yielded which relate to gravitation. But as the exposition of the entire group of considerations would be rather difficult to follow, only a few quite elementary reflexions will be given in the following pages, from which the reader will readily be able to inform himself as to the suppositions of the theory and its line of thought. The relations here deduced, even if the theoretical foundation is sound, are valid only to a first approximation.

## § 1. A Hypothesis as to the Physical Nature of the Gravitational Field

In a homogeneous gravitational field (acceleration of gravity $\gamma$) let there be a stationary system of co-ordinates K, orientated so that

---

*A. Einstein, Jahrbuch für Radioakt. und Elektronik, 4, 1907.

the lines of force of the gravitational field run in the negative direction of the axis of $z$. In a space free of gravitational fields let there be a second system of coordinates K', moving with uniform acceleration $(\gamma)$ in the positive direction of its axis of $z$. To avoid unnecessary complications, let us for the present disregard the theory of relativity, and regard both systems from the customary point of view of kinematics, and the movements occurring in them from that of ordinary mechanics.

Relatively to K, as well as relatively to K', material points which are not subjected to the action of other material points, move in keeping with the equations

$$\frac{d^2x}{dt^2} = 0, \, \frac{d^2y}{dt^2} = 0, \, \frac{d^2z}{dt^2} = -\gamma.$$

For the accelerated system K' this follows directly from Galileo's principle, but for the system K, at rest in a homogeneous gravitational field, from the experience that all bodies in such a field are equally and uniformly accelerated. This experience, of the equal falling of all bodies in the gravitational field, is one of the most universal which the observation of nature has yielded; but in spite of that the law has not found any place in the foundations of our edifice of the physical universe.

But we arrive at a very satisfactory interpretation of this law of experience, if we assume that the systems K and K' are physically exactly equivalent, that is, if we assume that we may just as well regard the system K as being in a space free from gravitational fields, if we then regard K as uniformly accelerated. This assumption of exact physical equivalence makes it impossible for us to speak of the absolute acceleration of the system of reference, just as the usual theory of relativity forbids us to talk of the absolute velocity of a system;* and it makes the equal falling of all bodies in a gravitational field seem a matter of course.

As long as we restrict ourselves to purely mechanical processes in the realm where Newton's mechanics holds sway, we are certain of the equivalence of the systems K and K'.

*Of course we cannot replace any arbitrary gravitational field by a state of motion of the system without a gravitational field, any more than, by a transformation of relativity, we can transform all points of a medium in any kind of motion to rest.

But this view of ours will not have any deeper significance unless the systems K and K′ are equivalent with respect to all physical processes, that is, unless the laws of nature with respect to K are in entire agreement with those with respect to K′. By assuming this to be so, we arrive at a principle which, if it is really true, has great heuristic importance. For by theoretical consideration of processes which take place relatively to a system of reference with uniform acceleration, we obtain information as to the career of processes in a homogeneous gravitational field. We shall now show, first of all, from the standpoint of the ordinary theory of relativity, what degree of probability is inherent in our hypothesis.

## § 2. On the Gravitation of Energy

One result yielded by the theory of relativity is that the inertia mass of a body increases with the energy it contains; if the increase of energy amounts to E, the increase in inertia mass is equal to $E/c^2$, when $c$ denotes the velocity of light. Now is there an increase of gravitating mass corresponding to this increase of inertia mass? If not, then a body would fall in the same gravitational field with varying acceleration according to the energy it contained. That highly satisfactory result of the theory of relativity by which the law of the conservation of mass is merged in the law of conservation of energy could not be maintained, because it would compel us to abandon the law of the conservation of mass in its old form for inertia mass, and maintain it for gravitating mass.

But this must be regarded as very improbable. On the other hand, the usual theory of relativity does not provide us with any argument from which to infer that the weight of a body depends on the energy contained in it. But we shall show that our hypothesis of the equivalence of the systems K and K′ gives us gravitation of energy as a necessary consequence.

Let the two material systems $S_1$ and $S_2$, provided with instruments, of measurement, be situated on the $z$-axis of K at the distance $h$ from each other,* so that the gravitation potential in $S_2$ is

*The dimensions of $S_1$ and $S_2$ are regarded as infinitely small in comparison with $h$.

greater than that in $S_1$ by $\gamma h$. Let a definite quantity of energy E be emitted from $S_2$ towards $S_1$. Let the quantities of energy in $S_1$ and $S_2$ be measured by contrivances which—brought to one place in the system $z$ and there compared—shall be perfectly alike. As to the process of this conveyance of energy by radiation we can make no *a priori* assertion, because we do not know the influence of the gravitational field on the radiation and the measuring instruments in $S_1$ and $S_2$.

But by our postulate of the equivalence of K and K' we are able, in place of the system K in a homogeneous gravitational field, to set the gravitation-free system K', which moves with uniform acceleration in the direction of positive $z$, and with the $z$-axis of which the material systems $S_1$ and $S_2$ are rigidly connected.

We judge of the process of the transference of energy by radiation from $S_2$ to $S_1$ from a system $K_0$, which is to be free from acceleration. At the moment when the radiation energy $E_2$ is emitted from $S_2$ toward $S_1$, let the velocity of K' relatively to $K_0$ be zero. The radiation will arrive at $S_1$ when the time $h/c$ has elapsed (to a first approximation). But at this moment the velocity of $S_1$ relatively to $K_0$ is $\gamma h/c = v$. Therefore by the ordinary theory of relativity the radiation arriving at $S_1$ does not possess the energy $E_2$, but a greater energy $E_1$, which is related to $E_2$ to a first approximation by the equation*

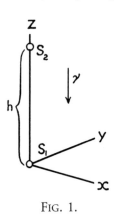

FIG. 1.

$$E_1 = E_2\left(1 + \frac{v}{c}\right) = E_2\left(1 + \gamma\frac{h}{c^2}\right) \qquad . \qquad . \ (1)$$

By our assumption exactly the same relation holds if the same process takes place in the system K, which is not accelerated, but is provided with a gravitational field. In this case we may replace $\gamma h$ by the potential $\Phi$ of the gravitation vector in $S_2$, if the

*See above, pp. 32–34.

arbitrary constant of $\Phi$ in $S_1$ is equated to zero. We then have the equation

$$E_1 = E_2 + \frac{E_2}{c^2} \Phi \qquad . \qquad . \qquad . \qquad (1a)$$

This equation expresses the law of energy for the process under observation. The energy $E_1$ arriving at $S_1$ is greater than the energy $E_2$, measured by the same means, which was emitted in $S_2$, the excess being the potential energy of the mass $E_2/c^2$ in the gravitational field. It thus proves that for the fulfilment of the principle of energy we have to ascribe to the energy E, before its emission in $S_2$, a potential energy due to gravity, which corresponds to the gravitational mass $E/c^2$. Our assumption of the equivalence of K and K' thus removes the difficulty mentioned at the beginning of this paragraph which is left unsolved by the ordinary theory of relativity.

The meaning of this result is shown particularly clearly if we consider the following cycle of operations:—

1. The energy E, as measured in $S_2$, is emitted in the form of radiation in $S_2$ towards $S_1$, where, by the result just obtained, the energy $E(1 + \gamma h/c^2)$, as measured in $S_1$, is absorbed.

2. A body W of mass M is lowered from $S_2$ to $S_1$, work $M\gamma h$ being done in the process.

3. The energy E is transferred from $S_1$ to the body W while W is in $S_1$. Let the gravitational mass M be thereby changed so that it acquires the value M'.

4. Let W be again raised to $S_2$, work $M'\gamma h$ being done in the process.

5. Let E be transferred from W back to $S_2$.

The effect of this cycle is simply that $S_1$ has undergone the increase of energy $E\gamma h/c^2$, and that the quantity of energy $M'\gamma h - M\gamma h$ has been conveyed to the system in the form of mechanical work. By the principle of energy, we must therefore have

$$E\gamma \frac{h}{c^2} = M'\gamma h - M\gamma h,$$

or

$$M' - M = E/c^2. \qquad . \qquad . \qquad . \qquad . \qquad (1b)$$

The increase in gravitational mass is thus equal to $E/c^2$, and therefore equal to the increase in inertia mass as given by the theory of relativity.

The result emerges still more directly from the equivalence of the systems K and K', according to which the gravitational mass in respect of K is exactly equal to the inertia mass in respect of K'; energy must therefore possess a gravitational mass which is equal to its inertia mass. If a mass $M_0$ be suspended on a spring balance in the system K', the balance will indicate the apparent weight $M_0\gamma$ on account of the inertia of $M_0$. If the quantity of energy E be transferred to $M_0$, the spring balance, by the law of the inertia of energy, will indicate $(M_0 + E/c^2)\gamma$. By reason of our fundamental assumption exactly the same thing must occur when the experiment is repeated in the system K, that is, in the gravitational field.

## § 3. TIME AND THE VELOCITY OF LIGHT IN THE GRAVITATIONAL FIELD

If the radiation emitted in the uniformly accelerated system K' in $S_2$ toward $S_1$ had the frequency $\nu_2$ relatively to the clock in $S_2$, then, relatively to $S_1$, at its arrival in $S_1$ it no longer has the frequency $\nu_2$ relatively to an identical clock in $S_1$, but a greater frequency $\nu_1$, such that to a first approximation

$$\nu_1 = \nu_2\left(1 + \gamma\frac{h}{c^2}\right) \qquad . \qquad . \qquad . \qquad . \qquad (2)$$

For if we again introduce the unaccelerated system of reference $K_0$, relatively to which, at the time of the emission of light, K' has no velocity, then $S_1$, at the time of arrival of the radiation at $S_1$, has, relatively to $K_0$, the velocity $\gamma h/c$, from which, by Doppler's principle, the relation as given results immediately.

In agreement with our assumption of the equivalence of the systems K' and K, this equation also holds for the stationary system of co-ordinates K, provided with a uniform gravitational field, if in it the transference by radiation takes place as described. It follows, then, that a ray of light emitted in $S_2$ with a definite gravitational potential, and possessing at its emission the frequency $v_2$—compared with a clock in $S_2$—will, at its arrival in $S_1$, possess a different frequency $v_1$—measured by an identical clock in $S_1$. For $\gamma h$ we substitute the gravitational potential $\Phi$ of $S_2$—that of $S_1$ being taken as zero—and assume that the relation which we have deduced for the homogeneous gravitational field also holds for other forms of field. Then

$$v_1 = v_2\left(1 + \frac{\Phi}{c^2}\right) \quad . \quad . \quad . \quad . \quad (2a)$$

This result (which by our deduction is valid to a first approximation) permits, in the first place, of the following application. Let $v_0$ be the vibration-number of an elementary light-generator, measured by a delicate clock at the same place. Let us imagine them both at a place on the surface of the Sun (where our $S_2$ is located). Of the light there emitted, a portion reaches the Earth ($S_1$), where we measure the frequency of the arriving light with a clock U in all respects resembling the one just mentioned. Then by (2a),

$$v = v_0\left(1 + \frac{\Phi}{c^2}\right),$$

where $\Phi$ is the (negative) difference of gravitational potential between the surface of the Sun and the Earth. Thus according to our view the spectral lines of sunlight, as compared with the corresponding spectral lines of terrestrial sources of light, must be somewhat displaced toward the red, in fact by the relative amount

$$\frac{v_0 - v}{v_0} = -\frac{\Phi}{c^2} = 2.10^{-6}$$

If the conditions under which the solar bands arise were exactly known, this shifting would be susceptible of measurement. But as other influences (pressure, temperature) affect the position of the centres of the spectral lines, it is difficult to discover whether the inferred influence of the gravitational potential really exists.*

*L. F. Jewell (Journ. de Phys., 6, 1897, p. 84) and particularly Ch. Fabry and H. Boisson (Comptes rendus, 148, 1909, pp. 688–690) have actually found such displacements of fine spectral lines toward the red end of the spectrum, of the order of magnitude here calculated, but have ascribed them to an effect of pressure in the absorbing layer.

On a superficial consideration equation (2), or (2a), respectively, seems to assert an absurdity. If there is constant transmission of light from $S_2$ to $S_1$, how can any other number of periods per second arrive in $S_1$ than is emitted in $S_2$? But the answer is simple. We cannot regard $\nu_2$ or respectively $\nu_1$ simply as frequencies (as the number of periods per second) since we have not yet determined the time in system K. What $\nu_2$ denotes is the number of periods with reference to the time-unit of the clock U in $S_2$, while $\nu_1$ denotes the number of periods per second with reference to the identical clock in $S_1$. Nothing compels us to assume that the clocks U in different gravitation potentials must be regarded as going at the same rate. On the contrary, we must certainly define the time in K in such a way that the number of wave crests and troughs between $S_2$ and $S_1$ is independent of the absolute value of time; for the process under observation is by nature a stationary one. If we did not satisfy this condition, we should arrive at a definition of time by the application of which time would merge explicitly into the laws of nature, and this would certainly be unnatural and unpractical. Therefore the two clocks in $S_1$ and $S_2$ do not both give the "time" correctly. If we measure time in $S_1$ with the lock U, then we must measure time in $S_2$ with a clock which goes $1 + \Phi/c^2$ times more slowly than the clock U when compared with U at one and the same place. For when measured by such a clock the frequency of the ray of light which is considered above is at its emission in $S_2$

$$\nu_2\left(1 + \frac{\Phi}{c^2}\right)$$

and is therefore, by (2a), equal to the frequency $\nu_1$ of the same ray of light on its arrival in $S_1$.

This has a consequence which is of fundamental importance for our theory. For if we measure the velocity of light at different places in the accelerated, gravitation-free system K', employing clocks U of identical constitution, we obtain the same magnitude at all these places. The same holds good, by our fundamental assumption, for the system K as well. But from what has just been said we must use

clocks of unlike constitution, for measuring time at places with differing gravitation potential. For measuring time at a place which,
relatively to the origin of the co-ordinates, has the gravitation potential $\Phi$, we must employ a clock which—when removed to the origin of co-ordinates—goes $(1 + \Phi/c^2)$ times more slowly than the
clock used for measuring time at the origin of co-ordinates. If we call
the velocity of light at the origin of co-ordinates $c_0$, then the velocity of light $c$ at a place with the gravitation potential $\Phi$ will be given
by the relation

$$c = c_0\left(1 + \frac{\Phi}{c^2}\right) \qquad . \qquad . \qquad . \qquad . \quad (3)$$

The principle of the constancy of the velocity of light holds good
according to this theory in a different form from that which usually
underlies the ordinary theory of relativity.

## § 4. BENDING OF LIGHT-RAYS IN THE GRAVITATIONAL FIELD

From the proposition which has just been proved, that the velocity of
light in the gravitational field is a function of the place, we may easily infer, by means of Huyghens's principle, that light-rays propagated
across a gravitational field undergo deflexion. For let E be a wave front
of a plane light-wave at the time $t$, and let $P_1$ and $P_2$ be two points
in that plane at unit distance from each other. $P_1$ and $P_2$ lie in the

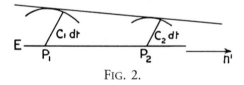

FIG. 2.

plane of the paper, which is chosen so that the differential coefficient
of $\Phi$, taken in the direction of the normal to the plane, vanishes, and
therefore also that of $c$. We obtain the corresponding wave front at
time $t + dt$, or, rather, its line of section with the plane of the paper,
by describing circles round the points $P_1$ and $P_2$ with radii $c_1 dt$ and
$c_2 dt$ respectively, where $c_1$ and $c_2$ denote the velocity of light at the
points $P_1$ and $P_2$ respectively, and by drawing the tangent to these

circles. The angle through which the light-ray is deflected in the path $cdt$ is therefore

$$(c_1 - c_2)\,dt = -\frac{\partial c}{\partial n'}\,dt,$$

if we calculate the angle positively when the ray is bent toward the side of increasing $n'$. The angle of deflexion per unit of path of the light-ray is thus

$$-\frac{1}{c}\frac{\partial c}{\partial n'}, \text{ or by (3)} \quad -\frac{1}{c^2}\frac{\partial \Phi}{\partial n'}.$$

Finally, we obtain for the deflexion which a light-ray experiences toward the side $n'$ on any path $(s)$ the expression

$$a = -\frac{1}{c^2}\int \frac{\partial \Phi}{\partial n'}\,ds \qquad . \qquad . \qquad . \qquad . \quad (4)$$

We might have obtained the same result by directly considering the propagation of a ray of light in the uniformly accelerated system $K'$, and transferring the result to the system K, and thence to the case of a gravitational field of any form.

By equation (4) a ray of light passing along by a heavenly body suffers a deflexion to the side of the diminishing gravitational potential, that is, on the side directed toward the heavenly body, of the magnitude

$$a = \frac{1}{c^2}\int_{\theta=-\frac{1}{2}\pi}^{\theta=\frac{1}{2}\pi} \frac{kM}{r^2}\cos\theta\,ds = 2\frac{kM}{c^2\Delta}$$

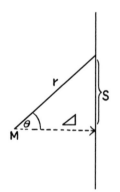

where $k$ denotes the constant of gravitation, M the mass of the heavenly body, $\Delta$ the distance of the ray from the centre of the body. A ray of light going past the Sun would accordingly undergo deflexion to the amount of $4.10^{-6} = .83$ seconds of arc. The angular distance of the star from the centre of the Sun appears to be increased by this amount. As the fixed stars in the parts of the sky near the Sun are visible during total eclipses of the Sun, this consequence of

FIG. 3.

the theory may be compared with experience. With the planet Jupiter the displacement to be expected reaches to about $\frac{1}{100}$ of the amount given. It would be a most desirable thing if astronomers would take up the question here raised. For apart from any theory there is the question whether it is possible with the equipment at present available to detect an influence of gravitational fields on the propagation of light.

# THE FOUNDATION OF THE GENERAL THEORY OF RELATIVITY

BY

A. EINSTEIN

*Translated from "Die Grundlage der allgemeinen Relativitätstheorie,"*
*Annalen der Physik, 49, 1916.*

## A. FUNDAMENTAL CONSIDERATIONS ON THE POSTULATE OF RELATIVITY

§ 1. OBSERVATIONS ON THE SPECIAL THEORY OF RELATIVITY

The special theory of relativity is based on the following postulate, which is also satisfied by the mechanics of Galileo and Newton.

If a system of co-ordinates K is chosen so that, in relation to it, physical laws hold good in their simplest form, the *same* laws also hold good in relation to any other system of co-ordinates K' moving in uniform translation relatively to K. This postulate we call the "special principle of relativity." The word "special" is meant to intimate that the principle is restricted to the case when K' has a motion of uniform translation relatively to K, but that the equivalence of K' and K does not extend to the case of non-uniform motion of K' relatively to K.

Thus the special theory of relativity does not depart from classical mechanics through the postulate of relativity, but through the postulate of the constancy of the velocity of light *in vacuo*, from which, in combination with the special principle of relativity, there follow, in the well-known way, the relativity of simultaneity, the Lorentzian transformation, and the related laws for the behaviour of moving bodies and clocks.

The modification to which the special theory of relativity has subjected the theory of space and time is indeed far-reaching, but one important point has remained unaffected.

For the laws of geometry, even according to the special theory of relativity, are to be interpreted directly as laws relating to the possible relative positions of solid bodies at rest; and, in a more general way, the laws of kinematics are to be interpreted as laws which describe the relations of measuring bodies and clocks. To two selected material points of a stationary rigid body there always corresponds a distance of quite definite length, which is independent of the locality and orientation of the body, and is also independent of the time. To two selected positions of the hands of a clock at rest relatively to the privileged system of reference there always corresponds an interval of time of a definite length, which is independent of place and time. We shall soon see that the general theory of relativity cannot adhere to this simple physical interpretation of space and time.

## § 2. The Need for an Extension of the Postulate of Relativity

In classical mechanics, and no less in the special theory of relativity, there is an inherent epistemological defect which was, perhaps for the first time, clearly pointed out by Ernst Mach. We will elucidate it by the following example:—Two fluid bodies of the same size and nature hover freely in space at so great a distance from each other and from all other masses that only those gravitational forces need be taken into account which arise from the interaction of different parts of the same body. Let the distance between the two bodies be invariable, and in neither of the bodies let there be any relative movements of the parts with respect to one another. But let either mass, as judged by an observer at rest relatively to the other mass, rotate with constant angular velocity about the line joining the masses. This is a verifiable relative motion of the two bodies. Now let us imagine that each of the bodies has been surveyed by means of measuring instruments at rest relatively to itself, and let the surface of $S_1$ prove to be a sphere, and that of $S_2$ an ellipsoid of revolution. Thereupon we put the question— What is the reason for this difference in the two bodies? No answer can be admitted as epistemologically satisfactory,* unless the reason given is an *observable fact of experience.* The law of causality has not the significance of a statement as to the world of experience, except when *observable facts* ultimately appear as causes and effects.

Newtonian mechanics does not give a satisfactory answer to this question. It pronounces as follows:—The laws of mechanics apply to the space $R_1$, in respect to which the body $S_1$ is at rest, but not to the space $R_2$, in respect to which the body $S_2$ is at rest. But the privileged space $R_1$ of Galileo, thus introduced, is a merely *factitious* cause, and not a thing that can be observed. It is therefore clear that Newton's mechanics does not really satisfy the requirement of causality in the case under consideration, but only apparently does so, since it makes the factitious cause $R_1$ responsible for the observable difference in the bodies $S_1$ and $S_2$.

The only satisfactory answer must be that the physical system consisting of $S_1$ and $S_2$ reveals within itself no imaginable cause to which the differing behaviour of $S_1$ and $S_2$ can be referred. The cause must therefore lie *outside* this system. We have to take it that the general laws of motion, which in particular determine the shapes of $S_1$ and $S_2$, must be such that the mechanical behaviour of $S_1$ and $S_2$ is partly conditioned, in quite essential respects, by distant masses which we have not included in the system under consideration. These distant masses and their motions relative to $S_1$ and $S_2$ must then be regarded as the seat of the causes (which must be susceptible to observation) of the different behaviour of our two bodies $S_1$ and $S_2$. They take over the rôle of the factitious cause $R_1$. Of all imaginable spaces $R_1$, $R_2$, etc., in any kind of motion relatively to one another, there is none which we may look upon as privileged *a priori* without reviving the above-mentioned epistemological objection. *The laws of physics must be of such a nature that they apply to systems of reference in any hind of motion.* Along this road we arrive at an extension of the postulate of relativity.

In addition to this weighty argument from the theory of knowledge, there is a well-known physical fact which favours an extension

*Of course an answer may be satisfactory from the point of view of epistemology, and yet be unsound physically, if it is in conflict with other experiences.

of the theory of relativity. Let K be a Galilean system of reference, i.e. a system relatively to which (at least in the four-dimensional region under consideration) a mass, sufficiently distant from other masses, is moving with uniform motion in a straight line. Let K' be a second system of reference which is moving relatively to K in *uniformly accelerated* translation. Then, relatively to K', a mass sufficiently distant from other masses would have an accelerated motion such that its acceleration and direction of acceleration are independent of the material composition and physical state of the mass.

Does this permit an observer at rest relatively to K' to infer that he is on a "really" accelerated system of reference? The answer is in the negative; for the above-mentioned relation of freely movable masses to K' may be interpreted equally well in the following way. The system of reference K' is unaccelerated, but the space-time territory in question is under the sway of a gravitational field, which generates the accelerated motion of the bodies relatively to K'.

This view is made possible for us by the teaching of experience as to the existence of a field of force, namely, the gravitational field, which possesses the remarkable property of imparting the same acceleration to all bodies.* The mechanical behaviour of bodies relatively to K' is the same as presents itself to experience in the case of systems which we are wont to regard as "stationary" or as "privileged." Therefore, from the physical standpoint, the assumption readily suggests itself that the systems K and K' may both with equal right be looked upon as "stationary," that is to say, they have an equal title as systems of reference for the physical description of phenomena.

It will be seen from these reflexions that in pursuing the general theory of relativity we shall be led to a theory of gravitation, since we are able to "produce" a gravitational field merely by changing the system of co-ordinates. It will also be obvious that the principle of the constancy of the velocity of light *in vacuo* must be modified, since we easily recognize that the path of a ray of light with respect to K' must

---

*Eötvös has proved experimentally that the gravitational field has this property in great accuracy.

in general be curvilinear, if with respect to K light is propagated in a straight line with a definite constant velocity.

## § 3. The Space-Time Continuum. Requirement of General Co-Variance for the Equations Expressing General Laws of Nature

In classical mechanics, as well as in the special theory of relativity, the co-ordinates of space and time have a direct physical meaning. To say that a point-event has the $X_1$ coordinate $x_1$ means that the projection of the point-event on the axis of $X_1$, determined by rigid rods and in accordance with the rules of Euclidean geometry, is obtained by measuring off a given rod (the unit of length) $x_1$ times from the origin of coordinates along the axis of $X_1$. To say that a point-event has the $X_4$ co-ordinate $x_4 = t$, means that a standard clock, made to measure time in a definite unit period, and which is stationary relatively to the system of co-ordinates and practically coincident in space with the point-event,* will have measured off $x_4 = t$ periods at the occurrence of the event.

This view of space and time has always been in the minds of physicists, even if, as a rule, they have been unconscious of it. This is clear from the part which these concepts play in physical measurements; it must also have underlain the reader's reflexions on the preceding paragraph (§ 2) for him to connect any meaning with what he there read. But we shall now show that we must put it aside and replace it by a more general view, in order to be able to carry through the postulate of general relativity, if the special theory of relativity applies to the special case of the absence of a gravitational field.

In a space which is free of gravitational fields we introduce a Galilean system of reference K ($x$, $y$, $z$, $t$), and also a system of co-ordinates K' ($x'$, $y'$, $z'$, $t'$) in uniform rotation relatively to K. Let the origins of both systems, as well as their axes of Z, permanently coincide. We shall show that for a space-time measurement in the system K' the above definition of the physical meaning of lengths and times

---

*We assume the possibility of verifying "simultaneity" for events immediately proximate in space, or—to speak more precisely—for immediate proximity or coincidence in space-time, without giving a definition of this fundamental concept.

cannot be maintained. For reasons of symmetry it is clear that a circle around the origin in the X, Y plane of K may at the same time be regarded as a circle in the X′, Y′ plane of K′. We suppose that the circumference and diameter of this circle have been measured with a unit measure infinitely small compared with the radius, and that we have the quotient of the two results. If this experiment were performed with a measuring-rod at rest relatively to the Galilean system K, the quotient would be $\pi$. With a measuring-rod at rest relatively to K′, the quotient would be greater than $\pi$. This is readily understood if we envisage the whole process of measuring from the "stationary" system K, and take into consideration that the measuring-rod applied to the periphery undergoes a Lorentzian contraction, while the one applied along the radius does not. Hence Euclidean geometry does not apply to K′. The notion of co-ordinates defined above, which pre-supposes the validity of Euclidean geometry, therefore breaks down in relation to the System K′. So, too, we are unable to introduce a time corresponding to physical requirements in K′, indicated by clocks at rest relatively to K′. To convince ourselves of this impossibility, let us imagine two clocks of identical constitution placed, one at the origin of co-ordinates, and the other at the circumference of the circle, and both envisaged from the "stationary" system K. By a familiar result of the special theory of relativity, the clock at the circumference—judged from K—goes more slowly than the other, because the former is in motion and the latter at rest. An observer at the common origin of co-ordinates, capable of observing the clock at the circumference by means of light, would therefore see it lagging behind the clock beside him. As he will not make up his mind to let the velocity of light along the path in question depend explicitly on the time, he will interpret his observations as showing that the clock at the circumference "really" goes more slowly than the clock at the origin. So he will be obliged to define time in such a way that the rate of a clock depends upon where the clock may be.

We therefore reach this result:—In the general theory of relativity, space and time cannot be defined in such a way that differences of the spatial co-ordinates can be directly measured by the unit measuring-rod, or differences in the time co-ordinate by a standard clock.

The method hitherto employed for laying co-ordinates into the space-time continuum in a definite manner thus breaks down, and there seems to be no other way which would allow us to adapt systems of co-ordinates to the four-dimensional universe so that we might expect from their application a particularly simple formulation of the laws of nature. So there is nothing for it but to regard all imaginable systems of co-ordinates, on principle, as equally suitable for the description of nature. This comes to requiring that:—

*The general laws of nature are to be expressed by equations which hold good for all systems of co-ordinates, that is, are co-variant with respect to any substitutions whatever (generally co-variant).*

It is clear that a physical theory which satisfies this postulate will also be suitable for the general postulate of relativity. For the sum of *all* substitutions in any case includes those which correspond to all relative motions of three-dimensional systems of co-ordinates. That this requirement of general co-variance, which takes away from space and time the last remnant of physical objectivity, is a natural one, will be seen from the following reflexion. All our space-time verifications invariably amount to a determination of space-time coincidences. If, for example, events consisted merely in the motion of material points, then ultimately nothing would be observable but the meetings of two or more of these points. Moreover, the results of our measurings are nothing but verifications of such meetings of the material points of our measuring instruments with other material points, coincidences between the hands of a clock and points on the clock dial, and observed point-events happening at the same place at the same time.

The introduction of a system of reference serves no other purpose than to facilitate the description of the totality of such coincidences. We allot to the universe four space-time variables $x_1$, $x_2$, $x_3$, $x_4$ in such a way that for every point-event there is a corresponding system of values of the variables $x_1 \ldots x_4$. To two coincident point-events there corresponds one system of values of the variables $x_1 \ldots x_4$, i.e. coincidence is characterized by the identity of the co-ordinates. If, in place of the variables $x_1 \ldots x_4$, we introduce functions of them, $x_1'$, $x_2'$, $x_3'$, $x_4'$, as a new system of co-ordinates, so that the systems of values are made to correspond to one another without ambiguity, the equality of all four co-ordinates in the new system will also serve as an expression for the space-time coincidence of the two point-events. As all our physical experience can be ultimately reduced to such coincidences, there is no immediate reason for preferring certain systems of co-ordinates to others, that is to say, we arrive at the requirement of general co-variance.

## § 4. THE RELATION OF THE FOUR CO-ORDINATES TO MEASUREMENT IN SPACE AND TIME

It is not my purpose in this discussion to represent the general theory of relativity as a system that is as simple and logical as possible, and with the minimum number of axioms; but my main object is to develop this theory in such a way that the reader will feel that the path we have entered upon is psychologically the natural one, and that the underlying assumptions will seem to have the highest possible degree of security. With this aim in view let it now be granted that:—

For infinitely small four-dimensional regions the theory of relativity in the restricted sense is appropriate, if the coordinates are suitably chosen.

For this purpose we must choose the acceleration of the infinitely small ("local") system of co-ordinates so that no gravitational field occurs; this is possible for an infinitely small region. Let $X_1$, $X_2$, $X_3$, be the co-ordinates of space, and $X_4$ the appertaining co-ordinate of time measured in the appropriate unit.* If a rigid rod is imagined to be given as the unit measure, the co-ordinates, with a given orientation of the system of co-ordinates, have a direct physical meaning in the sense of the special theory of relativity. By the special theory of relativity the expression

$$ds^2 = -dX_1^2 - dX_2^2 - dX_3^2 + dX_4^2 \quad \cdot \quad \cdot \quad \cdot \quad \cdot \ (1)$$

*The unit of time is to be chosen so that the velocity of light *in vacuo* as measured in the "local" system of co-ordinates is to be equal to unity.

then has a value which is independent of the orientation of the local system of co-ordinates, and is ascertainable by measurements of space and time. The magnitude of the linear element pertaining to points of the four-dimensional continuum in infinite proximity, we call $ds$. If the $ds$ belonging to the element $dX_1 \ldots dX_4$ is positive, we follow Minkowski in calling it time-like; if it is negative, we call it space-like.

To the "linear element" in question, or to the two infinitely proximate point-events, there will also correspond definite differentials $dx_1 \ldots dx_4$ of the four-dimensional co-ordinates of any chosen system of reference. If this system, as well as the "local" system, is given for the region under consideration, the $dX_\nu$ will allow themselves to be represented here by definite linear homogeneous expressions of the $dx_\sigma$:—

$$dX_\nu = \sum_\sigma \alpha_{\nu\sigma} \, dx_\sigma \, . \qquad . \qquad . \qquad . \qquad . \quad (2)$$

Inserting these expressions in (1), we obtain

$$ds^2 = \sum_{\tau\sigma} g_{\sigma\tau} \, dx_\sigma \, dx_\tau, \, . \qquad . \qquad . \qquad . \qquad . \quad (3)$$

where the $g_{\sigma\tau}$ will be functions of the $x_\sigma$. These can no longer be dependent on the orientation and the state of motion of the "local" system of co-ordinates, for $ds^2$ is a quantity ascertainable by rod-clock measurement of point-events infinitely proximate in space-time, and defined independently of any particular choice of co-ordinates. The $g_{\sigma\tau}$ are to be chosen here so that $g_{\sigma\tau} = g_{\tau\sigma}$; the summation is to extend over all values of $\sigma$ and $\tau$, so that the sum consists of $4 \times 4$ terms, of which twelve are equal in pairs.

The case of the ordinary theory of relativity arises out of the case here considered, if it is possible, by reason of the particular relations of the $g_{\sigma\tau}$ in a finite region, to choose the system of reference in the finite region in such a way that the $g_{\sigma\tau}$ assume the constant values

$$\left. \begin{array}{cccc} -1 & 0 & 0 & 0 \\ 0 & -1 & 0 & 0 \\ 0 & 0 & -1 & 0 \\ 0 & 0 & 0 & +1 \end{array} \right\} \quad . \qquad . \qquad . \quad (4)$$

We shall find hereafter that the choice of such co-ordinates is, in general, not possible for a finite region.

From the considerations of § 2 and § 3 it follows that the quantities $g_{\tau\sigma}$ are to be regarded from the physical standpoint as the quantities which describe the gravitational field in relation to the chosen system of reference. For, if we now assume the special theory of relativity to apply to a certain four-dimensional region with the co-ordinates properly chosen, then the $g_{\sigma\tau}$ have the values given in (4). A free material point then moves, relatively to this system, with uniform motion in a straight line. Then if we introduce new space-time co-ordinates $x_1$, $x_2$, $x_3$, $x_4$, by means of any substitution we choose, the $g^{\sigma\tau}$ in this new system will no longer be constants, but functions of space and time. At the same time the motion of the free material point will present itself in the new co-ordinates as a curvilinear non-uniform motion, and the law of this motion will be independent of the nature of the moving particle. We shall therefore interpret this motion as a motion under the influence of a gravitational field. We thus find the occurrence of a gravitational field connected with a space-time variability of the $g_{\sigma}$. So, too, in the general case, when we are no longer able by a suitable choice of co-ordinates to apply the special theory of relativity to a finite region, we shall hold fast to the view that the $g_{\sigma\tau}$ describe the gravitational field.

Thus, according to the general theory of relativity, gravitation occupies an exceptional position with regard to other forces, particularly the electromagnetic forces, since the ten functions representing the gravitational field at the same time define the metrical properties of the space measured.

# B. MATHEMATICAL AIDS TO THE FORMULATION OF GENERALLY COVARIANT EQUATIONS

Having seen in the foregoing that the general postulate of relativity leads to the requirement that the equations of physics shall be covariant in the face of any substitution of the co-ordinates $x_1 \ldots x_4$, we have to consider how such generally covariant equations can be found. We now turn to this purely mathematical task, and we shall find that in its solution a fundamental rôle is played by the invariant *ds* given in equation (3), which, borrowing from Gauss's theory of surfaces, we have called the "linear element."

The fundamental idea of this general theory of covariants is the following:—Let certain things ("tensors") be defined with respect to any system of co-ordinates by a number of functions of the co-ordinates, called the "components" of the tensor. There are then certain rules by which these components can be calculated for a new system of co-ordinates, if they are known for the original system of co-ordinates, and if the transformation connecting the two systems is known. The things hereafter called tensors are further characterized by the fact that the equations of transformation for their components are linear and homogeneous. Accordingly, all the components in the new system vanish, if they all vanish in the original system. If, therefore, a law of nature is expressed by equating all the components of a tensor to zero, it is generally covariant. By examining the laws of the formation of tensors, we acquire the means of formulating generally covariant laws.

## § 5. CONTRAVARIANT AND COVARIANT FOUR-VECTORS

*Contravariant Four-vectors.*—The linear element is defined by the four "components" $dx_\nu$, for which the law of transformation is expressed by the equation

$$dx'_\sigma = \sum_\nu \frac{\partial x'_\sigma}{\partial x_\nu} dx_\nu \qquad . \qquad . \qquad . \qquad . \quad (5)$$

The $dx'_\sigma$ are expressed as linear and homogeneous functions of the $dx_\nu$. Hence we may look upon these co-ordinate differentials as the components of a "tensor" of the particular kind which we call a contravariant four-vector. Any thing which is defined relatively to the system of co-ordinates by four quantities $A^\nu$, and which is transformed by the same law

$$A'^\sigma = \sum_\nu \frac{\partial x'_\sigma}{\partial x_\nu} A^\nu, \qquad . \qquad . \qquad . \quad (5a)$$

we also call a contravariant four-vector. From (5a) it follows at once that the sums $A^\sigma \pm B^\sigma$ are also components of a four-vector, if $A^\sigma$ and $B^\sigma$ are such. Corresponding relations hold for all "tensors" subsequently to be introduced. (Rule for the addition and subtraction of tensors.)

*Covariant Four-vectors.*—We call four quantities $A_\nu$ the components of a covariant four-vector, if for any arbitrary choice of the contravariant four-vector $B^\nu$

$$\sum_\nu A_\nu B^\nu = \text{Invariant} \qquad . \qquad . \qquad . \qquad (6)$$

The law of transformation of a covariant four-vector follows from this definition. For if we replace $B^\nu$ on the right-hand side of the equation

$$\sum_\sigma A'_\sigma B'^\sigma = \sum_\nu A_\nu B^\nu$$

by the expression resulting from the inversion of (5a),

$$\sum_\sigma \frac{\partial x_\nu}{\partial x'_\sigma} B'^\sigma,$$

we obtain

$$\sum_\sigma B'^\sigma \sum_\nu \frac{\partial x_\nu}{\partial x'_\sigma} A_\nu = \sum_\sigma B'^\sigma A'_\sigma.$$

Since this equation is true for arbitrary values of the $B'^\sigma$, it follows that the law of transformation is

$$A'_\sigma = \sum_\nu \frac{\partial x_\nu}{\partial x'_\sigma} A_\nu \qquad . \qquad . \qquad . \qquad . \qquad (7)$$

*Note on a Simplified Way of Writing the Expressions.*—A glance at the equations of this paragraph shows that there is always a summation with respect to the indices which occur twice under a sign of summation (e.g. the index $\nu$ in (5)), and only with respect to indices which occur twice. It is therefore possible, without loss of clearness, to omit the sign of summation. In its place we introduce the convention:—If an index occurs twice in one term of an expression, it is always to be summed unless the contrary is expressly stated.

The difference between covariant and contravariant four-vectors lies in the law of transformation ((7) or (5) respectively). Both forms are tensors in the sense of the general remark above. Therein lies their importance. Following Ricci and Levi-Civita, we denote the contravariant character by placing the index above, the covariant by placing it below.

## § 6. TENSORS OF THE SECOND AND HIGHER RANKS

*Contravariant Tensors.*—If we form all the sixteen products $A^{\mu\nu}$ of the components $A^{\mu}$ and $B^{\nu}$ of two contravariant four-vectors

$$A^{\mu\nu} = A^{\mu}B^{\nu} \qquad . \qquad . \qquad . \qquad . \qquad (8)$$

then by (8) and (5a) $A^{\mu\nu}$ satisfies the law of transformation

$$A'^{\sigma\tau} = \frac{\partial x'_{\sigma}}{\partial x_{\mu}} \frac{\partial x'_{\tau}}{\partial x_{\nu}} A^{\mu\nu} \qquad . \qquad . \qquad . \qquad . \qquad (9)$$

We call a thing which is described relatively to any system of reference by sixteen quantities; satisfying the law of transformation (9), a contravariant tensor of the second rank. Not every such tensor allows itself to be formed in accordance with (8) from two four-vectors, but it is easily shown that any given sixteen $A^{\mu\nu}$ can be represented as the sums of the $A^{\mu}B^{\nu}$ of four appropriately selected pairs of four-vectors. Hence we can prove nearly all the laws which apply to the tensor of the second rank defined by (9) in the simplest manner by demonstrating them for the special tensors of the type (8).

*Contravariant Tensors of Any Bank.*—It is clear that, on the lines of (8) and (9), contravariant tensors of the third and higher ranks may also be defined with $4^3$ components, and so on. In the same way it follows from (8) and (9) that the contravariant four-vector may be taken in this sense as a contravariant tensor of the first rank.

*Covariant Tensors.*—On the other hand, if we take the sixteen products $A_{\mu\nu}$ of two covariant four-vectors $A_{\mu}$ and $B_{\nu}$,

$$A_{\mu\nu} = A_{\mu}B_{\nu}, \qquad . \qquad . \qquad . \qquad . \qquad (10)$$

the law of transformation for these is

$$A'_{\sigma\tau} = \frac{\partial x_{\mu}}{\partial x'_{\sigma}} \frac{\partial x_{\nu}}{\partial x'_{\tau}} A_{\mu\nu} \qquad . \qquad . \qquad . \qquad . \qquad (11)$$

This law of transformation defines the covariant tensor of the second rank. All our previous remarks on contravariant tensors apply equally to covariant tensors.

NOTE.—It is convenient to treat the scalar (or invariant) both as a contravariant and a covariant tensor of zero rank.

*Mixed Tensors.*—We may also define a tensor of the second rank of the type

$$A_\mu^\nu = A_\mu B^\nu \qquad . \qquad . \qquad . \qquad . \qquad (12)$$

which is covariant with respect to the index $\mu$, and contravariant with respect to the index $\nu$. Its law of transformation is

$$A_\sigma'^\tau = \frac{\partial x_\tau'}{\partial x_\nu} \frac{\partial x_\mu}{\partial x_\sigma'} A_\mu^\nu \qquad . \qquad . \qquad . \qquad . \qquad (13)$$

Naturally there are mixed tensors with any number of indices of covariant character, and any number of indices of contravariant character. Covariant and contravariant tensors may be looked upon as special cases of mixed tensors.

*Symmetrical Tensors.*—A contravariant, or a covariant tensor, of the second or higher rank is said to be symmetrical if two components, which are obtained the one from the other by the interchange of two indices, are equal. The tensor $A^{\mu\nu}$, or the tensor $A_{\mu\nu}$, is thus symmetrical if for any combination of the indices $\mu$, $\nu$,

$$A^{\mu\nu} = A^{\nu\mu}, \qquad . \qquad . \qquad . \qquad . \qquad (14)$$

or respectively,

$$A_{\mu\nu} = A_{\nu\mu}. \qquad . \qquad . \qquad . \qquad . \qquad (14a)$$

It has to be proved that the symmetry thus defined is a property which is independent of the system of reference. It follows in fact from (9), when (14) is taken into consideration, that

$$A'^{\sigma\tau} = \frac{\partial x_\sigma'}{\partial x_\mu} \frac{\partial x_\tau'}{\partial x_\nu} A^{\mu\nu} = \frac{\partial x_\sigma'}{\partial x_\mu} \frac{\partial x_\tau'}{\partial x_\nu} A^{\nu\mu} = \frac{\partial x_\sigma'}{\partial x_\nu} \frac{\partial x_\tau'}{\partial x_\mu} A^{\mu\nu} = A'^{\tau\sigma}.$$

The last equation but one depends upon the interchange of the summation indices $\mu$ and $\nu$, i.e. merely on a change of notation.

*Antisymmetrical Tensors.*—A contravariant or a covariant tensor of the second, third, or fourth rank is said to be antisymmetrical if two components, which are obtained the one from the other by the inter-change of two indices, are equal and of opposite sign. The tensor $A^{\mu\nu}$, or the tensor $A_{\mu\nu}$, is therefore antisymmetrical, if always

$$A^{\mu\nu} = -A^{\nu\mu}, \quad \ldots \quad \ldots \quad (15)$$

or respectively,

$$A_{\mu\nu} = -A_{\nu\mu} \quad \ldots \quad \ldots \quad (15a)$$

Of the sixteen components $A^{\mu\nu}$, the four components $A^{\mu\mu}$ vanish; the rest are equal and of opposite sign in pairs, so that there are only six components numerically different (a six-vector). Similarly we see that the antisymmetrical tensor of the third rank $A^{\mu\nu\sigma}$ has only four numerically different components, while the antisymmetrical tensor $A^{\mu\nu\sigma\tau}$ has only one. There are no antisymmetrical tensors of higher rank than the fourth in a continuum of four dimensions.

## § 7. MULTIPLICATION OF TENSORS

*Outer Multiplication of Tensors.*—We obtain from the components of a tensor of rank $n$ and of a tensor of rank $m$ the components of a ten-sor of rank $n + m$ by multiplying each component of the one tensor by each component of the other. Thus, for example, the tensors T arise out of the tensors A and B of different kinds,

$$T_{\mu\nu\sigma} = A_{\mu\nu}B_{\sigma},$$
$$T^{\mu\nu\sigma\tau} = A^{\mu\nu}B^{\sigma\tau},$$
$$T_{\mu\nu}^{\sigma\tau} = A_{\mu\nu}B^{\sigma\nu}.$$

The proof of the tensor character of T is given directly by the rep-resentations (8), (10), (12), or by the laws of transformation (9), (11), (13). The equations (8), (10), (12) are themselves examples of outer multiplication of tensors of the first rank.

*"Contraction" of a Mixed Tensor.*—From any mixed tensor we may form a tensor whose rank is less by two, by equating an index of covari-ant with one of contravariant character, and summing with respect to this index ("contraction"). Thus, for example, from the mixed tensor of the fourth rank $A_{\mu\nu}^{\sigma\tau}$, we obtain the mixed tensor of the second rank,

$$A_{\nu}^{\tau} = A_{\mu\nu}^{\mu\tau}\left(=\sum_{\mu}A_{\mu\nu}^{\mu\tau}\right),$$

and from this, by a second contraction, the tensor of zero rank,

$$A = A_\nu^\nu = A_{\mu\nu}^{\mu\nu}.$$

The proof that the result of contraction really possesses the tensor character is given either by the representation of a tensor according to the generalization of (12) in combination with (6), or by the generalization of (13).

*Inner and Mixed Multiplication of Tensors.*—These consist in a combination of outer multiplication with contraction.

*Examples.*—From the covariant tensor of the second rank $A_{\mu\nu}$ and the contravariant tensor of the first rank $B^\sigma$ we form by outer multiplication the mixed tensor

$$D_{\mu\nu}^\sigma = A_{\mu\nu} B^\sigma.$$

On contraction with respect to the indices $\nu$ and $\sigma$, we obtain the covariant four-vector

$$D_\mu = D_{\mu\nu}^\nu = A_{\mu\nu} B^\nu.$$

This we call the inner product of the tensors $A_{\mu\nu}$ and $B^\sigma$. Analogously we form from the tensors $A_{\mu\nu}$ and $B^{\sigma\tau}$, by outer multiplication and double contraction, the inner product $A_{\mu\nu} B^{\mu\nu}$. By outer multiplication and one contraction, we obtain from $A_{\mu\nu}$ and $B^{\sigma\tau}$ the mixed tensor of the second rank $D_\mu^\tau = A_{\mu\nu} B^{\nu\tau}$. This operation may be aptly characterized as a mixed one, being "outer" with respect to the indices $\mu$ and $\tau$, and "inner" with respect to the indices $\nu$ and $\sigma$.

We now prove a proposition which is often useful as evidence of tensor character. From what has just been explained, $A_{\mu\nu} B^{\mu\nu}$ is a scalar if $A_{\mu\nu}$ and $B^{\sigma\tau}$ are tensors. But we may also make the following assertion: If $A_{\mu\nu} B^{\mu\nu}$ is a scalar *for any choice of the tensor* $B^{\mu\nu}$, then $A_{\mu\nu}$ has tensor character. For, by hypothesis, for any substitution,

$$A'_{\sigma\tau} B'^{\sigma\tau} = A_{\mu\nu} B^{\mu\nu}.$$

But by an inversion of (9)

$$B^{\mu\nu} = \frac{\partial x_\mu}{\partial x'_\sigma} \frac{\partial x_\nu}{\partial x'_\tau} B'^{\sigma\tau}.$$

This, inserted in the above equation, gives

$$\left( A'_{\sigma\tau} - \frac{\partial x_\mu}{\partial x'_\sigma} \frac{\partial x_\nu}{\partial x'_\tau} A_{\mu\nu} \right) B'^{\sigma\tau} = 0.$$

This can only be satisfied for arbitrary values of $B'^{\sigma\tau}$ if the bracket vanishes. The result then follows by equation (11). This rule applies correspondingly to tensors of any rank and character, and the proof is analogous in all cases.

The rule may also be demonstrated in this form: If $B^\mu$ and $C^\nu$ are any vectors, and if, for all values of these, the inner product $A_{\mu\nu}B^\mu C^\nu$ is a scalar, then $A_{\mu\nu}$ is a covariant tensor. This latter proposition also holds good even if only the more special assertion is correct, that with any choice of the four-vector $B^\mu$ the inner product $A_{\mu\nu}B^\mu B^\nu$ is a scalar, if in addition it is known that $A_{\mu\nu}$ satisfies the condition of symmetry $A_{\mu\nu} = A_{\nu\mu}$. For by the method given above we prove the tensor character of $(A_{\mu\nu} + A_{\nu\mu})$, and from this the tensor character of $A_{\mu\nu}$ follows on account of symmetry. This also can be easily generalized to the case of covariant and contravariant tensors of any rank.

Finally, there follows from what has been proved, this law, which may also be generalized for any tensors: If for any choice of the four-vector $B^\nu$ the quantities $A_{\mu\nu}B^\nu$ form a tensor of the first rank, then $A_{\mu\nu}$ is a tensor of the second rank. For, if $C^\mu$ is any four-vector, then on account of the tensor character of $A_{\mu\nu}B^\nu$, the inner product $A_{\mu\nu}B^\nu C^\mu$ is a scalar for any choice of the two four-vectors $B^\nu$ and $C^\mu$. From which the proposition follows.

## § 8. SOME ASPECTS OF THE FUNDAMENTAL TENSOR $g_{\mu\nu}$

*The Covariant Fundamental Tensor.*—In the invariant expression for the square of the linear element,

$$ds^2 = g_{\mu\nu}\,dx_\mu\,dx_\nu,$$

the part played by the $dx_\mu$ is that of a contravariant vector which may be chosen at will. Since further, $g_{\mu\nu} = g_{\nu\mu}$, it follows from the considerations of the preceding paragraph that $g_{\mu\nu}$ is a covariant tensor of the second rank. We call it the "fundamental tensor." In what follows we deduce some properties of this tensor which, it is true, apply to any tensor of the second rank. But as the fundamental tensor plays a special part in our theory, which has its physical basis in the peculiar effects of gravitation, it so happens that the relations to be developed are of importance to us only in the case of the fundamental tensor.

*The Contravariant Fundamental Tensor.*—If in the determinant formed by the elements $g_{\mu\nu}$, we take the co-factor of each of the $g_{\mu\nu}$ and divide it by the determinant $g = |g_{\mu\nu}|$, we obtain certain quantities $g^{\mu\nu}(=g^{\nu\mu})$ which, as we shall demonstrate, form a contravariant tensor.

By a known property of determinants

$$g_{\mu\sigma} g^{\nu\sigma} = \delta_\mu^\nu \qquad \cdot \qquad \cdot \qquad \cdot \qquad \cdot \quad (16)$$

where the symbol $\delta_\mu^\nu$ denotes 1 or 0, according as $\mu = \nu$ or $\mu \neq \nu$.

Instead of the above expression for $ds^2$ we may thus write

$$g_{\mu\sigma} \delta_\nu^\sigma dx_\mu dx_\nu$$

or, by (16)

$$g_{\mu\sigma} g_{\nu\tau} g^{\sigma\tau} dx_\mu dx_\nu.$$

But, by the multiplication rules of the preceding paragraphs, the quantities

$$d\xi_\sigma = g_{\mu\sigma} dx_\mu$$

form a covariant four-vector, and in fact an arbitrary vector, since the $dx_\mu$ are arbitrary. By introducing this into our expression we obtain

$$ds^2 = g^{\sigma\tau} d\xi_\sigma d\xi_\tau.$$

Since this, with the arbitrary choice of the vector $d\xi_\sigma$, is a scalar, and $g^{\sigma\tau}$ by its definition is symmetrical in the indices $\sigma$ and $\tau$, it follows from the results of the preceding paragraph that $g^{\sigma\tau}$ is a contravariant tensor.

It further follows from (16) that $\delta_\mu$ is also a tensor, which we may call the mixed fundamental tensor.

*The Determinant of the Fundamental Tensor.*—By the rule for the multiplication of determinants

$$|g_{\mu\alpha} g^{\alpha\nu}| = |g_{\mu\alpha}| \times |g^{\alpha\nu}|.$$

On the other hand

$$|g_{\mu\alpha} g^{\alpha\nu}| = |\delta_\mu^\nu| = 1.$$

It therefore follows that

$$|g_{\mu\nu}| \times |g^{\mu\nu}| = 1 \qquad \cdot \qquad \cdot \qquad \cdot \qquad \cdot \quad (17)$$

*The Volume Scalar.*—We seek first the law of transformation of the determinant $g = |g_{\mu\nu}|$. In accordance with (11)

$$g' = \left| \frac{\partial x_\mu}{\partial x'_\sigma} \frac{\partial x}{\partial x'_\tau} g_{\mu\nu} \right|.$$

Hence, by a double application of the rule for the multiplication of determinants, it follows that

$$g' = \left| \frac{\partial x_\mu}{\partial x'_\sigma} \right| \cdot \left| \frac{\partial x_\nu}{\partial x'_\tau} \right| \cdot |g_{\mu\nu}| = \left| \frac{\partial x_\mu}{\partial x'_\sigma} \right|^2 g,$$

or

$$\sqrt{g'} = \left| \frac{\partial x_\mu}{\partial x'_\sigma} \right| \sqrt{g}.$$

On the other hand, the law of transformation of the element of volume

$$d\tau = \int dx_1 \, dx_2 \, dx_3 \, dx_4$$

is, in accordance with the theorem of Jacobi,

$$d\tau' = \left| \frac{\partial x'_\sigma}{\partial x_\mu} \right| d\tau.$$

By multiplication of the last two equations, we obtain

$$\sqrt{g'} \, d\tau' = \sqrt{g} \, d\tau \qquad . \qquad . \qquad . \qquad (18)$$

Instead of $\sqrt{g}$, we introduce in what follows the quantity $\sqrt{-g}$, which is always real on account of the hyperbolic character of the space-time continuum. The invariant $\sqrt{-g} \, d\tau$ is equal to the magnitude of the four-dimensional element of volume in the "local" system of reference, as measured with rigid rods and clocks in the sense of the special theory of relativity.

*Note on the Character of the Space-time Continuum.*—Our assumption that the special theory of relativity can always be applied to an infinitely small region, implies that $ds^2$ can always be expressed in accordance with (1) by means of real quantities $dX_1 \ldots dX_4$. If we denote by $d\tau_0$ the "natural" element of volume $dX_1, dX_2, dX_3, dX_4$, then

$$d\tau_0 = \sqrt{-g} \, d\tau \qquad . \qquad . \qquad . \qquad (18a)$$

If $\sqrt{-g}$ were to vanish at a point of the four-dimensional continuum, it would mean that at this point an infinitely small "natural"
volume would correspond to a finite volume in the co-ordinates. Let
us assume that this is never the case. Then $g$ cannot change sign. We
will assume that, in the sense of the special theory of relativity, $g$ always
has a finite negative value. This is a hypothesis as to the physical nature
of the continuum under consideration, and at the same time a convention as to the choice of co-ordinates.

But if $-g$ is always finite and positive, it is natural to settle the
choice of co-ordinates *a posteriori* in such a way that this quantity is
always equal to unity. We shall see later that by such a restriction of
the choice of co-ordinates it is possible to achieve an important simplification of the laws of nature.

In place of (18), we then have simply $d\tau' = d\tau$, from which, in
view of Jacobi's theorem, it follows that

$$\left| \frac{\partial x'_\sigma}{\partial x_\mu} \right| = 1 \qquad \cdot \qquad \cdot \qquad \cdot \qquad \cdot \quad (19)$$

Thus, with this choice of co-ordinates, only substitutions for which
the determinant is unity are permissible.

But it would be erroneous to believe that this step indicates a partial abandonment of the general postulate of relativity. We do not ask
"What are the laws of nature which are co-variant in face of all substitutions for which the determinant is unity?" but our question is
"What are the generally co-variant laws of nature?" It is not until we
have formulated these that we simplify their expression by a particular choice of the system of reference.

*The Formation of New Tensors by Means of the Fundamental Tensor.*
—Inner, outer, and mixed multiplication of a tensor by the fundamental tensor give tensors of different character and rank. For
example,

$$A^\mu = g^{\mu\sigma} A_\sigma,$$
$$A = g_{\mu\nu} A^{\mu\nu}.$$

The following forms may be specially noted:—

$$A^{\mu\nu} = g^{\mu\alpha} g^{\nu\beta} A_{\alpha\beta},$$

$$A_{\mu\nu} = g_{\mu\alpha} g_{\nu\beta} A^{\alpha\beta}$$

(the "complements" of covariant and contravariant tensors respectively), and

$$B_{\mu\nu} = g_{\mu\nu} g^{\alpha\beta} A_{\alpha\beta}.$$

We call $B_{\mu\nu}$ the reduced tensor associated with $A_{\mu\nu}$. Similarly,

$$B^{\mu\nu} = g^{\mu\nu} g_{\alpha\beta} A^{\alpha\beta}.$$

It may be noted that $g^{\mu\nu}$ is nothing more than the complement of $g_{\mu\nu}$, since

$$g^{\mu\alpha} g^{\nu\beta} g_{\alpha\beta} = g^{\mu\alpha} \delta^{\nu}_{\alpha} = g^{\mu\nu}.$$

## § 9. THE EQUATION OF THE GEODETIC LINE. THE MOTION OF A PARTICLE

As the linear element $ds$ is defined independently of the system of co-ordinates, the line drawn between two points P and P' of the four-dimensional continuum in such a way that $\int ds$ is stationary—a geodetic line—has a meaning which also is independent of the choice of co-ordinates. Its equation is

$$\delta \int_{P}^{P'} ds = 0 \qquad \cdot \qquad \cdot \qquad \cdot \qquad \cdot \qquad (20)$$

Carrying out the variation in the usual way, we obtain from this equation four differential equations which define the geodetic line; this operation will be inserted here for the sake of completeness. Let $\lambda$ be a function of the co-ordinates $x_{\nu}$, and let this define a family of surfaces which intersect the required geodetic line as well as all the lines in immediate proximity to it which are drawn through the points P and P'. Any such line may then be supposed to be given by expressing its co-ordinates $x_{\nu}$ as functions of $\lambda$. Let the symbol $\delta$ indicate the transition from a point of the required geodetic to the point corresponding to the same $\lambda$ on a neighbouring line. Then for (20) we may substitute

$$\left.\begin{array}{c}\displaystyle\int_{\lambda_1}^{\lambda_2}\delta w d\lambda = 0\\[2mm]\displaystyle w^2 = g_{\mu\nu}\frac{dx_\mu}{d\lambda}\frac{dx_\nu}{d\lambda}\end{array}\right\}\quad\cdot\quad\cdot\quad\cdot\quad(20a)$$

But since

$$\delta w = \frac{1}{w}\left\{\frac{1}{2}\frac{\partial g_{\mu\nu}}{\partial x_\sigma}\frac{dx_\mu}{d\lambda}\frac{dx_\nu}{d\lambda}\delta x_\sigma + g_{\mu\nu}\frac{dx_\mu}{d\lambda}\delta\left(\frac{dx_\nu}{d\lambda}\right)\right\},$$

and

$$\delta\left(\frac{dx_\nu}{d\lambda}\right) = \frac{d}{d\lambda}(\delta x_\nu),$$

we obtain from (20a), after a partial integration,

$$\int_{\lambda_1}^{\lambda_2}\kappa_\sigma\delta x_\sigma d\lambda = 0,$$

where

$$\kappa_\sigma = \frac{d}{d\lambda}\left\{\frac{g_{\mu\nu}}{w}\frac{dx_\mu}{d\lambda}\right\} - \frac{1}{2w}\frac{\partial g_{\mu\nu}}{\partial x_\sigma}\frac{dx_\mu}{d\lambda}\frac{dx_\nu}{d\lambda}\quad\cdot\quad(20b)$$

Since the values of $\delta x_\sigma$ are arbitrary, it follows from this that

$$\kappa_\sigma = 0\quad\cdot\quad\cdot\quad\cdot\quad\cdot\quad(20c)$$

are the equations of the geodetic line.

If $ds$ does not vanish along the geodetic line we may choose the "length of the arc" $s$, measured along the geodetic line, for the parameter $\lambda$. Then $w = 1$, and in place of (20c) we obtain

$$g_{\mu\nu}\frac{d^2x_\mu}{ds^2} + \frac{\partial g_{\mu\nu}}{\partial x_\sigma}\frac{dx_\sigma}{ds}\frac{dx_\mu}{ds} - \frac{1}{2}\frac{\partial g_{\mu\nu}}{\partial x_\sigma}\frac{dx_\mu}{ds}\frac{dx_\nu}{ds} = 0$$

or, by a mere change of notation,

$$g_{\alpha\sigma}\frac{d^2x_\alpha}{ds^2} + [\mu\nu,\sigma]\frac{dx_\mu}{ds}\frac{dx_\nu}{ds} = 0\quad\cdot\quad\cdot\quad(20d)$$

where, following Christoffel, we have written

$$[\mu\nu,\sigma] = \frac{1}{2}\left(\frac{\partial g_{\mu\sigma}}{\partial x_\nu} + \frac{\partial g_{\nu\sigma}}{\partial x_\mu} - \frac{\partial g_{\mu\nu}}{\partial x_\sigma}\right)\quad\cdot\quad\cdot\quad(21)$$

Finally, if we multiply (20d) by $g^{\sigma\tau}$ (outer multiplication with respect to $\tau$, inner with respect to $\sigma$), we obtain the equations of the geodetic line in the form

$$\frac{d^2x_\tau}{ds^2} + \{\mu\nu, \tau\} \frac{dx_\mu}{ds} \frac{dx_\nu}{ds} = 0. \qquad . \qquad . \qquad (22)$$

where, following Christoffel, we have set

$$\{\mu\nu, \tau\} = g^{\tau\alpha}[\mu\nu, \alpha] \qquad . \qquad . \qquad . \qquad (23)$$

## § 10. THE FORMATION OF TENSORS BY DIFFERENTIATION

With the help of the equation of the geodetic line we can now easily deduce the laws by which new tensors can be formed from old by differentiation. By this means we are able for the first time to formulate generally covariant differential equations. We reach this goal by repeated application of the following simple law:—

If in our continuum a curve is given, the points of which are specified by the arcual distance $s$ measured from a fixed point on the curve, and if, further, $\phi$ is an invariant function of space, then $d\phi/ds$ is also an invariant. The proof lies in this, that $ds$ is an invariant as well as $d\phi$.

As

$$\frac{d\phi}{ds} = \frac{\partial\phi}{\partial x_\mu} \frac{dx_\mu}{ds}$$

therefore

$$\psi = \frac{\partial\phi}{dx_\mu} \frac{dx_\mu}{ds}$$

is also an invariant, and an invariant for all curves starting from a point of the continuum, that is, for any choice of the vector $dx_\mu$. Hence it immediately follows that

$$A_\mu = \frac{\partial\phi}{\partial x_\mu} \qquad . \qquad . \qquad . \qquad . \qquad (24)$$

is a covariant four-vector—the "gradient" of $\phi$.

According to our rule, the differential quotient

$$\chi = \frac{d\psi}{ds}$$

taken on a curve, is similarly an invariant. Inserting the value of $\psi$, we obtain in the first place

$$\chi = \frac{\partial^2 \phi}{\partial x_\mu \partial x_\nu} \frac{dx_\mu}{ds} \frac{dx_\nu}{ds} + \frac{\partial \phi}{\partial x_\mu} \frac{d^2 x_\mu}{ds^2}.$$

The existence of a tensor cannot be deduced from this forthwith. But if we may take the curve along which we have differentiated to be a geodetic, we obtain on substitution for $d^2 x_\nu / ds^2$ from (22),

$$\chi = \left( \frac{\partial^2 \phi}{\partial x_\mu \partial x_\nu} - \{\mu\nu, \tau\} \frac{\partial \phi}{\partial x_\tau} \right) \frac{dx_\mu}{ds} \frac{dx_\nu}{ds}.$$

Since we may interchange the order of the differentiations, and since by (23) and (21) $\{\mu\nu, \tau\}$ is symmetrical in $\mu$ and $\nu$, it follows that the expression in brackets is symmetrical in $\mu$ and $\nu$. Since a geodetic line can be drawn in any direction from a point of the continuum, and therefore $dx_\mu / ds$ is a four-vector with the ratio of its components arbitrary, it follows from the results of § 7 that

$$A_{\mu\nu} = \frac{\partial^2 \phi}{\partial x_\mu \partial x_\nu} - \{\mu\nu, \tau\} \frac{\partial \phi}{\partial x_\tau} \qquad \cdot \qquad \cdot \qquad \cdot \quad (25)$$

is a covariant tensor of the second rank. We have therefore come to this result: from the covariant tensor of the first rank

$$A_\mu = \frac{\partial \phi}{\partial x_\mu}$$

we can, by differentiation, form a covariant tensor of the second rank

$$A_{\mu\nu} = \frac{\partial A_\mu}{\partial x_\nu} - \{\mu\nu, \tau\} A_\tau \qquad \cdot \qquad \cdot \qquad \cdot \quad (26)$$

We call the tensor $A_{\mu\nu}$ the "extension" (covariant derivative) of the tensor $A_\mu$. In the first place we can readily show that the operation leads to a tensor, even if the vector $A_\mu$ cannot be represented as a gradient. To see this, we first observe that

$$\psi \frac{\partial \phi}{\partial x_\mu}$$

is a covariant vector, if $\psi$ and $\phi$ are scalars. The sum of four such terms

$$S_\mu = \psi^{(1)} \frac{\phi \partial^{(1)}}{\partial x_\mu} + \cdot + \cdot + \psi^{(4)} \frac{\partial \phi^{(4)}}{\partial x_\mu},$$

THE PRINCIPLE OF RELATIVITY

is also a covariant vector, if $\psi^{(1)}$, $\phi^{(1)}$ ... $\psi^{(4)}$, $\phi^{(4)}$ are scalars. But it is clear that any covariant vector can be represented in the form $S_\mu$. For, if $A_\mu$ is a vector whose components are any given functions of the $x_\nu$, we have only to put (in terms of the selected system of co-ordinates)

$$\psi^{(1)} = A_1, \quad \phi^{(1)} = x_1,$$
$$\psi^{(2)} = A_2, \quad \phi^{(2)} = x_2,$$
$$\psi^{(3)} = A_3, \quad \phi^{(3)} = x_3,$$
$$\psi^{(4)} = A_4, \quad \phi^{(4)} = x_4,$$

in order to ensure that $S_\mu$ shall be equal to $A_\mu$.

Therefore, in order to demonstrate that $A_{\mu\nu}$ is a tensor if *any* covariant vector is inserted on the right-hand side for $A_\mu$, we only need show that this is so for the vector $S_\mu$. But for this latter purpose it is sufficient, as a glance at the right-hand side of (26) teaches us, to furnish the proof for the case

$$A_\mu = \psi \frac{\partial \phi}{\partial x_\mu}.$$

Now the right-hand side of (25) multiplied by $\psi$,

$$\psi \frac{\partial^2 \phi}{\partial x_\mu \partial x_\nu} - \{\mu\nu, \tau\} \psi \frac{\partial \phi}{\partial x_\tau}$$

is a tensor. Similarly

$$\frac{\partial \psi}{\partial x_\mu} \frac{\partial \phi}{\partial x_\nu}$$

being the outer product of two vectors, is a tensor. By addition, there follows the tensor character of

$$\frac{\partial}{\partial x_\nu} \left( \psi \frac{\partial \phi}{\partial x_\mu} \right) - \{\mu\nu, \tau\} \left( \psi \frac{\partial \phi}{\partial x_\tau} \right).$$

As a glance at (26) will show, this completes the demonstration for the vector

$$\psi \frac{\partial \phi}{\partial x_\mu}$$

and consequently, from what has already been proved, for any vector $A_\mu$.

By means of the extension of the vector, we may easily define the "extension" of a covariant tensor of any rank. This operation is a generalization of the extension of a vector. We restrict ourselves to the case of a tensor of the second rank, since this suffices to give a clear idea of the law of formation.

As has already been observed, any covariant tensor of the second rank can be represented* as the sum of tensors of the type $A_\mu B_\nu$. It will therefore be sufficient to deduce the expression for the extension of a tensor of this special type. By (26) the expressions

$$\frac{\partial A_\mu}{\partial x_\sigma} - \{\sigma\mu, \tau\}A_\tau,$$

$$\frac{\partial B_\nu}{\partial x_\sigma} - \{\sigma\nu, \tau\}B_\tau,$$

are tensors. On outer multiplication of the first by $B_\nu$, and of the second by $A_\mu$, we obtain in each case a tensor of the third rank. By adding these, we have the tensor of the third rank

$$A_{\mu\nu\sigma} = \frac{\partial A_{\mu\nu}}{\partial x_\sigma} - \{\sigma\mu, \tau\}A_{\tau\nu} - \{\sigma\nu, \tau\}A_{\mu\tau}. \quad \cdot \quad \cdot (27)$$

where we have put $A_{\mu\nu} = A_\mu B_\nu$. As the right-hand side of (27) is linear and homogeneous in the $A_{\mu\nu}$ and their first derivatives, this law of formation leads to a tensor, not only in the case of a tensor of the type $A_\mu B_\nu$, but also in the case of a sum of such tensors, i.e. in the case of any covariant tensor of the second rank. We call $A_{\mu\nu\sigma}$ the extension of the tensor $A_{\mu\nu}$.

It is clear that (26) and (24) concern only special cases of extension (the extension of the tensors of rank one and zero respectively).

In general, all special laws of formation of tensors are included in (27) in combination with the multiplication of tensors.

## § 11. SOME CASES OF SPECIAL IMPORTANCE

*The Fundamental Tensor.*—We will first prove some lemmas which will be useful hereafter. By the rule for the differentiation of determinants

$$dg = g^{\mu\nu} g \, dg_{\mu\nu} = -g_{\mu\nu} g \, dg^{\mu\nu} \quad \cdot \quad \cdot (28)$$

---

*By outer multiplication of the vector with arbitrary components $A_{11}$, $A_{12}$, $A_{13}$, $A_{14}$ by the vector with components 1, 0, 0, 0, we produce a tensor with components

$$\begin{matrix} A_{11} & A_{12} & A_{13} & A_{14} \\ 0 & 0 & 0 & 0 \\ 0 & 0 & 0 & 0 \\ 0 & 0 & 0 & 0 \end{matrix}$$

By the addition of four tensors of this type, we obtain the tensor $A_{\mu\nu}$ with any assigned components.

The last member is obtained from the last but one, if we bear in mind that $g_{\mu\nu}g^{\mu'\nu} = \delta_\mu^{\mu'}$, so that $g_{\mu\nu}g^{\mu\nu} = 4$, and consequently

$$g_{\mu\nu}\,dg^{\mu\nu} + g^{\mu\nu}dg_{\mu\nu} = 0.$$

From (28), it follows that

$$\frac{1}{\sqrt{-g}}\frac{\partial\sqrt{-g}}{\partial x_\sigma} = \frac{1}{2}\frac{\partial\log(-g)}{\partial x_\sigma} = \frac{1}{2}g^{\mu\nu}\frac{\partial g_{\mu\nu}}{\partial x_\sigma} = \frac{1}{2}g_{\mu\nu}\frac{\partial g^{\mu\nu}}{\partial x_\sigma}. \quad (29)$$

Further, from $g_{\mu\sigma}g^{\nu\sigma} = \delta_\mu^\nu$, it follows on differentiation that

$$\left.\begin{array}{r}g_{\mu\sigma}dg^{\nu\sigma} = -g^{\nu\sigma}dg_{\mu\sigma}\\[1mm]g_{\mu\sigma}\dfrac{\partial g^{\nu\sigma}}{\partial x_\lambda} = -g^{\nu\sigma}\dfrac{\partial g_{\mu\sigma}}{\partial x_\lambda}\end{array}\right\} \quad\cdot\quad\cdot\quad (30)$$

From these, by mixed multiplication by $g^{\sigma\tau}$ and $g_{\nu\lambda}$ respectively, and a change of notation for the indices, we have

$$\left.\begin{array}{r}dg^{\mu\nu} = -g^{\mu\alpha}g^{\nu\beta}dg_{\alpha\beta}\\[1mm]\dfrac{\partial g^{\mu\nu}}{\partial x_\sigma} = -g^{\mu\alpha}g^{\nu\beta}\dfrac{\partial g_{\alpha\beta}}{\partial x_\sigma}\end{array}\right\} \quad\cdot\quad\cdot\quad (31)$$

and

$$\left.\begin{array}{r}dg_{\mu\nu} = -g_{\mu\alpha}g_{\nu\beta}dg^{\alpha\beta}\\[1mm]\dfrac{\partial g_{\mu\nu}}{\partial x_\sigma} = -g_{\mu\alpha}g_{\nu\beta}\dfrac{\partial g^{\alpha\beta}}{\partial x_\sigma}\end{array}\right\} \quad\cdot\quad\cdot\quad (32)$$

The relation (31) admits of a transformation, of which we also have frequently to make use. From (21)

$$\frac{\partial g_{\alpha\beta}}{\partial x_\sigma} = [\alpha\sigma, \beta] + [\beta\sigma, \alpha] \quad\cdot\quad\cdot\quad\cdot\quad (33)$$

Inserting this in the second formula of (31), we obtain, in view of (23)

$$\frac{\partial g^{\mu\nu}}{\partial x_\sigma} = -g^{\mu\tau}\{\tau\sigma, \nu\} - g^{\nu\tau}\{\tau\sigma, \mu\} \quad\cdot\quad\cdot\quad (34)$$

Substituting the right-hand side of (34) in (29), we have

$$\frac{1}{\sqrt{-g}}\frac{\partial\sqrt{-g}}{\partial x_\sigma} = \{\mu\sigma, \mu\} \quad\cdot\quad\cdot\quad (29a)$$

*The "Divergence" of a Contravariant Vector.*—If we take the inner product of (26) by the contravariant fundamental tensor $g^{\mu\nu}$, the right-hand side, after a transformation of the first term, assumes the form

$$\frac{\partial}{\partial x_\nu}(g^{\mu\nu}A_\mu) - A_\mu\frac{\partial g^{\mu\nu}}{\partial x_\nu} - \tfrac{1}{2}g^{\tau\alpha}\left(\frac{\partial g_{\mu\alpha}}{\partial x_\nu} + \frac{\partial g_{\nu\alpha}}{\partial x_\mu} - \frac{\partial g_{\mu\nu}}{\partial x_\alpha}\right)g^{\mu\nu}A_\tau.$$

In accordance with (31) and (29), the last term of this expression may be written

$$\tfrac{1}{2}\frac{\partial g^{\tau\nu}}{\partial x_\nu}A_\tau + \tfrac{1}{2}\frac{\partial g^{\tau\mu}}{\partial x_\mu}A_\tau + \frac{1}{\sqrt{-g}}\frac{\partial\sqrt{-g}}{\partial x_\alpha}g^{\mu\nu}A_\tau.$$

As the symbols of the indices of summation are immaterial, the first two terms of this expression cancel the second of the one above. If we then write $g^{\mu\nu}A_\mu = A^\nu$, so that $A^\nu$ like $A_\mu$ is an arbitrary vector, we finally obtain

$$\Phi = \frac{1}{\sqrt{-g}}\frac{\partial}{\partial x_\nu}(\sqrt{-g}A^\nu) \quad \cdot \quad \cdot \quad \cdot \quad (35)$$

This scalar is the *divergence* of the contravariant vector $A^\nu$.

*The "Curl" of a Covariant Vector.*—The second term in (26) is symmetrical in the indices $\mu$ and $\nu$. Therefore $A_{\mu\nu} - A_{\nu\mu}$ is a particularly simply constructed antisymmetrical tensor. We obtain

$$B_{\mu\nu} = \frac{\partial A_\mu}{\partial x_\nu} - \frac{\partial A_\nu}{\partial x_\mu} \quad \cdot \quad \cdot \quad \cdot \quad (36)$$

*Antisymmetrical Extension of a Six-vector.*—Applying (27) to an antisymmetrical tensor of the second rank $A_{\mu\nu}$, forming in addition the two equations which arise through cyclic permutations of the indices, and adding these three equations, we obtain the tensor of the third rank

$$B_{\mu\nu\sigma} = A_{\mu\nu\sigma} + A_{\nu\sigma\mu} + A_{\sigma\mu\nu} = \frac{\partial A_{\mu\nu}}{\partial x_\sigma} + \frac{\partial A_{\nu\sigma}}{\partial x_\mu} + \frac{\partial A_{\sigma\mu}}{\partial x_\nu} \quad (37)$$

which it is easy to prove is antisymmetrical.

*The Divergence of a Six-vector.*—Taking the mixed product of (27) by $g^{\mu\alpha}g^{\nu\beta}$, we also obtain a tensor. The first term on the right-hand side of (27) may be written in the form

$$\frac{\partial}{\partial x_\sigma}(g^{\mu\alpha}g^{\nu\beta}A_{\mu\nu}) - g^{\mu\alpha}\frac{\partial g^{\nu\beta}}{\partial x_\sigma}A_{\mu\nu} - g^{\nu\beta}\frac{\partial g^{\mu\alpha}}{\partial x_\sigma}A_{\mu\nu}.$$

If we write $A_\sigma^{\alpha\beta}$ for $g^{\mu\alpha}g^{\nu\beta}A_{\mu\nu\sigma}$ and $A^{\alpha\beta}$ for $g^{\mu\alpha}g^{\nu\beta}A_{\mu\nu}$, and in the transformed first term replace

$$\frac{\partial g^{\nu\beta}}{\partial x_\sigma} \text{ and } \frac{\partial g^{\mu\alpha}}{\partial x_\sigma}$$

by their values as given by (34), there results from the right-hand side of (27) an expression consisting of seven terms, of which four cancel, and there remains

$$A_\sigma^{\alpha\beta} = \frac{\partial A^{\alpha\beta}}{\partial x_\sigma} + \{\sigma\gamma, \alpha\}A^{\gamma\beta} + \{\sigma\gamma, \beta\}A^{\alpha\gamma} \quad . \quad . \quad (38)$$

This is the expression for the extension of a contravariant tensor of the second rank, and corresponding expressions for the extension of contravariant tensors of higher and lower rank may also be formed.

We note that in an analogous way we may also form the extension of a mixed tensor:—

$$A_{\mu\sigma}^\alpha = \frac{\partial A_\mu^\alpha}{\partial x_\sigma} - \{\sigma\mu, \tau\}A_\tau^\alpha + \{\sigma\tau, \alpha\}A_\mu^\tau \quad . \quad . \quad (39)$$

On contracting (38) with respect to the indices $\beta$ and $\sigma$ (inner multiplication by $\delta_\beta^\sigma$), we obtain the vector

$$A^\alpha = \frac{\partial A^{\alpha\beta}}{\partial x_\beta} + \{\beta\gamma, \beta\}A^{\alpha\gamma} + \{\beta\gamma, \alpha\}A^{\gamma\beta}.$$

On account of the symmetry of $\{\beta\gamma, \alpha\}$ with respect to the indices $\beta$ and $\gamma$, the third term on the right-hand side vanishes, if $A^{\alpha\beta}$ is, as we will assume, an antisymmetrical tensor. The second term allows itself to be transformed in accordance with (29a). Thus we obtain

$$A^\alpha = \frac{1}{\sqrt{-g}}\frac{\partial(\sqrt{-g}A^{\alpha\beta})}{\partial x_\beta} \quad . \quad . \quad . \quad (40)$$

This is the expression for the divergence of a contravariant six-vector.

*The Divergence of a Mixed Tensor of the Second Rank.*—Contracting (39) with respect to the indices $\alpha$ and $\sigma$, and taking (29a) into consideration, we obtain

$$\sqrt{-g}A_\mu = \frac{\partial(\sqrt{-g}A_\mu^\sigma)}{\partial x_\sigma} - \{\sigma\mu, \tau\}\sqrt{-g}A_\tau^\sigma \quad . \quad (41)$$

If we introduce the contravariant tensor $A^{\rho\sigma} = g^{\rho\tau}A_\tau^\sigma$ in the last term, it assumes the form

$$-[\sigma\mu, \rho]\sqrt{-g}A^{\rho\sigma}.$$

If, further, the tensor $A^{\rho\sigma}$ is symmetrical, this reduces to

$$-\tfrac{1}{2}\sqrt{-g}\frac{\partial g_{\rho\sigma}}{\partial x_\mu}A^{\rho\sigma}.$$

Had we introduced, instead of $A^{\rho\sigma}$, the covariant tensor $A_{\rho\sigma} = g_{\rho\alpha}g_{\sigma\beta}A^{\alpha\beta}$, which is also symmetrical, the last term, by virtue of (31), would assume the form

$$\tfrac{1}{2}\sqrt{-g}\frac{\partial g^{\rho\sigma}}{\partial x_\mu}A_{\rho\sigma}.$$

In the case of symmetry in question, (41) may therefore be replaced by the two forms

$$\sqrt{-g}A_\mu = \frac{\partial(\sqrt{-g}A_\mu^\sigma)}{\partial x_\sigma} - \tfrac{1}{2}\frac{\partial g_{\rho\sigma}}{\partial x_\mu}\sqrt{-g}A^{\rho\sigma} \quad . \quad (41a)$$

$$\sqrt{-g}A_\mu = \frac{\partial(\sqrt{-g}A_\mu^\sigma)}{\partial x_\sigma} + \tfrac{1}{2}\frac{\partial g^{\rho\sigma}}{\partial x_\mu}\sqrt{-g}A_{\rho\sigma} \quad . \quad (41b)$$

which we have to employ later on.

## § 12. THE RIEMANN-CHRISTOFFEL TENSOR

We now seek the tensor which can be obtained from the fundamental tensor *alone*, by differentiation. At first sight the solution seems obvious. We place the fundamental tensor of the $g_{\mu\nu}$ in (27) instead of any given tensor $A_{\mu\nu}$, and thus have a new tensor, namely, the extension of the fundamental tensor. But we easily convince ourselves that this extension vanishes identically. We reach our goal, however, in the following way. In (27) place

$$A_{\mu\nu} = \frac{\partial A_\mu}{\partial x_\nu} - \{\mu\nu, \rho\}A_\rho,$$

i.e. the extension of the four-vector $A_\mu$. Then (with a somewhat different naming of the indices) we get the tensor of the third rank

$$A_{\mu\sigma\tau} = \frac{\partial^2 A_\mu}{\partial x_\sigma \partial x_\tau} - \{\mu\sigma, \rho\}\frac{\partial A_\rho}{\partial x_\tau} - \{\mu\tau, \rho\}\frac{\partial A_\rho}{\partial x_\sigma} - \{\sigma\tau, \rho\}\frac{\partial A_\mu}{\partial x_\rho}$$

$$+ \left[ -\frac{\partial}{\partial x_\tau}\{\mu\sigma, \rho\} + \{\mu\tau, \alpha\}\{\alpha\sigma, \rho\} + \{\sigma\tau, \alpha\}\{\alpha\mu, \rho\} \right]A_\rho.$$

This expression suggests forming the tensor $A_{\mu\sigma\tau} - A_{\mu\tau\sigma}$. For, if we do so, the following terms of the expression for $A_{\mu\sigma\tau}$ cancel those of $A_{\mu\tau\sigma}$, the first, the fourth, and the member corresponding to the last term in square brackets; because all these are symmetrical in $\sigma$ and $\tau$. The same holds good for the sum of the second and third terms. Thus we obtain

$$A_{\mu\sigma\tau} - A_{\mu\tau\sigma} = B^\rho_{\mu\sigma\tau}A_\rho \qquad . \qquad . \qquad . \qquad (42)$$

where

$$B^\rho_{\mu\sigma\tau} = -\frac{\partial}{\partial x_\tau}\{\mu\sigma, \rho\} + \frac{\partial}{\partial x_\sigma}\{\mu\tau, \rho\} - \{\mu\sigma, \alpha\}\{\alpha\tau, \rho\}$$

$$+ \{\mu\tau, \alpha\}\{\alpha\sigma, \rho\} \qquad (43)$$

The essential feature of the result is that on the right side of (42) the $A_\rho$ occur alone, without their derivatives. From the tensor character of $A_{\mu\sigma\tau} - A_{\mu\tau\sigma}$ in conjunction with the fact that $A_\rho$ is an arbitrary vector, it follows, by reason of § 7, that $B^\rho_{\mu\sigma\tau}$ is a tensor (the Riemann-Christoffel tensor).

The mathematical importance of this tensor is as follows: If the continuum is of such a nature that there is a co-ordinate system with reference to which the $g_{\mu\nu}$ are constants, then all the $B^\rho_{\mu\sigma\tau}$ vanish. If we choose any new system of coordinates in place of the original ones, the $g_{\mu\nu}$ referred thereto will not be constants, but in consequence of its tensor nature, the transformed components of $B^\rho_{\mu\sigma\tau}$ will still vanish in the new system. Thus the vanishing of the Riemann tensor is a necessary condition that, by an appropriate choice of the system of reference, the $g_{\mu\nu}$ may be constants. In our problem this corresponds to the case in which,* with a suitable choice of the system of reference,

---

*The mathematicians have proved that this is also a *sufficient* condition.

the special theory of relativity holds good for a *finite* region of the continuum.

Contracting (43) with respect to the indices $\tau$ and $\rho$ we obtain the covariant tensor of second rank

$$G_{\mu\nu} = B^{\rho}_{\mu\nu\rho} = R_{\mu\nu} + S_{\mu\nu}$$

where

$$\left. \begin{array}{l} R_{\mu\nu} = -\dfrac{\partial}{\partial x_\alpha}\{\mu\nu, \alpha\} + \{\mu\alpha, \beta\}\{\nu\beta, \alpha\} \\[2em] S_{\mu\nu} = \dfrac{\partial^2 \log \sqrt{-g}}{\partial x_\mu \partial x_\nu} - \{\mu\nu, \alpha\}\dfrac{\partial \log \sqrt{-g}}{\partial x_\alpha} \end{array} \right\} \qquad (44)$$

*Note on the Choice of Co-ordinates.*—It has already been observed in § 8, in connexion with equation (18a), that the choice of co-ordinates may with advantage be made so that $\sqrt{-g} = 1$. A glance at the equations obtained in the last two sections shows that by such a choice the laws of formation of tensors undergo an important simplification. This applies particularly to $G_{\mu\nu}$, the tensor just developed, which plays a fundamental part in the theory to be set forth. For this specialization of the choice of co-ordinates brings about the vanishing of $S_{\mu\nu}$, so that the tensor $G_{\mu\nu}$ reduces to $R_{\mu\nu}$.

On this account I shall hereafter give all relations in the simplified form which this specialization of the choice of co-ordinates brings with it. It will then be an easy matter to revert to the *generally* covariant equations, if this seems desirable in a special case.

# C. THEORY OF THE GRAVITATIONAL FIELD

### § 13. Equations of Motion of a Material Point in the Gravitational Field. Expression for the Field-components of Gravitation

A freely movable body not subjected to external forces moves, according to the special theory of relativity, in a straight line and uniformly. This is also the case, according to the general theory of relativity, for a part of four-dimensional space in which the system of co-ordinates

$K_0$, may be, and is, so chosen that they have the special constant values given in (4).

If we consider precisely this movement from any chosen system of co-ordinates $K_1$, the body, observed from $K_1$, moves, according to the considerations in § 2, in a gravitational field. The law of motion with respect to $K_1$ results without difficulty from the following consideration. With respect to $K_0$ the law of motion corresponds to a four-dimensional straight line, i.e. to a geodetic line. Now since the geodetic line is defined independently of the system of reference, its equations will also be the equation of motion of the material point with respect to $K_1$. If we set

$$\Gamma^\tau_{\mu\nu} = -\{\mu\nu, \tau\} \qquad \cdot \quad \cdot \quad \cdot \quad \cdot \quad (45)$$

the equation of the motion of the point with respect to $K_1$, becomes

$$\frac{d^2 x_\tau}{ds^2} = \Gamma^\tau_{\mu\nu} \frac{dx_\mu}{ds} \frac{dx_\nu}{ds} \qquad \cdot \quad \cdot \quad \cdot \quad (46)$$

We now make the assumption, which readily suggests itself, that this covariant system of equations also defines the motion of the point in the gravitational field in the case when there is no system of reference $K_0$, with respect to which the special theory of relativity holds good in a finite region. We have all the more justification for this assumption as (46) contains only *first* derivatives of the $g_{\mu\nu}$, between which even in the special case of the existence of $K_0$, no relations subsist.*

If the $\Gamma^\tau_{\mu\nu}$ vanish, then the point moves uniformly in a straight line. These quantities therefore condition the deviation of the motion from uniformity. They are the components of the gravitational field.

## § 14. The Field Equations of Gravitation
### in the Absence of Matter

We make a distinction hereafter between "gravitational field" and "matter" in this way, that we denote everything but the gravitational field as "matter." Our use of the word therefore includes not only matter in the ordinary sense, but the electromagnetic field as well.

---

*It is only between the second (and first) derivatives that, by § 12, the relations $B^\rho_{\mu\sigma\tau} = 0$ subsist.

Our next task is to find the field equations of gravitation in the absence of matter. Here we again apply the method employed in the preceding paragraph in formulating the equations of motion of the material point. A special case in which the required equations must in any case be satisfied is that of the special theory of relativity, in which the $g_{\mu\nu}$ have certain constant values. Let this be the case in a certain finite space in relation to a definite system of co-ordinates $K_0$. Relatively to this system all the components of the Riemann tensor $B^{\rho}_{\mu\sigma\tau}$, defined in (43), vanish. For the space under consideration they then vanish, also in any other system of co-ordinates.

Thus the required equations of the matter-free gravitational field must in any case be satisfied if all $B^{\rho}_{\mu\sigma\tau}$ vanish. But this condition goes too far. For it is clear that, e.g., the gravitational field generated by a material point in its environment certainly cannot be "transformed away" by any choice of the system of co-ordinates, i.e. it cannot be transformed to the case of constant $g_{\mu\nu}$.

This prompts us to require for the matter-free gravitational field that the symmetrical tensor $G_{\mu\nu}$, derived from the tensor $B^{\rho}_{\mu\nu\tau}$, shall vanish. Thus we obtain ten equations for the ten quantities $g_{\mu\nu}$, which are satisfied in the special case of the vanishing of all $B^{\rho}_{\mu\nu\tau}$. With the choice which we have made of a system of co-ordinates, and taking (44) into consideration, the equations for the matter-free field are

$$\left.\begin{array}{l} \dfrac{\partial \Gamma^{\alpha}_{\mu\nu}}{\partial x_{\alpha}} + \Gamma^{\alpha}_{\mu\beta}\Gamma^{\beta}_{\nu\alpha} = 0 \\[2mm] \sqrt{-g} = 1 \end{array}\right\} \qquad . \qquad . \qquad . \quad (47)$$

It must be pointed out that there is only a minimum of arbitrariness in the choice of these equations. For besides $G_{\mu\nu}$ there is no tensor of second rank which is formed from the $g_{\mu\nu}$ and its derivatives, contains no derivations higher than second, and is linear in these derivatives.*

These equations, which proceed, by the method of pure mathematics, from the requirement of the general theory of relativity, give

---

*Properly speaking, this can be affirmed only of the tensor
$$G_{\mu\nu} + \lambda g_{\mu\nu} \, g^{\alpha\beta}G_{\alpha\beta},$$
where $\lambda$ is a constant. If, however, we set this tensor $= 0$, we come back again to the equations $G_{\mu\nu} = 0$.

us, in combination with the equations of motion (46), to a first approximation Newton's law of attraction, and to a second approximation the explanation of the motion of the perihelion of the planet Mercury discovered by Leverrier (as it remains after corrections for perturbation have been made). These facts must, in my opinion, be taken as a convincing proof of the correctness of the theory.

## § 15. THE HAMILTONIAN FUNCTION FOR THE GRAVITATIONAL FIELD. LAWS OF MOMENTUM AND ENERGY

To show that the field equations correspond to the laws of momentum and energy, it is most convenient to write them in the following Hamiltonian form:—

$$\left.\begin{array}{l} \delta \int H d\tau = 0 \\ H = g^{\mu\nu}\Gamma^{\alpha}_{\mu\beta}\Gamma^{\beta}_{\nu\alpha} \\ \sqrt{-g} = 1 \end{array}\right\} \qquad \cdot \qquad \cdot \qquad \cdot \qquad (47a)$$

where, on the boundary of the finite four-dimensional region of integration which we have in view, the variations vanish.

We first have to show that the form (47a) is equivalent to the equations (47). For this purpose we regard H as a function of the $g^{\mu\nu}$ and the $g^{\mu\nu}_\sigma (= \partial g^{\mu\nu}/\partial x_\sigma)$.

Then in the first place

$$\delta H = \Gamma^{\alpha}_{\mu\beta}\Gamma^{\beta}_{\nu\alpha}\delta g^{\mu\nu} + 2g^{\mu\nu}\Gamma^{\alpha}_{\mu\beta}\delta\Gamma^{\beta}_{\nu\alpha}$$
$$= -\Gamma^{\alpha}_{\mu\beta}\Gamma^{\beta}_{\nu\alpha}\delta g^{\mu\nu} + 2\Gamma^{\alpha}_{\mu\beta}\delta(g^{\mu\nu}\Gamma^{\beta}_{\nu\alpha}).$$

But

$$\delta(g^{\mu\nu}\Gamma^{\beta}_{\nu\alpha}) = -\frac{1}{2}\delta\left[g^{\mu\nu}g^{\beta\lambda}\left(\frac{\partial g_{\nu\lambda}}{\partial x_\alpha} + \frac{\partial g_{\alpha\lambda}}{\partial x_\nu} - \frac{\partial g_{\alpha\nu}}{\partial x_\lambda}\right)\right].$$

The terms arising from the last two terms in round brackets are of different sign, and result from each other (since the denomination of the summation indices is immaterial) through interchange of the indices $\mu$ and $\beta$. They cancel each other in the expression for $\delta H$, because they are multiplied by the quantity $\Gamma^{\alpha}_{\mu\beta}$, which is symmetrical with respect to the indices $\mu$ and $\beta$. Thus there remains only the first term in round brackets to be considered, so that, taking (31) into account, we obtain

$$\delta H = -\Gamma^{\alpha}_{\mu\beta}\Gamma^{\beta}_{\nu\alpha}\delta g^{\mu\nu} + \Gamma^{\alpha}_{\mu\beta}\delta g^{\mu\beta}_{\alpha}.$$

Thus

$$\left.\begin{array}{c} \dfrac{\partial H}{\partial g^{\mu\nu}} = -\Gamma^{\alpha}_{\mu\beta}\Gamma^{\beta}_{\nu\alpha} \\[4mm] \dfrac{\partial H}{\partial g^{\mu\nu}_{\sigma}} = \Gamma^{\sigma}_{\mu\nu} \end{array}\right\} \qquad . \quad . \quad . \quad (48)$$

Carrying out the variation in (47a), we get in the first place

$$\frac{\partial}{\partial x_{\alpha}}\left(\frac{\partial H}{\partial g^{\mu\nu}_{\alpha}}\right) - \frac{\partial H}{\partial g^{\mu\nu}} = 0, \qquad . \quad . \quad . \quad (47b)$$

which, on account of (48), agrees with (47), as was to be proved.

If we multiply (47b) by $g^{\mu\nu}_{\sigma}$, then because

$$\frac{\partial g^{\mu\nu}_{\sigma}}{\partial x_{\alpha}} = \frac{\partial g^{\mu\nu}_{\alpha}}{\partial x_{\sigma}}$$

and, consequently,

$$g^{\mu\nu}_{\sigma}\frac{\partial}{\partial x_{\alpha}}\left(\frac{\partial H}{\partial g^{\mu\nu}_{\alpha}}\right) = \frac{\partial}{\partial x_{\alpha}}\left(g^{\mu\nu}_{\sigma}\frac{\partial H}{\partial g^{\mu\nu}_{\alpha}}\right) - \frac{\partial H}{\partial g^{\mu\nu}_{\alpha}}\frac{\partial g^{\mu\nu}_{\alpha}}{\partial x_{\sigma}},$$

we obtain the equation

$$\frac{\partial}{\partial x_{\alpha}}\left(g^{\mu\nu}_{\sigma}\frac{\partial H}{\partial g^{\mu\nu}_{\alpha}}\right) - \frac{\partial H}{\partial x_{\sigma}} = 0$$

or*

$$\left.\begin{array}{c} \dfrac{\partial t^{\alpha}_{\sigma}}{\partial x_{\alpha}} = 0 \\[4mm] -2\kappa t^{\alpha}_{\sigma} = g^{\mu\nu}_{\sigma}\dfrac{\partial H}{\partial g^{\mu\nu}_{\alpha}} - \delta^{\alpha}_{\sigma}H \end{array}\right\} \qquad . \quad . \quad . \quad (49)$$

where, on account of (48), the second equation of (47), and (34)

$$\kappa t^{\alpha}_{\sigma} = \tfrac{1}{2}\delta^{\alpha}_{\sigma}g^{\mu\nu}\Gamma^{\lambda}_{\mu\beta}\Gamma^{\beta}_{\nu\lambda} - g^{\mu\nu}\Gamma^{\alpha}_{\mu\beta}\Gamma^{\beta}_{\nu\sigma} \qquad . \quad . \quad (50)$$

It is to be noticed that $t^{\alpha}_{\sigma}$ is not a tensor; on the other hand (49) applies to all systems of co-ordinates for which $\sqrt{-g} = 1$. This equation expresses the law of conservation of momentum and of energy

*The reason for the introduction of the factor $-2\kappa$ will be apparent later.

for the gravitational field. Actually the integration of this equation over a three-dimensional volume V yields the four equations

$$\frac{d}{dx_4} \int t^4_\sigma \, dV = \int (lt^1_\sigma + mt^2_\sigma + nt^3_\sigma) \, dS. \qquad (49a)$$

where $l$, $m$, $n$ denote the direction-cosines of direction of the inward drawn normal at the element $dS$ of the bounding surface (in the sense of Euclidean geometry). We recognize in this the expression of the laws of conservation in their usual form. The quantities $t^\alpha_\sigma$ we call the "energy components" of the gravitational field.

I will now give equations (47) in a third form, which is particularly useful for a vivid grasp of our subject. By multiplication of the field equations (47) by $g^{\nu\sigma}$ these are obtained in the "mixed" form. Note that

$$g^{\nu\sigma} \frac{\partial \Gamma^\alpha_{\mu\nu}}{\partial x_\alpha} = \frac{\partial}{\partial x_\alpha}(g^{\nu\sigma}\Gamma^\alpha_{\mu\nu}) - \frac{\partial g^{\nu\sigma}}{\partial x_\alpha}\Gamma^\alpha_{\mu\nu},$$

which quantity, by reason of (34), is equal to

$$\frac{\partial}{\partial x_\alpha}(g^{\nu\sigma}\Gamma^\alpha_{\mu\nu}) - g^{\nu\beta}\Gamma^\sigma_{\alpha\beta}\Gamma^\alpha_{\mu\nu} - g^{\sigma\beta}\Gamma^\nu_{\beta\alpha}\Gamma^\alpha_{\mu\nu},$$

or (with different symbols for the summation indices)

$$\frac{\partial}{\partial x_\alpha}(g^{\sigma\beta}\Gamma^\alpha_{\mu\beta}) - g^{\gamma\delta}\Gamma^\alpha_{\gamma\beta}\Gamma^\beta_{\delta\mu} - g^{\nu\sigma}\Gamma^\alpha_{\mu\beta}\Gamma^\beta_{\nu\alpha}.$$

The third term of this expression cancels with the one arising from the second term of the field equations (47); using relation (50), the second term may be written

$$\kappa(t^\sigma_\mu - \tfrac{1}{2}\delta^\sigma_\mu t),$$

where $t = t^\alpha_\alpha$. Thus instead of equations (47) we obtain

$$\left.\begin{array}{c} \dfrac{\partial}{\partial x_\alpha}(g^{\sigma\beta}\Gamma^\alpha_{\mu\beta}) = -\kappa(t^\sigma_\mu - \tfrac{1}{2}\delta^\sigma_\mu t) \\[2mm] \sqrt{-g} = 1 \end{array}\right\} \qquad . \quad . \quad (51)$$

## § 16. THE GENERAL FORM OF THE FIELD EQUATIONS OF GRAVITATION

The field equations for matter-free space formulated in § 15 are to be compared with the field equation

$$\nabla^2 \phi = 0$$

of Newton's theory. We require the equation corresponding to Poisson's equation

$$\nabla^2\phi = 4\pi\kappa\rho,$$

where $\rho$ denotes the density of matter.

The special theory of relativity has led to the conclusion that inert mass is nothing more or less than energy, which finds its complete mathematical expression in a symmetrical tensor of second rank, the energy-tensor. Thus in the general theory of relativity we must introduce a corresponding energy-tensor of matter $T_\sigma^\alpha$, which, like the energy-components $t_\sigma$ [equations (49) and (50)] of the gravitational field, will have mixed character, but will pertain to a symmetrical covariant tensor.*

The system of equation (51) shows how this energy-tensor (corresponding to the density $\rho$ in Poisson's equation) is to be introduced into the field equations of gravitation. For if we consider a complete system (e.g. the solar system), the total mass of the system, and therefore its total gravitating action as well, will depend on the total energy of the system, and therefore on the ponderable energy together with the gravitational energy. This will allow itself to be expressed by introducing into (51), in place of the energy-components of the gravitational field alone, the sums $t_\mu^\sigma + T_\mu^\sigma$ of the energy-components of matter and of gravitational field. Thus instead of (51) we obtain the tensor equation

$$\left. \begin{aligned} \frac{\partial}{\partial x_\alpha}(g^{\sigma\beta}T_{\mu\beta}) &= -\kappa[(t_\mu^\sigma + T_\mu^\sigma) - \tfrac{1}{2}\delta_\mu^\sigma(t + T)], \\ \sqrt{-g} &= 1 \end{aligned} \right\} \quad . \quad (52)$$

where we have set $T = T_\mu^\mu$ (Laue's scalar). These are the required general field equations of gravitation in mixed form. Working back from these, we have in place of (47)

$$\left. \begin{aligned} \frac{\partial}{\partial x_\alpha}\Gamma_{\mu\nu}^\alpha + \Gamma_{\mu\beta}^\alpha \Gamma_{\nu\alpha}^\beta &= -\kappa(T_{\mu\nu} - \tfrac{1}{2}g_{\mu\nu}T), \\ \sqrt{-g} &= 1 \end{aligned} \right\} \quad . \quad (53)$$

---

*$g_{\sigma\tau}T_\sigma^\alpha = T_{\sigma\tau}$ and $g^{\sigma\beta}T_\sigma^\alpha = T^{\alpha\beta}$ are to be symmetrical tensors.

It must be admitted that this introduction of the energy-tensor of matter is not justified by the relativity postulate alone. For this reason we have here deduced it from the requirement that the energy of the gravitational field shall act gravitatively in the same way as any other kind of energy. But the strongest reason for the choice of these equations lies in their consequence, that the equations of conservation of momentum and energy, corresponding exactly to equations (49) and (49a), hold good for the components of the total energy. This will be shown in § 17.

### § 17. THE LAWS OF CONSERVATION IN THE GENERAL CASE

Equation (52) may readily be transformed so that the second term on the right-hand side vanishes. Contract (52) with respect to the indices $\mu$ and $\sigma$, and after multiplying the resulting equation by $\frac{1}{2}\delta^\sigma_\mu$, subtract it from equation (52). This gives

$$\frac{\partial}{\partial x_\alpha}\left(g^{\sigma\beta}\Gamma^\alpha_{\mu\beta} - \tfrac{1}{2}\delta^\sigma_\mu g^{\lambda\beta}\Gamma^\alpha_{\lambda\beta}\right) = -\kappa(t^\sigma_\mu + T^\sigma_\mu). \qquad (52a)$$

On this equation we perform the operation $\partial/\partial x_\sigma$. We have

$$\frac{\partial^2}{\partial x_\alpha \partial x_\sigma}\left(g^\sigma \Gamma^\alpha_{\beta\mu}\right) = -\frac{1}{2}\frac{\partial^2}{\partial x_\alpha \partial x_\sigma}\left[g^{\sigma\beta}g^{\alpha\lambda}\left(\frac{\partial g_{\mu\lambda}}{\partial x_\beta} + \frac{\partial g_{\beta\lambda}}{\partial x_\mu} - \frac{\partial g_{\mu\beta}}{\partial x_\lambda}\right)\right].$$

The first and third terms of the round brackets yield contributions which cancel one another, as may be seen by interchanging, in the contribution of the third term, the summation indices $\alpha$ and $\sigma$ on the one hand, and $\beta$ and $\lambda$ on the other. The second term may be remodelled by (31), so that we have

$$\frac{\partial^2}{\partial x_\alpha \partial x_\sigma}\left(g^{\sigma\beta}\Gamma^\alpha_{\mu\beta}\right) = \tfrac{1}{2}\frac{\partial^3 g^{\alpha\beta}}{\partial x_\alpha \partial x_\beta \partial x_\mu} \qquad . \qquad . \qquad (54)$$

The second term on the left-hand side of (52a) yields in the first place

$$-\tfrac{1}{2}\frac{\partial^2}{\partial x_\alpha \partial x_\mu}\left(g^{\lambda\beta}\Gamma^\alpha_{\lambda\beta}\right)$$

or

$$\tfrac{1}{4}\frac{\partial^2}{\partial x_\alpha \partial x_\mu}\left[g^{\lambda\beta}g^{\alpha\delta}\left(\frac{\partial g_{\delta\lambda}}{\partial x_\beta} + \frac{\partial g_{\delta\beta}}{\partial x_\lambda} - \frac{\partial g_{\lambda\beta}}{\partial x_\delta}\right)\right].$$

With the choice of co-ordinates which we have made, the term deriving from the last term in round brackets disappears by reason of (29). The other two may be combined, and together, by (31), they give

$$-\tfrac{1}{2}\frac{\partial^3 g^{\alpha\beta}}{\partial x_\alpha \partial x_\beta \partial x_\mu},$$

so that in consideration of (54), we have the identity

$$\frac{\partial^2}{\partial x_\alpha \partial x_\sigma}(g^{\rho\beta}\Gamma_{\mu\beta} - \tfrac{1}{2}\delta_\mu^\sigma g^{\lambda\beta}\Gamma_{\lambda\beta}^\alpha) \equiv 0 \qquad \cdot \quad \cdot \qquad (55)$$

From (55) and (52a), it follows that

$$\frac{\partial(t_\mu^\sigma + T_\mu^\sigma)}{\partial x_\sigma} = 0 \qquad \cdot \qquad \cdot \qquad \cdot \qquad (56)$$

Thus it results from our field equations of gravitation that the laws of conservation of momentum and energy are satisfied. This may be seen most easily from the consideration which leads to equation (49a); except that here, instead of the energy components $t^\sigma$ of the gravitational field, we have to introduce the totality of the energy components of matter and gravitational field.

## § 18. THE LAWS OF MOMENTUM AND ENERGY FOR MATTER, AS A CONSEQUENCE OF THE FIELD EQUATIONS

Multiplying (53) by $\partial g^{\mu\nu}/\partial x_\sigma$, we obtain, by the method adopted in § 15, in view of the vanishing of

$$g_{\mu\nu}\frac{\partial g^{\mu\nu}}{\partial x_\sigma},$$

the equation

$$\frac{\partial t_\sigma^\alpha}{\partial x_\alpha} + \tfrac{1}{2}\frac{\partial g^{\mu\nu}}{\partial x_\sigma}T_{\mu\nu} = 0,$$

or, in view of (56),

$$\frac{\partial T_\sigma^\alpha}{\partial x_\alpha} + \tfrac{1}{2}\frac{\partial g^{\mu\nu}}{\partial x_\sigma}T_{\mu\nu} = 0 \qquad \cdot \quad \cdot \quad \cdot (57)$$

Comparison with (41b) shows that with the choice of system of co-ordinates which we have made, this equation predicates nothing more or less than the vanishing of divergence of the material energy-tensor. Physically, the occurrence of the second term on the left-hand side shows that laws of conservation of momentum and energy do not apply in the strict sense for matter alone, or else that they apply only when the $g^{\mu\nu}$ are constant, i.e. when the field intensities of gravitation vanish. This second term is an expression for momentum, and for energy, as transferred per unit of volume and time from the gravitational field to matter. This is brought out still more clearly by re-writing (57) in the sense of (41) as

$$\frac{\partial T_\sigma^\alpha}{\partial x_\alpha} = -\Gamma_{\alpha\sigma}^\beta T_\beta^\alpha \qquad \cdot \qquad \cdot \qquad \cdot \qquad (57a)$$

The right side expresses the energetic effect of the gravitational field on matter.

Thus the field equations of gravitation contain four conditions which govern the course of material phenomena. They give the equations of material phenomena completely, if the latter is capable of being characterized by four differential equations independent of one another.*

# D. MATERIAL PHENOMENA

The mathematical aids developed in part B enable us forthwith to generalize the physical laws of matter (hydrodynamics, Maxwell's electrodynamics), as they are formulated in the special theory of relativity, so that they will fit in with the general theory of relativity. When this is done, the general principle of relativity does not indeed afford us a further limitation of possibilities; but it makes us acquainted with the influence of the gravitational field on all processes, without our having to introduce any new hypothesis whatever.

Hence it comes about that it is not necessary to introduce definite assumptions as to the physical nature of matter (in the narrower sense). In particular it may remain an open question whether the theory of the electromagnetic field in conjunction with that of the gravitational field furnishes a sufficient basis for the theory of matter or not. The general postulate of relativity is unable on principle to tell

*On this question cf. H. Hilbert, Nachr. d. K. Gesellsch. d. Wiss. zu Göttingen, Math.-phys. Klasse, 1915, p. 3.

us anything about this. It must remain to be seen, during the working out of the theory, whether electromagnetics and the doctrine of
gravitation are able in collaboration to perform what the former by
itself is unable to do.

## § 19. Euler's Equations for a Frictionless Adiabatic Fluid

Let $p$ and $\rho$ be two scalars, the former of which we call the "pressure,"
the latter the "density" of a fluid; and let an equation subsist between
them. Let the contravariant symmetrical tensor

$$T^{\alpha\beta} = -g^{\alpha\beta}p + \rho\frac{dx_\alpha}{ds}\frac{dx_\beta}{ds} \qquad \cdot \quad \cdot \quad \cdot \quad (58)$$

be the contravariant energy-tensor of the fluid. To it belongs the
covariant tensor

$$T_{\mu\nu} = -g_{\mu\nu}p + g_{\mu\alpha}\,g_{\mu\beta}\frac{dx_\alpha}{ds}\frac{dx_\beta}{ds}\rho, \qquad \cdot \quad \cdot \quad (58a)$$

as well as the mixed tensor*

$$T_\sigma^\alpha = -\delta_\sigma^\alpha p + g_{\sigma\beta}\frac{dx_\beta}{ds}\frac{dx_\alpha}{ds}\rho \qquad \cdot \quad \cdot \quad \cdot \quad (58b)$$

Inserting the right-hand side of (58b) in (57a), we obtain the Euler-
ian hydrodynamical equations of the general theory of relativity. They
give, in theory, a complete solution of the problem of motion, since
the four equations (57a), together with the given equation between $p$
and $\rho$, and the equation

$$g_{\alpha\beta}\frac{dx_\alpha}{ds}\frac{dx_\beta}{ds} = 1,$$

are sufficient, $g_{\alpha\beta}$ being given, to define the six unknowns

$$p, \rho, \frac{dx_1}{ds}, \frac{dx_2}{ds}, \frac{dx_3}{ds}, \frac{dx_4}{ds}.$$

If the $g_{\mu\nu}$ are also unknown, the equations (53) are brought in. These
are eleven equations for defining the ten functions $g_{\mu\nu}$, so that these

---

*For an observer using a system of reference in the sense of the special theory of relativity for an infinitely small region, and moving
with it, the density of energy $T_4^4$ equals $\rho - p$. This gives the definition of $\rho$. Thus $\rho$ is not constant for an incompressible fluid.

functions appear over-defined. We must remember, however, that the equations (57a) are already contained in the equations (53), so that the latter represent only seven independent equations. There is good reason for this lack of definition, in that the wide freedom of the choice of co-ordinates causes the problem to remain mathematically undefined to such a degree that three of the functions of space may be chosen at will.*

## § 20. MAXWELL'S ELECTROMAGNETIC FIELD EQUATIONS FOR FREE SPACE

Let $\phi_\nu$ be the components of a covariant vector—the electromagnetic potential vector. From them we form, in accordance with (36), the components $F_{\rho\sigma}$ of the covariant six-vector of the electromagnetic field, in accordance with the system of equations

$$F_{\rho\sigma} = \frac{\partial \phi_\rho}{\partial x_\sigma} - \frac{\partial \phi_\sigma}{\partial x_\rho} \quad . \quad . \quad . \quad (59)$$

It follows from (59) that the system of equations

$$\frac{\partial F_{\rho\sigma}}{\partial x_\tau} + \frac{\partial F_{\sigma\tau}}{\partial x_\rho} + \frac{\partial F_{\tau\rho}}{\partial x_\sigma} = 0 \quad . \quad . \quad . \quad (60)$$

is satisfied, its left side being, by (37), an antisymmetrical tensor of the third rank. System (60) thus contains essentially four equations which are written out as follows:—

$$\left.\begin{array}{c} \dfrac{\partial F_{23}}{\partial x_4} + \dfrac{\partial F_{34}}{\partial x_2} + \dfrac{\partial F_{42}}{\partial x_3} = 0 \\[2ex] \dfrac{\partial F_{34}}{\partial x_1} + \dfrac{\partial F_{41}}{\partial x_3} + \dfrac{\partial F_{13}}{\partial x_4} = 0 \\[2ex] \dfrac{\partial F_{41}}{\partial x_2} + \dfrac{\partial F_{12}}{\partial x_4} + \dfrac{\partial F_{24}}{\partial x_1} = 0 \\[2ex] \dfrac{\partial F_{12}}{\partial x_3} + \dfrac{\partial F_{23}}{\partial x_1} + \dfrac{\partial F_{31}}{\partial x_2} = 0 \end{array}\right\} \quad . \quad . \quad . \quad (60a)$$

*On the abandonment of the choice of co-ordinates with $g = -1$, there remain *four* functions of space with liberty of choice, corresponding to the four arbitrary functions at our disposal in the choice of co-ordinates.

This system corresponds to the second of Maxwell's systems of equations. We recognize this at once by setting

$$\left.\begin{array}{l} F_{23} = H_x, F_{14} = E_x \\ F_{31} = H_y, F_{24} = E_y \\ F_{12} = H_z, F_{34} = E_z \end{array}\right\} \quad \cdot \quad \cdot \quad \cdot \quad (61)$$

Then in place of (60a) we may set, in the usual notation of three-dimensional vector analysis,

$$\left.\begin{array}{l} -\dfrac{\partial H}{\partial t} = \text{curl } E \\ \\ \text{div } H = 0 \end{array}\right\} \quad \cdot \quad \cdot \quad \cdot \quad (60b)$$

We obtain Maxwell's first system by generalizing the form given by Minkowski. We introduce the contravariant six-vector associated with $F^{\alpha\beta}$

$$F^{\mu\nu} = g^{\mu\alpha} g^{\nu\beta} F_{\alpha\beta} \quad \cdot \quad \cdot \quad \cdot \quad \cdot \quad (62)$$

and also the contravariant vector $J^\mu$ of the density of the electric current. Then, taking (40) into consideration, the following equations will be invariant for any substitution whose invariant is unity (in agreement with the chosen coordinates):—

$$\frac{\partial}{\partial x_\nu} F^{\mu\nu} = J^\mu \quad \cdot \quad \cdot \quad \cdot \quad \cdot \quad (63)$$

Let

$$\left.\begin{array}{l} F^{23} = H'_x, F^{14} = -E'_x \\ F^{31} = H'_y, F^{24} = -E'_y \\ F^{12} = H'_z, F^{34} = -E'_z \end{array}\right\} \quad \cdot \quad \cdot \quad (64)$$

which quantities are equal to the quantities $H_x \ldots E_z$ in the special case of the restricted theory of relativity; and in addition

$$J^1 = j_x, J^2 = j_y, J^3 = j_z, J^4 = \rho,$$

we obtain in place of (63)

$$\left.\begin{array}{l} \dfrac{\partial E'}{\partial t} + j = \text{curl } H' \\ \\ \text{div } E' = \rho \end{array}\right\} \quad \cdot \quad \cdot \quad \cdot \quad (63a)$$

The equations (60), (62), and (63) thus form the generalization of Maxwell's field equations for free space, with the convention which we have established with respect to the choice of co-ordinates.

*The Energy-components of the Electromagnetic Field.*—We form the inner product

$$\kappa_\sigma = F_{\sigma\mu}J^\mu \qquad . \qquad . \qquad . \qquad (65)$$

By (61) its components, written in the three-dimensional manner, are

$$\left.\begin{array}{c} \kappa_1 = \rho E_x + [j\cdot H]^x \\ \cdot \qquad \cdot \qquad \cdot \qquad \cdot \\ \cdot \qquad \cdot \qquad \cdot \qquad \cdot \\ \kappa_4 = -(jE) \end{array}\right\} \qquad . \qquad . \qquad . \ (65a)$$

$\kappa_\sigma$ is a covariant vector the components of which are equal to the negative momentum, or, respectively, the energy, which is transferred from the electric masses to the electromagnetic field per unit of time and volume. If the electric masses are free, that is, under the sole influence of the electromagnetic field, the covariant vector $\kappa_\sigma$ will vanish.

To obtain the energy-components $T^\nu_\sigma$ of the electromagnetic field, we need only give to equation $\kappa_\sigma = 0$ the form of equation (57). From (63) and (65) we have in the first place

$$\kappa_\sigma = F_{\sigma\mu}\frac{\partial F^{\mu\nu}}{\partial x_\nu} = \frac{\partial}{\partial x_\nu}(F_{\sigma\mu}F^{\mu\nu}) - F^{\mu\rho}\frac{\partial F_{\sigma\mu}}{\partial x_\nu}.$$

The second term of the right-hand side, by reason of (60), permits the transformation

$$F^{\mu\nu}\frac{\partial F_{\sigma\mu}}{\partial x_\nu} = -\tfrac{1}{2}F^{\mu\nu}\frac{\partial F_{\mu\nu}}{\partial x_\sigma} = -\tfrac{1}{2}g^{\mu\alpha}g^{\nu\beta}F_{\alpha\beta}\frac{\partial F_{\mu\nu}}{\partial x_\sigma},$$

which latter expression may, for reasons of symmetry, also be written in the form

$$-\tfrac{1}{4}\left[g^{\mu\alpha}g^{\nu\beta}F_{\alpha\beta}\frac{\partial F_{\mu\nu}}{\partial x_\sigma} + g^{\mu\alpha}g^{\nu\beta}\frac{\partial F_{\alpha\beta}}{\partial x_\sigma}F_{\mu\nu}\right].$$

But for this we may set

$$-\tfrac{1}{4}\frac{\partial}{\partial x_\sigma}(g^{\mu\alpha}g^{\nu\beta}F_{\alpha\beta}F_{\mu\nu}) + \tfrac{1}{4}F_{\alpha\beta}F_{\mu\nu}\frac{\partial}{\partial x_\sigma}(g^{\mu\alpha}g^{\nu\beta}).$$

The first of these terms is written more briefly

$$-\tfrac{1}{4}\frac{\partial}{\partial x_\sigma}(F^{\mu\nu}F_{\mu\nu});$$

the second, after the differentiation is carried out, and after some reduction, results in

$$-\tfrac{1}{2}F^{\mu\tau}F_{\mu\nu}\,g^{\nu\rho}\frac{\partial g_{\sigma\tau}}{\partial x_\sigma}.$$

Taking all three terms together we obtain the relation

$$\kappa_\sigma = \frac{\partial T_\sigma^\nu}{\partial x_\nu} - \tfrac{1}{2}g^{\tau\mu}\frac{\partial g_{\mu\nu}}{\partial x_\sigma}T_\tau^\nu \qquad \cdot \quad \cdot \quad \cdot \quad (66)$$

where

$$T_\sigma^\nu = -F_{\sigma\alpha}F^{\nu\alpha} + \tfrac{1}{4}\delta_\sigma^\nu F_{\alpha\beta}F^{\alpha\beta}.$$

Equation (66), if $\kappa_\sigma$ vanishes, is, on account of (30), equivalent to (57) or (57a) respectively. Therefore the $T_\sigma^\nu$ are the energy-components of the electromagnetic field. With the help of (61) and (64), it is easy to show that these energy-components of the electromagnetic field in the case of the special theory of relativity give the well-known Maxwell-Poynting expressions.

We have now deduced the general laws which are satisfied by the gravitational field and matter, by consistently using a system of co-ordinates for which $\sqrt{-g} = 1$. We have thereby achieved a considerable simplification of formulae and calculations, without failing to comply with the requirement of general covariance; for we have drawn our equations from generally covariant equations by specializing the system of co-ordinates.

Still the question is not without a formal interest, whether with a correspondingly generalized definition of the energy-components of gravitational field and matter, even without specializing the system of co-ordinates, it is possible to formulate laws of conservation in the form of equation (56), and field equations of gravitation of the same nature as (52) or (52a), in such a manner that on the left we have a divergence (in the ordinary sense), and on the right the sum of the energy-components of matter and gravitation. I have found that in both cases this is actually so. But I do not think that the communication of my somewhat extensive reflexions on this subject would be worth while, because after all they do not give us anything that is materially new.

# E

## § 21. NEWTON'S THEORY AS A FIRST APPROXIMATION

As has already been mentioned more than once, the special theory of relativity as a special case of the general theory is characterized by the $g_{\mu\nu}$ having the constant values (4). From what has already been said, this means complete neglect of the effects of gravitation. We arrive at a closer approximation to reality by considering the case where the $g_{\mu\nu}$ differ from the values of (4) by quantities which are small compared with 1, and neglecting small quantities of second and higher order. (First point of view of approximation.)

It is further to be assumed that in the space-time territory under consideration the $g_{\mu\nu}$ at spatial infinity, with a suitable choice of co-ordinates, tend toward the values (4); i.e. we are considering gravitational fields which may be regarded as generated exclusively by matter in the finite region.

It might be thought that these approximations must lead us to Newton's theory. But to that end we still need to approximate the fundamental equations from a second point of view. We give our attention to the motion of a material point in accordance with the equations (16). In the case of the special theory of relativity the components

$$\frac{dx_1}{ds}, \frac{dx_2}{ds}, \frac{dx_3}{ds}$$

may take on any values. This signifies that any velocity

$$v = \sqrt{\left(\frac{dx_1}{dx_4}\right)^2 + \left(\frac{dx_2}{dx_4}\right)^2 + \left(\frac{dx_3}{dx_4}\right)^2}$$

may occur, which is less than the velocity of light *in vacuo*. If we restrict ourselves to the case which almost exclusively offers itself to our experience, of $v$ being small as compared with the velocity of light, this denotes that the components

$$\frac{dx_1}{ds}, \frac{dx_2}{ds}, \frac{dx_3}{ds}$$

are to be treated as small quantities, while $dx_4/ds$, to the second order of small quantities, is equal to one. (Second point of view of approximation.)

Now we remark that from the first point of view of approximation the magnitudes $\Gamma^\tau_{\mu\nu}$ are all small magnitudes of at least the first order. A glance at (46) thus shows that in this equation, from the second point of view of approximation, we have to consider only terms for which $\mu = \nu = 4$. Restricting ourselves to terms of lowest order we first obtain in place of (46) the equations

$$\frac{d^2 x_\tau}{dt^2} = \Gamma^\tau_{44}$$

where we have set $ds = dx_4 = dt$; or with restriction to terms which from the first point of view of approximation are of first order:—

$$\frac{d^2 x_\tau}{dt^2} = [44, \tau]\,(\tau = 1, 2, 3)$$

$$\frac{d^2 x_4}{dt^2} = -[44, 4].$$

If in addition we suppose the gravitational field to be a quasi-static field, by confining ourselves to the case where the motion of the matter generating the gravitational field is but slow (in comparison with the velocity of the propagation of light), we may neglect on the right-hand side differentiations with respect to the time in comparison with those with respect to the space co-ordinates, so that we have

$$\frac{d^2 x_\tau}{dt^2} = -\frac{1}{2}\frac{\partial g_{44}}{\partial x_\tau} \quad (\tau = 1, 2, 3) \qquad . \qquad . \qquad (67)$$

This is the equation of motion of the material point according to Newton's theory, in which $\frac{1}{2} g_{44}$ plays the part of the gravitational potential. What is remarkable in this result is that the component $g_{44}$ of the fundamental tensor alone defines, to a first approximation, the motion of the material point.

We now turn to the field equations (53). Here we have to take into consideration that the energy-tensor of "matter" is almost exclusively defined by the density of matter in the narrower sense, i.e. by the second term of the right-hand side of (58) [or, respectively, (58a) or (58b)]. If we form the approximation in question, all the components vanish with the one exception of $T_{44} = \rho = T$. On the left-hand side of (53) the second term is a small quantity of second order; the first yields, to the approximation in question,

$$\frac{\partial}{\partial x_1}[\mu\nu, 1] + \frac{\partial}{\partial x_2}[\mu\nu, 2] + \frac{\partial}{\partial x_3}[\mu\nu, 3] - \frac{\partial}{\partial x_4}[\mu\nu, 4].$$

For $\mu = \nu = 4$, this gives, with the omission of terms differentiated with respect to time,

$$-\tfrac{1}{2}\left(\frac{\partial^2 g_{44}}{\partial x_1^2} + \frac{\partial^2 g_{44}}{\partial x_2^2} + \frac{\partial^2 g_{44}}{\partial x_3^2}\right) = -\tfrac{1}{2}\nabla^2 g_{44}.$$

The last of equations (53) thus yields

$$\nabla^2 g_{44} = \kappa\rho \qquad \cdot \qquad \cdot \qquad \cdot \qquad (68)$$

The equations (67) and (68) together are equivalent to Newton's law of gravitation.

By (67) and (68) the expression for the gravitational potential becomes

$$-\frac{\kappa}{8\pi}\int \frac{\rho d\tau}{r} \qquad \cdot \qquad \cdot \qquad \cdot \qquad (68a)$$

while Newton's theory, with the unit of time which we have chosen, gives

$$-\frac{K}{c^2}\int \frac{\rho d\tau}{r}$$

in which K denotes the constant $6.7 \times 10^{-8}$, usually called the constant of gravitation. By comparison we obtain

$$\kappa = \frac{8\pi K}{c^2} = 1.87 \times 10^{-27} \qquad \cdot \qquad \cdot \qquad \cdot \qquad (69)$$

## § 22. BEHAVIOUR OF RODS AND CLOCKS IN THE STATIC GRAVITATIONAL FIELD. BENDING OF LIGHT-RAYS. MOTION OF THE PERIHELION OF A PLANETARY ORBIT

To arrive at Newton's theory as a first approximation we had to calculate only one component, $g_{44}$, of the ten $g_{\mu\nu}$ of the gravitational field, since this component alone enters into the first approximation, (67), of the equation for the motion of the material point in the gravitational field. From this, however, it is already apparent that other components of the $g_{\mu\nu}$ must differ from the values given in (4) by small quantities of the first order. This is required by the condition $g = -1$.

For a field-producing point mass at the origin of co-ordinates, we obtain, to the first approximation, the radially symmetrical solution

$$
\left.
\begin{aligned}
g_{\rho\sigma} &= -\delta_{\rho\sigma} - \alpha \frac{x_\rho x_\sigma}{r^3}(\rho, \sigma = 1, 2, 3) \\
g_{\rho 4} &= g_{4\rho} = 0 \qquad (\rho = 1, 2, 3) \\
g_{44} &= 1 - \frac{\alpha}{r}
\end{aligned}
\right\} \qquad \cdot \quad \cdot \quad (70)
$$

where $\delta_{\rho\sigma}$ is 1 or 0, respectively, accordingly as $\rho = \sigma$ or $\rho \neq \sigma$, and $r$ is the quantity $+\sqrt{x_1^2 + x_2^2 + x_3^2}$. On account of (68a)

$$
\alpha = \frac{\kappa M}{4\pi}, \qquad \cdot \quad \cdot \quad \cdot \quad \cdot \quad (70a)
$$

if M denotes the field-producing mass. It is easy to verify that the field equations (outside the mass) are satisfied to the first order of small quantities.

We now examine the influence exerted by the field of the mass M upon the metrical properties of space. The relation

$$
ds^2 = g_{\mu\nu}\, dx_\mu\, dx_\nu.
$$

always holds between the "locally" (§ 4) measured lengths and times $ds$ on the one hand, and the differences of co-ordinates $dx_\nu$ on the other hand.

For a unit-measure of length laid "parallel" to the axis of $x$, for example, we should have to set $ds^2 = -1$; $dx_2 = dx_3 = dx_4 = 0$. Therefore $-1 = g_{11} dx_1^2$. If, in addition, the unit-measure lies on the axis of $x$, the first of equations (70) gives

$$
g_{11} = -\left(1 + \frac{\alpha}{r}\right).
$$

From these two relations it follows that, correct to a first order of small quantities,

$$
dx = 1 - \frac{\alpha}{2r} \qquad \cdot \quad \cdot \quad \cdot \quad \cdot \quad (71)
$$

The unit measuring-rod thus appears a little shortened in relation to the system of co-ordinates by the presence of the gravitational field, if the rod is laid along a radius.

In an analogous manner we obtain the length of co-ordinates in tangential direction if, for example, we set

$$ds^2 = -1; dx_1 = dx_3 = dx_4 = 0; x_1 = r, x_2 = x_3 = 0.$$

The result is

$$-1 = g_{22}\,dx_2^2 = -dx_2^2 \quad \cdot \quad \cdot \quad \cdot \quad \cdot \quad (71a)$$

With the tangential position, therefore, the gravitational field of the point of mass has no influence on the length of a rod.

Thus Euclidean geometry does not hold even to a first approximation in the gravitational field, if we wish to take one and the same rod, independently of its place and orientation, as a realization of the same interval; although, to be sure, a glance at (70a) and (69) shows that the deviations to be expected are much too slight to be noticeable in measurements of the earth's surface.

Further, let us examine the rate of a unit clock, which is arranged to be at rest in a static gravitational field. Here we have for a clock period $ds = 1; dx_1 = dx_2 = dx_3 = 0$

Therefore

$$1 = g_{44}\,dx_4^2;$$

$$dx_4 = \frac{1}{\sqrt{g_{44}}} = \frac{1}{\sqrt{(1 + (g_{44} - 1))}} = 1 - \tfrac{1}{2}(g_{44} - 1)$$

or

$$dx_4 = 1 + \frac{\kappa}{8\pi}\int \rho\,\frac{d\tau}{r} \quad \cdot \quad \cdot \quad \cdot \quad \cdot \quad (72)$$

Thus the clock goes more slowly if set up in the neighbourhood of ponderable masses. From this it follows that the spectral lines of light reaching us from the surface of large stars must appear displaced towards the red end of the spectrum.*

---

*According to E. Freundlich, spectroscopical observations on fixed stars of certain types indicate the existence of an effect of this kind, but a crucial test of this consequence has not yet been made.

We now examine the course of light-rays in the static gravitational field. By the special theory of relativity the velocity of light is given by the equation

$$-dx_1^2 - dx_2 - dx_3^2 + dx_4^2 = 0$$

and therefore by the general theory of relativity by the equation

$$ds^2 = g_{\mu\nu} dx_\mu dx_\nu = 0 \quad \cdot \quad \cdot \quad \cdot \quad \cdot \quad (73)$$

If the direction, i.e. the ratio $dx_1 : dx_2 : dx_3$ is given, equation (73) gives the quantities

$$\frac{dx_1}{dx_4}, \frac{dx_2}{dx_4}, \frac{dx_3}{dx_4}$$

and accordingly the velocity

$$\sqrt{\left(\frac{dx_1}{dx_4}\right)^2 + \left(\frac{dx_2}{dx_4}\right)^2 + \left(\frac{dx_3}{dx_4}\right)^2} = \gamma$$

defined in the sense of Euclidean geometry. We easily recognize that the course of the light-rays must be bent with regard to the system of co-ordinates, if the $g_{\mu\nu}$ are not constant. If $n$ is a direction perpendicular to the propagation of light, the Huyghens principle shows that the light-ray, envisaged in the plane $(\gamma, n)$, has the curvature $-\partial\gamma/\partial n$.

We examine the curvature undergone by a ray of light passing by a mass M at the distance $\Delta$. If we choose the system of co-ordinates in agreement with the accompanying diagram, the total bending of the ray (calculated positively if concave towards the origin) is given in sufficient approximation by

$$B = \int_{-\infty}^{+\infty} \frac{\partial\gamma}{\partial x_1} dx_2,$$

while (73) and (70) give

$$\gamma = \sqrt{\left(-\frac{g_{44}}{g_{22}}\right)} = 1 - \frac{\alpha}{2r}\left(1 + \frac{x_2^2}{r^2}\right).$$

Carrying out the calculation, this gives

$$B = \frac{2\alpha}{\Delta} = \frac{\kappa M}{2\pi\Delta}. \quad \cdot \quad \cdot \quad \cdot \quad (74)$$

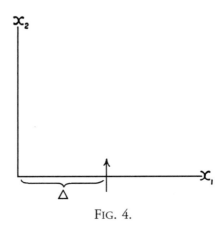

FIG. 4.

According to this, a ray of light going past the sun undergoes a deflexion of 1.7″; and a ray going past the planet Jupiter a deflexion of about .02″.

If we calculate the gravitational field to a higher degree of approximation, and likewise with corresponding accuracy the orbital motion of a material point of relatively infinitely small mass, we find a deviation of the following kind from the Kepler-Newton laws of planetary motion. The orbital ellipse of a planet undergoes a slow rotation, in the direction of motion, of amount

$$\varepsilon = 24\pi^3 \frac{\alpha^2}{T^2 c^2 (1 - e^2)} \qquad . \qquad . \qquad . \qquad (75)$$

per revolution. In this formula $a$ denotes the major semi-axis, $c$ the velocity of light in the usual measurement, $e$ the eccentricity, T the time of revolution in seconds.[*]

Calculation gives for the planet Mercury a rotation of the orbit of 43″ per century, corresopnding exactly to astronomical observation (Leverrier); for the astronomers have discovered in the motion of the perihelion of his planet, after allowing for disturbances by other planets, an inexplicable remainder of this magnitude.

---

[*]For the calculation I refer to the original papers: A. Einstein, Sitzungsber. d. Preuss. Akad. d. Wiss., 1915, p. 831; K. Schwarzschild, *ibid.*, 1916, p. 189.

# HAMILTON'S PRINCIPLE
# AND THE GENERAL THEORY
# OF RELATIVITY

BY

A. Einstein

*Translated from "Hamiltonsches Princip und allgemeine Relativitätstheorie,"*
*Sitzungsberichte der Preussischen Akad. d. Wissenschaften, 1916.*

THE general theory of relativity has recently been given in a particularly clear form by H. A. Lorentz and D. Hilbert,* who have deduced its equations from one single principle of variation. The same thing will be done in the present paper. But my purpose here is to present the fundamental connexions in as perspicuous a manner as possible, and in as general terms as is permissible from the point of view of the general theory of relativity. In particular we shall make as few specializing assumptions as possible, in marked contrast to Hilbert's treatment of the subject. On the other hand, in antithesis to my own most recent treatment of the subject, there is to be complete liberty in the choice of the system of co-ordinates.

§ 1. THE PRINCIPLE OF VARIATION AND THE FIELD-EQUATIONS
OF GRAVITATION AND MATTER

Let the gravitational field be described as usual by the tensor[†] of the $g_{\mu\nu}$ (or the $g^{\mu\nu}$); and matter, including the electromagnetic field, by any number of space-time functions $q_{(\rho)}$. How these functions may be characterized in the theory of invariants does not concern us. Further, let $\mathfrak{H}$ be a function of the

$$g^{\mu\nu}, g_\sigma^{\mu\nu}\left(=\frac{\partial g^{\mu\nu}}{\partial x_\sigma}\right) \text{ and } g_{\sigma\tau}^{\mu\nu}\left(=\frac{\partial^2 g^{\mu\nu}}{\partial x_\sigma \partial x_\tau}\right), \text{ the } q_{(\rho)} \text{ and } q_{(\rho)\alpha}\left(=\frac{\partial q_{(\rho)}}{\partial x_\alpha}\right).$$

---

*Four papers by Lorentz in the Publications of the Koninkl. Akad. yan Wetensch. te Amsterdam, 1915 and 1916; D. Hilbert, Göttinger Nachr., 1915, Part 3.
[†]No use is made for the present of the tensor character of the $g\mu\nu$.

The principle of variation

$$\delta \int \mathfrak{H} \, d\tau = 0 \qquad . \qquad . \qquad . \qquad . \qquad (1)$$

then gives us as many differential equations as there are functions $g_{\mu\nu}$ and $q_{(\rho)}$ to be defined, if the $g^{\mu\nu}$ and $q_{(\rho)}$ are varied independently of one another, and in such a way that at the limits of integration the $\delta q_{(\rho)}$, $\delta g^{\mu\nu}$, and $\dfrac{\partial}{\partial x_\sigma}(\delta g_{\mu\nu})$ all vanish.

We will now assume that $\mathfrak{H}$ is linear in the $g_{\sigma\tau}$, and that the coefficients of the $g_{\sigma\tau}^{\mu\nu}$ depend only on the $g^{\mu\nu}$. We may then replace the principle of variation (1) by one which is more convenient for us. For by appropriate partial integration we obtain

$$\int \mathfrak{H} \, d\tau = \int \mathfrak{H}^* \, d\tau + \mathrm{F} \qquad . \qquad . \qquad . \qquad (2)$$

where F denotes an integral over the boundary of the domain in question, and $\mathfrak{H}^*$ depends only on the $g^{\mu\nu}$, $g_\sigma^{\mu\nu}$, $q_{(\rho)}$, $q_{(\rho)\alpha}$, and no longer on the $g_{\sigma\tau}^{\mu\nu}$. From (2) we obtain, for such variations as are of interest to us,

$$\delta \int \mathfrak{H} \, d\tau = \delta \int \mathfrak{H}^* \, d\tau, \qquad . \qquad . \qquad . \qquad (3)$$

so that we may replace our principle of variation (1) by the more convenient form

$$\delta \int \mathfrak{H}^* \, d\tau = 0. \qquad . \qquad . \qquad . \qquad (1a)$$

By carrying out the variation of the $g^{\mu\nu}$ and the $q_{(\rho)}$ we obtain, as field-equations of gravitation and matter, the equations[†]

$$\frac{\partial}{\partial x_\alpha}\left(\frac{\partial \mathfrak{H}^*}{\partial g_\alpha^{\mu\nu}}\right) - \frac{\partial \mathfrak{H}^*}{\partial g^{\mu\nu}} = 0 \qquad . \qquad . \qquad . \qquad (4)$$

$$\frac{\partial}{\partial x_\alpha}\left(\frac{\partial \mathfrak{H}^*}{\partial q_{(\rho)\alpha}}\right) - \frac{\partial \mathfrak{H}^*}{\partial q_{(\rho)}} = 0 \qquad . \qquad . \qquad . \qquad (5)$$

---

[†]For brevity the summation symbols are omitted in the formulæ. Indices occurring twice in a term are always to be taken as summed. Thus in (4), for example, $\dfrac{\partial}{\partial x_\alpha}\left(\dfrac{\partial \mathfrak{H}^*}{\partial g_\alpha^{\mu\nu}}\right)$ denotes the term $\displaystyle\sum_\alpha \dfrac{\partial}{\partial x_\alpha}\left(\dfrac{\partial \mathfrak{H}^*}{\partial g_\alpha^{\mu\nu}}\right)$.

## § 2. Separate Existence of the Gravitational Field

If we make no restrictive assumption as to the manner in which $\mathfrak{H}$ depends on the $g^{\mu\nu}$, $g^{\mu\nu}_\sigma$, $g^{\mu\nu}_{\sigma\tau}$, $q_{(\rho)}$, $q_{(\rho)\alpha}$, the energy-components cannot be divided into two parts, one belonging to the gravitational field, the other to matter. To ensure this feature of the theory, we make the following assumption

$$\mathfrak{H} = \mathfrak{G} + \mathfrak{M} \qquad . \qquad . \qquad . \qquad (6)$$

where $\mathfrak{G}$ is to depend only on the $g^{\mu\nu}$, $g^{\mu\nu}_\sigma$, $g^{\mu\nu}_{\sigma\tau}$, and $\mathfrak{M}$ only on $g^{\mu\nu}$, $q_{(\rho)}$, $q_{(\rho)\alpha}$. Equations (4), (4a) then assume the form

$$\frac{\partial}{\partial x_\alpha}\left(\frac{\partial \mathfrak{G}^*}{\partial g^{\mu\nu}_\alpha}\right) - \frac{\partial \mathfrak{G}^*}{\partial g^{\mu\nu}} = \frac{\partial \mathfrak{M}}{\partial g^{\mu\nu}} \qquad . \qquad . \qquad . \qquad (7)$$

$$\frac{\partial}{\partial x_\alpha}\left(\frac{\partial \mathfrak{M}}{\partial q_{(\rho)\alpha}}\right) - \frac{\partial \mathfrak{M}}{\partial q_{(\rho)}} = 0 \qquad . \qquad . \qquad . \qquad (8)$$

Here $\mathfrak{G}^*$ stands in the same relation to $\mathfrak{G}$ as $\mathfrak{H}^*$ to $\mathfrak{H}$.

It is to be noted carefully that equations (8) or (5) would have to give way to others, if we were to assume $\mathfrak{M}$ or $\mathfrak{H}$ to be also dependent on derivatives of the $q_{(\rho)}$ of order higher than the first. Likewise it might be imaginable that the $q_{(\rho)}$ would have to be taken, not as independent of one another, but as connected by conditional equations. All this is of no importance for the following developments, as these are based solely on the equations (7), which have been found by varying our integral with respect to the $g^{\mu\nu}$.

## § 3. Properties of the Field Equations of Gravitation Conditioned by the Theory of Invariants

We now introduce the assumption that

$$ds^2 = g_{\mu\nu}\, dx_\mu dx_\nu \qquad . \qquad . \qquad . \qquad (9)$$

is an invariant. This determines the transformational character of the $g_{\mu\nu}$. As to the transformational character of the $q_{(\rho)}$, which describe matter, we make no supposition. On the other hand, let the functions $H = \dfrac{\mathfrak{H}}{\sqrt{-g}}$, as well as $G = \dfrac{\mathfrak{G}}{\sqrt{-g}}$, and $M = \dfrac{\mathfrak{M}}{\sqrt{-g}}$, be invariants in relation to any substitutions of space-time co-ordinates. From these assumptions follows the general covariance of the equations (7) and (8), deduced from (1). It further follows that G (apart from a constant factor) must be equal to the scalar of Riemann's tensor of curvature; because there is no other invariant with the properties required for G.[†] Thereby $\mathfrak{G}^*$ is also perfectly determined, and consequently the left-hand side of field equation (7) as well.[‡]

From the general postulate of relativity there follow certain properties of the function $\mathfrak{G}^*$ which we shall now deduce. For this purpose we carry through an infinitesimal transformation of the co-ordinates, by setting

$$x'_\nu = x_\nu + \Delta x_\nu \quad . \quad . \quad . \quad . \quad (10)$$

where the $\Delta x_\nu$ are arbitrary, infinitely small functions of the co-ordinates, and $x'_\nu$ are the co-ordinates, in the new system, of the world-point having the co-ordinates $x_\nu$ in the original system. As for the co-ordinates, so too for any other magnitude $\psi$, a law of transformation holds good, of the type

$$\psi' = \psi + \Delta\psi,$$

where $\Delta\psi$ must always be expressible by the $\Delta x_\nu$. From the covariant property of the $g^{\mu\nu}$ we easily deduce for the $g^{\mu\nu}$ and $g_\sigma^{\mu\nu}$ the laws of transformation

$$\Delta g^{\mu\nu} = g^{\mu\alpha}\frac{\partial(\Delta x_\nu)}{\partial x_\alpha} + g^{\nu\alpha}\frac{\partial(\Delta x_\mu)}{\partial x_\alpha} \quad . \quad . \quad (11)$$

$$\Delta g_\sigma^{\mu\nu} = \frac{\partial(\Delta g^{\mu\nu})}{\partial x_\sigma} - g_\alpha^{\mu\nu}\frac{\partial(\Delta x_\alpha)}{\partial x_\sigma} \quad . \quad . \quad (12)$$

Since $\mathfrak{G}^*$ depends only on the $g^{\mu\nu}$ and $g_\sigma^{\mu\nu}$, it is possible, with the help of (11) and (12), to calculate $\Delta\mathfrak{G}^*$. We thus obtain the equation

$$\sqrt{-g}\,\Delta\left(\frac{\mathfrak{G}^*}{\sqrt{-g}}\right) = S_\sigma^\nu\frac{\partial(\Delta x_\sigma)}{\partial x_\nu} + 2\frac{\partial\mathfrak{G}^*}{\partial g_\alpha^{\mu\sigma}}g^{\mu\nu}\frac{\partial^2\Delta x_\sigma}{\partial x_\nu\partial x_\alpha}, \quad (13)$$

where for brevity we have set

$$S_\sigma^\nu = 2\frac{\partial\mathfrak{G}^*}{\partial g^{\mu\sigma}}g^{\mu\nu} + 2\frac{\partial\mathfrak{G}^*}{\partial g_\alpha^{\mu\sigma}}g_\alpha^{\mu\nu} + \mathfrak{G}^*\delta_\sigma^\nu - \frac{\partial\mathfrak{G}^*}{\partial g_\nu^{\mu\alpha}}g_\sigma^{\mu\alpha}. \quad (14)$$

---

[†]Herein is to be found the reason why the general postulate of relativity leads to a very definite theory of gravitation.
[‡]By performing partial integration we obtain
$\mathfrak{G}^* = \sqrt{-g}\,g^{\mu\nu}[\{\mu\alpha,\beta\}\{\nu\beta,\alpha\} - \{\mu\nu,\alpha\}\{\alpha\beta,\beta\}].$

From these two equations we draw two inferences which are important for what follows. We know that $\dfrac{\mathfrak{G}}{\sqrt{-g}}$ is an invariant with respect to any substitution, but we do not know this of $\dfrac{\mathfrak{G}^*}{\sqrt{-g}}$. It is easy to demonstrate, however, that the latter quantity is an invariant with respect to any *linear* substitutions of the co-ordinates. Hence it follows that the right side of (13) must always vanish if all $\dfrac{\partial^2 \Delta x_\sigma}{\partial x_\nu \partial x_\alpha}$ vanish. Consequently $\mathfrak{G}^*$ must satisfy the identity

$$S_\sigma^\nu \equiv 0 \qquad . \qquad . \qquad . \qquad . \qquad (15)$$

If, further, we choose the $\Delta x_\nu$ so that they differ from zero only in the interior of a given domain, but in infinitesimal proximity to the boundary they vanish, then, with the transformation in question, the value of the boundary integral occurring in equation (2) does not change. Therefore $\Delta F = 0$, and, in consequence,[†]

$$\Delta \int \mathfrak{G} d\tau = \Delta \int \mathfrak{G}^* d\tau.$$

But the left-hand side of the equation must vanish, since both $\dfrac{\mathfrak{G}}{\sqrt{-g}}$ and $\sqrt{-g}\, d\tau$ are invariants. Consequently the right-hand side also vanishes. Thus, taking (14), (15), and (16) into consideration, we obtain, in the first place, the equation

$$\int \frac{\partial \mathfrak{G}^*}{\partial g_\alpha^{\mu\sigma}} g^{\mu\nu} \frac{\partial^2 (\Delta x_\sigma)}{\partial x_\nu \partial x_\alpha} d\tau = 0 \qquad . \qquad . \qquad . \qquad (16)$$

Transforming this equation by two partial integrations, and having regard to the liberty of choice of the $\Delta x_\sigma$, we obtain the identity

$$\frac{\partial^2}{\partial x_\nu \partial x_\alpha} \left( g^{\mu\nu} \frac{\partial \mathfrak{G}^*}{\partial g_\alpha^{\mu\sigma}} \right) \equiv 0 \qquad . \qquad . \qquad . \qquad (17)$$

From the two identities (16) and (17), which result from the invariance of $\dfrac{\mathfrak{G}}{\sqrt{-g}}$, and therefore from the postulate of general relativity, we now have to draw conclusions.

[†]By the introduction of the quantities $\mathfrak{G}$ and $\mathfrak{G}^*$ instead of $\mathfrak{H}$ and $\mathfrak{H}^*$.

We first transform the field equations (7) of gravitation by mixed multiplication by $g^{\mu\sigma}$. We then obtain (by interchanging the indices $\sigma$ and $\nu$), as equivalents of the field equations (7), the equations

$$\frac{\partial}{\partial x_\alpha}\left(g^{\mu\nu}\frac{\partial\mathfrak{G}^*}{\partial g^{\mu\sigma}_\alpha}\right) = -(\mathfrak{T}^\nu_\sigma + t^\nu_\sigma) \qquad . \qquad . \qquad (18)$$

where we have set

$$\mathfrak{T}^\nu_\sigma = -\frac{\partial\mathfrak{M}}{\partial g^{\mu\sigma}}g^{\mu\nu} \qquad . \qquad . \qquad . \qquad . \qquad . \qquad . \qquad (19)$$

$$t^\nu_\sigma = -\left(\frac{\partial\mathfrak{G}^*}{\partial g^{\mu\sigma}_\alpha}g^{\mu\nu}_\alpha + \frac{\partial\mathfrak{G}^*}{\partial g^{\mu\sigma}}g^{\mu\nu}\right) = \frac{1}{2}\left(\mathfrak{G}^*\delta^\nu_\sigma - \frac{\partial\mathfrak{G}^*}{\partial g^{\mu\alpha}_\nu}g^{\mu\alpha}_\sigma\right) \qquad (20)$$

The last expression for $t^\nu_\sigma$ is vindicated by (14) and (15). By differentiation of (18) with respect to $x_\nu$, and summation for $\nu$, there follows, in view of (17),

$$\frac{\partial}{\partial x_\nu}(\mathfrak{T}^\nu_\sigma + t^\nu_\sigma) = 0 \qquad . \qquad . \qquad . \qquad (21)$$

Equation (21) expresses the conservation of momentum and energy. We call $\mathfrak{T}^\nu_\sigma$ the components of the energy of matter, $t^\nu_\sigma$ the components of the energy of the gravitational field.

Having regard to (20), there follows from the field equations (7) of gravitation, by multiplication by $g^{\mu\nu}_\sigma$ and summation with respect to $\mu$ and $\nu$,

$$\frac{\partial t^\nu_\sigma}{\partial x_\nu} + \frac{1}{2}g^{\mu\nu}_\sigma\frac{\partial\mathfrak{M}}{\partial g^{\mu\nu}} = 0,$$

or, in view of (19) and (21),

$$\frac{\partial\mathfrak{T}^\nu_\sigma}{\partial x_\nu} + \frac{1}{2}g^{\mu\nu}_\sigma\mathfrak{T}_{\mu\nu} = 0 \qquad . \qquad . \qquad . \qquad (22)$$

where $\mathfrak{T}_{\mu\nu}$ denotes the quantities $g_{\nu\sigma}\mathfrak{T}^\sigma_\mu$. These are four equations which the energy-components of matter have to satisfy.

It is to be emphasized that the (generally covariant) laws of conservation (21) and (22) are deduced from the field equations (7) of gravitation, in combination with the postulate of general covariance (relativity) *alone*, without using the field equations (8) for material phenomena.

# COSMOLOGICAL CONSIDERATIONS ON THE GENERAL THEORY OF RELATIVITY

BY

A. EINSTEIN

*Translated from "Kosmologische Betrachtungen zur allgemeinen Relativitätstheorie,"
Sitzungsberichte der Preussischen Akad. d. Wissenschaften, 1917.*

IT is well known that Poisson's equation

$$\nabla^2 \phi = 4\pi K \rho \qquad \cdot \qquad \cdot \qquad \cdot \qquad \cdot \quad (1)$$

in combination with the equations of motion of a material point is not as yet a perfect substitute for Newton's theory of action at a distance. There is still to be taken into account the condition that at spatial infinity the potential $\phi$ tends toward a fixed limiting value. There is an analogous state of things in the theory of gravitation in general relativity. Here, too, we must supplement the differential equations by limiting conditions at spatial infinity, if we really have to regard the universe as being of infinite spatial extent.

In my treatment of the planetary problem I chose these limiting conditions in the form of the following assumption: it is possible to select a system of reference so that at spatial infinity all the gravitational potentials $g_{\mu\nu}$ become constant. But it is by no means evident *a priori* that we may lay down the same limiting conditions when we wish to take larger portions of the physical universe into consideration. In the following pages the reflexions will be given which, up to the present, I have made on this fundamentally important question.

## § 1. THE NEWTONIAN THEORY

It is well known that Newton's limiting condition of the constant limit for $\phi$ at spatial infinity leads to the view that the density of matter becomes zero at infinity. For we imagine that there may be a place in universal space round about which the gravitational field of matter, viewed on a large scale, possesses spherical symmetry. It then follows from Poisson's equation that, in order that $\phi$ may tend to a limit at infinity, the mean density $\rho$ must decrease toward zero more rapidly than $1/r^2$ as the distance $r$ from the centre increases.* In this sense, therefore, the universe according to Newton is finite, although it may possess an infinitely great total mass.

From this it follows in the first place that the radiation emitted by the heavenly bodies will, in part, leave the Newtonian system of the universe, passing radially outwards, to become ineffective and lost in the infinite. May not entire heavenly bodies fare likewise? It is hardly possible to give a negative answer to this question. For it follows from the assumption of a finite limit for $\phi$ at spatial infinity that a heavenly body with finite kinetic energy is able to reach spatial infinity by overcoming the Newtonian forces of attraction. By statistical mechanics this case must occur from time to time, as long as the total energy of the stellar system—transferred to one single star—is great enough to send that star on its journey to infinity, whence it never can return.

We might try to avoid this peculiar difficulty by assuming a very high value for the limiting potential at infinity. That would be a possible way, if the value of the gravitational potential were not itself necessarily conditioned by the heavenly bodies. The truth is that we are compelled to regard the occurrence of any great differences of potential of the gravitational field as contradicting the facts. These differences must really be of so low an order of magnitude that the stellar velocities generated by them do not exceed the velocities actually observed.

If we apply Boltzmann's law of distribution for gas molecules to the stars, by comparing the stellar system with a gas in thermal equilibrium, we find that the Newtonian stellar system cannot exist at all.

*$\rho$ is the mean density of matter, calculated for a region which is large as compared with the distance between neighbouring fixed stars, but small in comparison with the dimensions of the whole stellar system.

For there is a finite ratio of densities corresponding to the finite difference of potential between the centre and spatial infinity. A vanishing of the density at infinity thus implies a vanishing of the density at the centre.

It seems hardly possible to surmount these difficulties on the basis of the Newtonian theory. We may ask ourselves the question whether they can be removed by a modification of the Newtonian theory. First of all we will indicate a method which does not in itself claim to be taken seriously; it merely serves as a foil for what is to follow. In place of Poisson's equation we write

$$\nabla^2 \phi - \lambda \phi = 4\pi\kappa\rho \qquad \cdot \qquad \cdot \qquad \cdot \qquad (2)$$

where $\lambda$ denotes a universal constant. If $\rho_0$ be the uniform density of a distribution of mass, then

$$\phi = -\frac{4\pi\kappa}{\lambda}\rho_0 \qquad \cdot \qquad \cdot \qquad \cdot \qquad \cdot \qquad (3)$$

is a solution of equation (2). This solution would correspond to the case in which the matter of the fixed stars was distributed uniformly through space, if the density $\rho_0$ is equal to the actual mean density of the matter in the universe. The solution then corresponds to an infinite extension of the central space, filled uniformly with matter. If, without making any change in the mean density, we imagine matter to be non-uniformly distributed locally, there will be, over and above the $\phi$ with the constant value of equation (3), an additional $\phi$, which in the neighbourhood of denser masses will so much the more resemble the Newtonian field as $\lambda\phi$ is smaller in comparison with $4\pi\kappa\rho$.

A universe so constituted would have, with respect to its gravitational field, no centre. A decrease of density in spatial infinity would not have to be assumed, but both the mean potential and mean density would remain constant to infinity. The conflict with statistical mechanics which we found in the case of the Newtonian theory is not repeated. With a definite but extremely small density, matter is in equilibrium, without any internal material forces (pressures) being required to maintain equilibrium.

## § 2. THE BOUNDARY CONDITIONS ACCORDING TO THE GENERAL THEORY OF RELATIVITY

In the present paragraph I shall conduct the reader over the road that I have myself travelled, rather a rough and winding road, because otherwise I cannot hope that he will take much interest in the result at the end of the journey. The conclusion I shall arrive at is that the field equations of gravitation which I have championed hitherto still need a slight modification, so that on the basis of the general theory of relativity those fundamental difficulties may be avoided which have been set forth in § 1 as confronting the Newtonian theory. This modification corresponds perfectly to the transition from Poisson's equation (1) to equation (2) of § 1. We finally infer that boundary conditions in spatial infinity fall away altogether, because the universal continuum in respect of its spatial dimensions is to be viewed as a self-contained continuum of finite spatial (three-dimensional) volume.

The opinion which I entertained until recently, as to the limiting conditions to be laid down in spatial infinity, took its stand on the following considerations. In a consistent theory of relativity there can be no inertia *relatively to "space,"* but only an inertia of masses *relatively to one another.* If, therefore, I have a mass at a sufficient distance from all other masses in the universe, its inertia must fall to zero. We will try to formulate this condition mathematically.

According to the general theory of relativity the negative momentum is given by the first three components, the energy by the last component of the covariant tensor multiplied by $\sqrt{-g}$

$$m\sqrt{-g}\, g_{\mu\alpha} \frac{dx_\alpha}{ds} \qquad \cdot \qquad \cdot \qquad \cdot \qquad \cdot \ (4)$$

where, as always, we set

$$ds^2 = g_{\mu\nu}\, dx_\mu dx_\nu \qquad \cdot \qquad \cdot \qquad \cdot \ (5)$$

In the particularly perspicuous case of the possibility of choosing the system of co-ordinates so that the gravitational field at every point is spatially isotropic, we have more simply

$$ds^2 = -A\big(dx_1^2 + dx_2^2 + dx_3^2\big) + B dx_4^2.$$

If, moreover, at the same time

$$\sqrt{-g} = 1 = \sqrt{A^3B}$$

we obtain from (4), to a first approximation for small velocities,

$$m\frac{A}{\sqrt{B}}\frac{dx_1}{dx_4}, \quad m\frac{A}{\sqrt{B}}\frac{dx_2}{dx_4}, \quad m\frac{A}{\sqrt{B}}\frac{dx_3}{dx_4}$$

for the components of momentum, and for the energy (in the static case)

$$m\sqrt{B}.$$

From the expressions for the momentum, it follows that $m\dfrac{A}{\sqrt{B}}$ plays the part of the rest mass. As $m$ is a constant peculiar to the point of mass, independently of its position, this expression, if we retain the condition $\sqrt{g^-} = 1$ at spatial infinity, can vanish only when A diminishes to zero, while B increases to infinity. It seems, therefore, that such a degeneration of the co-efficients $g_{\mu\nu}$ is required by the postulate of relativity of all inertia. This requirement implies that the potential energy $m\sqrt{B}$ becomes infinitely great at infinity. Thus a point of mass can never leave the system; and a more detailed investigation shows that the same thing applies to light-rays. A system of the universe with such behaviour of the gravitational potentials at infinity would not therefore run the risk of wasting away which was mooted just now in connexion with the Newtonian theory.

I wish to point out that the simplifying assumptions as to the gravitational potentials on which this reasoning is based, have been introduced merely for the sake of lucidity. It is possible to find general formulations for the behaviour of the $g_{\mu\nu}$ at infinity which express the essentials of the question without further restrictive assumptions.

At this stage, with the kind assistance of the mathematician J. Grommer, I investigated centrally symmetrical, static gravitational fields, degenerating at infinity in the way mentioned. The gravitational potentials $g_{\mu\nu}$ were applied, and from them the energy-tensor $T_{\mu\nu}$ of matter was calculated on the basis of the field equations of gravitation. But here it proved that for the system of the fixed stars no boundary conditions of the kind can come into question at all, as was also rightly emphasized by the astronomer de Sitter recently.

For the contravariant energy-tensor $T^{\mu\nu}$ of ponderable matter is given by

$$T^{\mu\nu} = \rho \frac{dx_\mu}{ds} \frac{dx_\nu}{ds},$$

where $\rho$ is the density of matter in natural measure. With an appropriate choice of the system of co-ordinates the stellar velocities are very small in comparison with that of light. We may, therefore, substitute $\sqrt{g_{44}}\, dx_4$ for $ds$. This shows us that all components of $T^{\mu\nu}$ must be very small in comparison with the last component $T^{44}$. But it was quite impossible to reconcile this condition with the chosen boundary conditions. In the retrospect this result does not appear astonishing. The fact of the small velocities of the stars allows the conclusion that wherever there are fixed stars, the gravitational potential (in our case $\sqrt{B}$) can never be much greater than here on earth. This follows from statistical reasoning, exactly as in the case of the Newtonian theory. At any rate, our calculations have convinced me that such conditions of degeneration for the $g_{\mu\nu}$ in spatial infinity may not be postulated.

After the failure of this attempt, two possibilities next present themselves.

(a) We may require, as in the problem of the planets, that, with a suitable choice of the system of reference, the $g_{\mu\nu}$ in spatial infinity approximate to the values

$$
\begin{array}{cccc}
-1 & 0 & 0 & 0 \\
0 & -1 & 0 & 0 \\
0 & 0 & -1 & 0 \\
0 & 0 & 0 & 1
\end{array}
$$

(b) We may refrain entirely from laying down boundary conditions for spatial infinity claiming general validity; but at the spatial limit of the domain under consideration we have to give the $g_{\mu\nu}$ separately in each individual case, as hitherto we were accustomed to give the initial conditions for time separately.

The possibility (b) holds out no hope of solving the problem, but amounts to giving it up. This is an incontestable position, which is taken up at the present time by de Sitter.* But I must confess that such a complete resignation in this fundamental question is for me a difficult thing. I should not make up my mind to it until every effort to make headway toward a satisfactory view had proved to be vain.

Possibility (*a*) is unsatisfactory in more respects than one. In the first place those boundary conditions pre-suppose a definite choice of the system of reference, which is contrary to the spirit of the relativity principle. Secondly, if we adopt this view, we fail to comply with the requirement of the relativity of inertia. For the inertia of a material point of mass *m* (in natural measure) depends upon the $g_{\mu\nu}$; but these differ but little from their postulated values, as given above, for spatial infinity. Thus inertia would indeed be *influenced*, but would not be *conditioned* by matter (present in finite space). If only one single point of mass were present, according to this view, it would possess inertia, and in fact an inertia almost as great as when it is surrounded by the other masses of the actual universe. Finally, those statistical objections must be raised against this view which were mentioned in respect of the Newtonian theory.

From what has now been said it will be seen that I have not succeeded in formulating boundary conditions for spatial infinity. Nevertheless, there is still a possible way out, without resigning as suggested under (*b*). For if it were possible to regard the universe as a continuum which is *finite (closed) with respect to its spatial dimensions*, we should have no need at all of any such boundary conditions. We shall proceed to show that both the general postulate of relativity and the fact of the small stellar velocities are compatible with the hypothesis of a spatially finite universe; though certainly, in order to carry through this idea, we need a generalizing modification of the field equations of gravitation.

## § 3. The Spatially Finite Universe with a Uniform Distribution of Matter

According to the general theory of relativity the metrical character (curvature) of the four-dimensional space-time continuum is defined

*de Sitter, Akad. van Wetensch, te Amsterdam, 8 Nov., 1916.

at every point by the matter at that point and the state of that matter. Therefore, on account of the lack of uniformity in the distribution of matter, the metrical structure of this continuum must necessarily be extremely complicated. But if we are concerned with the structure only on a large scale, we may represent matter to ourselves as being uniformly distributed over enormous spaces, so that its density of distribution is a variable function which varies extremely slowly. Thus our procedure will somewhat resemble that of the geodesists who, by means of an ellipsoid, approximate to the shape of the earth's surface, which on a small scale is extremely complicated.

The most important fact that we draw from experience as to the distribution of matter is that the relative velocities of the stars are very small as compared with the velocity of light. So I think that for the present we may base our reasoning upon the following approximative assumption. There is a system of reference relatively to which matter may be looked upon as being permanently at rest. With respect to this system, therefore, the contravariant energy-tensor $T^{\mu\nu}$ of matter is, by reason of (5), of the simple form

$$\left.\begin{array}{cccc} 0 & 0 & 0 & 0 \\ 0 & 0 & 0 & 0 \\ 0 & 0 & 0 & 0 \\ 0 & 0 & 0 & \rho \end{array}\right\} \qquad \cdot \qquad \cdot \qquad \cdot \qquad \cdot \;(6)$$

The scalar $\rho$ of the (mean) density of distribution may be *a priori* a function of the space co-ordinates. But if we assume the universe to be spatially finite, we are prompted to the hypothesis that $\rho$ is to be independent of locality. On this hypothesis we base the following considerations.

As concerns the gravitational field, it follows from the equation of motion of the material point

$$\frac{d^2 x_\nu}{ds^2} + \{\alpha\beta, \nu\}\frac{dx_\alpha}{ds}\frac{dx_\beta}{ds} = 0$$

that a material point in a static gravitational field can remain at rest only when $g_{44}$ is independent of locality. Since, further, we presuppose

112

independence of the time co-ordinate $x_4$ for all magnitudes, we may demand for the required solution that, for all $x_\nu$,

$$g_{44} = 1 \qquad \cdot \qquad \cdot \qquad \cdot \qquad \cdot \qquad (7)$$

Further, as always with static problems, we shall have to set

$$g_{14} = g_{24} = g_{34} = 0 \qquad \cdot \qquad \cdot \qquad \cdot \qquad (8)$$

It remains now to determine those components of the gravitational potential which define the purely spatial-geometrical relations of our continuum $(g_{11}, g_{12}, \ldots g_{33})$. From our assumption as to the uniformity of distribution of the masses generating the field, it follows that the curvature of the required space must be constant. With this distribution of mass, therefore, the required finite continuum of the $x_1$, $x_2$, $x_3$, with constant $x_4$, will be a spherical space.

We arrive at such a space, for example, in the following way. We start from a Euclidean space of four dimensions, $\xi_1$, $\xi_2$, $\xi_3$, $\xi_4$, with a linear element $d\sigma$; let, therefore,

$$d\sigma^2 = d\xi_1^2 + d\xi_2^2 + d\xi_3^2 + d\xi_4^2 \qquad \cdot \qquad \cdot \qquad \cdot \qquad (9)$$

In this space we consider the hyper-surface

$$R^2 = \xi_1^2 + \xi_2^2 + \xi_3^2 + \xi_4^2, \qquad \cdot \qquad \cdot \qquad \cdot \qquad (10)$$

where R denotes a constant. The points of this hyper-surface form a three-dimensional continuum, a spherical space of radius of curvature R.

The four-dimensional Euclidean space with which we started serves only for a convenient definition of our hyper-surface. Only those points of the hyper-surface are of interest to us which have metrical properties in agreement with those of physical space with a uniform distribution of matter. For the description of this three-dimensional continuum we may employ the co-ordinates $\xi_1$, $\xi_2$, $\xi_3$ (the projection upon the hyper-plane $\xi_4 = 0$) since, by reason of (10), $\xi_4$ can be expressed in terms of $\xi_1$, $\xi_2$, $\xi_3$. Eliminating $\xi_4$ from (9), we obtain for the linear element of the spherical space the expression

$$\left. \begin{aligned} d\sigma^2 &= \gamma_{\mu\nu} \, d\xi_\mu d\xi_\nu \\ \gamma_{\mu\nu} &= \delta_{\mu\nu} + \frac{\xi_\mu \xi_\nu}{R^2 - \rho^2} \end{aligned} \right\} \qquad \cdot \qquad \cdot \qquad \cdot \qquad (11)$$

where $\delta_{\mu\nu} = 1$, if $\mu = \nu$; $\delta_{\mu\nu} = 0$, if $\mu \neq \nu$, and $\rho^2 = \xi_1^2 + \xi_2^2 + \xi_3^2$. The co-ordinates chosen are convenient when it is a question of examining the environment of one of the two points $\xi_1 = \xi_2 = \xi_3 = 0$.

Now the linear element of the required four-dimensional space-time universe is also given us. For the potential $g_{\mu\nu}$, both indices of which differ from 4, we have to set

$$g_{\mu\nu} = -\left(\delta_{\mu\nu} + \frac{x_\mu x_\nu}{R^2 - (x_1^2 + x_2^2 + x_3^2)}\right) \quad . \quad . \quad (12)$$

which equation, in combination with (7) and (8), perfectly defines the behaviour of measuring-rods, clocks, and light-rays.

## § 4. ON AN ADDITIONAL TERM FOR
### THE FIELD EQUATIONS OF GRAVITATION

My proposed field equations of gravitation for any chosen system of co-ordinates run as follows:—

$$\left.\begin{array}{l} G_{\mu\nu} = -\kappa(T_{\mu\nu} - \tfrac{1}{2} g_{\mu\nu}T), \\[2mm] G_{\mu\nu} = -\dfrac{\partial}{\partial x_\alpha}\{\mu\nu, \alpha\} + \{\mu\alpha, \beta\}\{\nu\beta, \alpha\} \\[3mm] \qquad + \dfrac{\partial^2 \log\sqrt{-g}}{\partial x_\mu \partial x_\nu} - \{\mu\nu, \alpha\}\dfrac{\partial \log\sqrt{-g}}{\partial x_\alpha} \end{array}\right\} \quad (13)$$

The system of equations (13) is by no means satisfied when we insert for the $g_{\mu\nu}$ the values given in (7), (8), and (12), and for the (contravariant) energy-tensor of matter the values indicated in (6). It will be shown in the next paragraph how this calculation may conveniently be made. So that, if it were certain that the field equations (13) which I have hitherto employed were the only ones compatible with the postulate of general relativity, we should probably have to conclude that the theory of relativity does not admit the hypothesis of a spatially finite universe.

However, the system of equations (14) allows a readily suggested extension which is compatible with the relativity postulate, and is perfectly analogous to the extension of Poisson's equation given by equation (2). For on the left-hand side of field equation (13) we may add the fundamental tensor $g_{\mu\nu}$, multiplied by a universal constant, $-\lambda$, at present unknown, without destroying the general covariance. In place of field equation (13) we write

$$G_{\mu\nu} - \lambda g_{\mu\nu} = -\kappa(T_{\mu\nu} - \tfrac{1}{2} g_{\mu\nu} T) \quad \cdot \quad \cdot \quad (13a)$$

This field equation, with $\lambda$ sufficiently small, is in any case also compatible with the facts of experience derived from the solar system. It also satisfies laws of conservation of momentum and energy, because we arrive at (13a) in place of (13) by introducing into Hamilton's principle, instead of the scalar of Riemann's tensor, this scalar increased by a universal constant; and Hamilton's principle, of course, guarantees the validity of laws of conservation. It will be shown in § 5 that field equation (13a) is compatible with our conjectures on field and matter.

## § 5. CALCULATION AND RESULT

Since all points of our continuum are on an equal footing, it is sufficient to carry through the calculation for *one* point, e.g. for one of the two points with the co-ordinates

$$x_1 = x_2 = x_3 = x_4 = 0.$$

Then for the $g_{\mu\nu}$ in (13a) we have to insert the values

$$\begin{matrix} -1 & 0 & 0 & 0 \\ 0 & -1 & 0 & 0 \\ 0 & 0 & -1 & 0 \\ 0 & 0 & 0 & 1 \end{matrix}$$

wherever they appear differentiated only once or not at all. We thus obtain in the first place

$$G_{\mu\nu} = \frac{\partial}{\partial x_1}[\mu\nu, 1] + \frac{\partial}{\partial x_2}[\mu\nu, 2] + \frac{\partial}{\partial x_3}[\mu\nu, 3] + \frac{\partial^2 \log\sqrt{-g}}{\partial x_\mu \partial x_\nu}.$$

From this we readily discover, taking (7), (8), and (13) into account, that all equations (13*a*) are satisfied if the two relations

$$-\frac{2}{R^2} + \lambda = -\frac{\kappa\rho}{2}, \quad -\lambda = -\frac{\kappa\rho}{2},$$

or

$$\lambda = \frac{\kappa\rho}{2} = \frac{1}{R^2} \quad \cdot \quad \cdot \quad \cdot \quad \cdot \quad (14)$$

are fulfilled.

Thus the newly introduced universal constant $\lambda$ defines both the mean density of distribution $\rho$ which can remain in equilibrium and also the radius R and the volume $2\pi^2R^3$ of spherical space. The total mass M of the universe, according to our view, is finite, and is in fact

$$M = \rho \cdot 2\pi^2R^3 = 4\pi^2\frac{R}{\kappa} = \pi^2\sqrt{\frac{32}{\kappa^3\rho}} \qquad \cdot \quad \cdot \quad (15)$$

Thus the theoretical view of the actual universe, if it is in correspondence with our reasoning, is the following. The curvature of space is variable in time and place, according to the distribution of matter, but we may roughly approximate to it by means of a spherical space. At any rate, this view is logically consistent, and from the standpoint of the general theory of relativity lies nearest at hand; whether, from the standpoint of present astronomical knowledge, it is tenable, will not here be discussed. In order to arrive at this consistent view, we admittedly had to introduce an extension of the field equations of gravitation which is not justified by our actual knowledge of gravitation. It is to be emphasized, however, that a positive curvature of space is given by our results, even if the supplementary term is not introduced. That term is necessary only for the purpose of making possible a quasi-static distribution of matter, as required by the fact of the small velocities of the stars.

# DO GRAVITATIONAL FIELDS PLAY AN ESSENTIAL PART IN THE STRUCTURE OF THE ELEMENTARY PARTICLES OF MATTER?

BY

A. EINSTEIN

*Translated from "Spielen Gravitationsfelder im Aufber der materiellen Elementarteilchen eine wesentliche Rolle?" Sitzungsberichte der Preussischen Akad. d. Wissenschaften, 1919.*

NEITHER the Newtonian nor the relativistic theory of gravitation has so far led to any advance in the theory of the constitution of matter. In view of this fact it will be shown in the following pages that there are reasons for thinking that the elementary formations which go to make up the atom are held together by gravitational forces.

§ 1. DEFECTS OF THE PRESENT VIEW

Great pains have been taken to elaborate a theory which will account for the equilibrium of the electricity constituting the electron. G. Mie, in particular, has devoted deep researches to this question. His theory, which has found considerable support among theoretical physicists, is based mainly on the introduction into the energy-tensor of supplementary terms depending on the components of the electro-dynamic potential, in addition to the energy terms of the Maxwell-Lorentz theory. These new terms, which in outside space are unimportant, are nevertheless effective in the interior of the electrons in maintaining equilibrium against the electric forces of repulsion. In spite of the beauty of the formal structure of this theory, as erected by Mie, Hilbert, and Weyl, its physical results have hitherto been unsatisfactory. On the one hand the multiplicity of possibilities is discouraging, and on the other hand those additional terms have not as yet allowed themselves to be framed in such a simple form that the solution could be satisfactory.

So far the general theory of relativity has made no change in this state of the question. If we for the moment disregard the additional cosmological term, the field equations take the form

$$G_{\mu\nu} - \tfrac{1}{2} g_{\mu\nu} G = -\kappa T_{\mu\nu} \qquad \cdot \qquad \cdot \qquad \cdot \quad (1)$$

where $G_{\mu\nu}$ denotes the contracted Riemann tensor of curvature, G the scalar of curvature formed by repeated contraction, and $T_{\mu\nu}$ the energy-tensor of "matter." The assumption that the $T_{\mu\nu}$ do *not* depend on the derivatives of the $g_{\mu\nu}$ is in keeping with the historical development of these equations. For these quantities are, of course, the energy-components in the sense of the special theory of relativity, in which variable $g_{\mu\nu}$ do not occur. The second term on the left-hand side of the equation is so chosen that the divergence of the left-hand side of (1) vanishes identically, so that taking the divergence of (1), we obtain the equation

$$\frac{\partial \mathfrak{T}_\mu^\sigma}{\partial x_\sigma} + \tfrac{1}{2} g_\mu^{\sigma\tau} \mathfrak{T}_{\sigma\tau} = 0 \qquad \cdot \qquad \cdot \qquad \cdot \qquad \cdot \quad (2)$$

which in the limiting case of the special theory of relativity gives the complete equations of conservation

$$\frac{\partial T_{\mu\nu}}{\partial x_\nu} = 0.$$

Therein lies the physical foundation for the second term of the left-hand side of (1). It is by no means settled *a priori* that a limiting transition of this kind has any possible meaning. For if gravitational fields do play an essential part in the structure of the particles of matter, the transition to the limiting case of constant $g_{\mu\nu}$ would, for them, lose its justification, for indeed, with constant $g_{\mu\nu}$ there could not be any particles of matter. So if we wish to contemplate the possibility that gravitation may take part in the structure of the fields which constitute the corpuscles, we cannot regard equation (1) as confirmed.

Placing in (1) the Maxwell-Lorentz energy-components of the electromagnetic field $\phi_{\mu\nu}$,

$$T_{\mu\nu} = \tfrac{1}{4} g_{\mu\nu} \phi_{\sigma\tau} \phi^{\sigma\tau} - \phi_{\mu\sigma} \phi_{\nu\tau} g^{\sigma\tau}, \quad \cdot \quad \cdot \quad \cdot \ (3)$$

we obtain for (2), by taking the divergence, and after some reduction,*

$$\phi_{\mu\sigma} \Im^\sigma = 0 \quad \cdot \quad \cdot \quad \cdot \quad \cdot \ (4)$$

where, for brevity, we have set

$$\frac{\partial}{\partial x_r} \left( \sqrt{-g}\, \phi_{\mu\nu} g^{\mu\sigma} g^{\nu\tau} \right) = \frac{\partial \mathfrak{f}^{\sigma\tau}}{\partial x_\tau} = \Im^\sigma \quad \cdot \quad \cdot \ (5)$$

In the calculation we have employed the second of Maxwell's systems of equations

$$\frac{\partial \phi_{\mu\nu}}{\partial x_\rho} + \frac{\partial \phi_{\nu\rho}}{\partial x_\mu} + \frac{\partial \phi_{\rho\mu}}{\partial x_\nu} = 0 \quad \cdot \quad \cdot \quad \cdot \ (6)$$

We see from (4) that the current-density $\Im^\sigma$ must everywhere vanish. Therefore, by equation (1), we cannot arrive at a theory of the electron by restricting ourselves to the electro-magnetic components of the Maxwell-Lorentz theory, as has long been known. Thus if we hold to (1) we are driven on to the path of Mie's theory.[†]

Not only the problem of matter, but the cosmological problem as well, leads to doubt as to equation (1). As I have shown in the previous paper, the general theory of relativity requires that the universe be spatially finite. But this view of the universe necessitated an extension of equations (1), with the introduction of a new universal constant $\lambda$, standing in a fixed relation to the total mass of the universe (or, respectively, to the equilibrium density of matter). This is gravely detrimental to the formal beauty of the theory.

## § 2. The Field Equations Freed of Scalars

The difficulties set forth above are removed by setting in place of field equations (1) the field equations

$$G_{\mu\nu} - \tfrac{1}{4} g_{\mu\nu} G = -\kappa T_{\mu\nu} \quad \cdot \quad \cdot \quad \cdot \ (1a)$$

where $T_{\mu\nu}$ denotes the energy-tensor of the electromagnetic field given by (3).

*Cf. e.g. A. Einstein, Sitzungsber. d. Preuss. Akad. d. Wiss., 1916, pp. 187, 188.
[†]Cf. D. Hilbert, Göttinger Nachr., 20 Nov., 1915.

The formal justification for the factor $-\frac{1}{4}$ in the second term of this equation lies in its causing the scalar of the left-hand side,

$$g^{\mu\nu}(G_{\mu\nu} - \tfrac{1}{4} g_{\mu\nu} G),$$

to vanish identically, as the scalar $g^{\nu\mu} T_{\mu\nu}$ of the right-hand side does by reason of (3). If we had reasoned on the basis of equations (1) instead of (1a), we should, on the contrary, have obtained the condition $G = 0$, which would have to hold good everywhere for the $g_{\mu\nu}$, independently of the electric field. It is clear that the system of equations [(1a), (3)] is a consequence of the system [(1), (3)], but not conversely.

We might at first sight feel doubtful whether (1a) together with (6) sufficiently define the entire field. In a generally relativistic theory we need $n - 4$ differential equations, independent of one another, for the definition of $n$ independent variables, since in the solution, on account of the liberty of choice of the co-ordinates, four quite arbitrary functions of all co-ordinates must naturally occur. Thus to define the sixteen independent quantities $g_{\mu\nu}$ and $\phi_{\mu\nu}$ we require twelve equations, all independent of one another. But as it happens, nine of the equations (1a), and three of the equations (6) are independent of one another.

Forming the divergence of (1a), and taking into account that the divergence of $G_{\mu\nu} - \frac{1}{2} g_{\mu\nu} G$ vanishes, we obtain

$$\phi_{\sigma\alpha} J^\alpha + \frac{1}{4\kappa} \frac{\partial G}{\partial x_\sigma} = 0 \qquad . \quad . \quad . \qquad (4a)$$

From this we recognize first of all that the scalar of curvature G in the four-dimensional domains in which the density of electricity vanishes, is constant. If we assume that all these parts of space are connected, and therefore that the density of electricity differs from zero only in separate "world-threads," then the scalar of curvature, everywhere outside these world-threads, possesses a constant value $G_0$. But equation (4a) also allows an important conclusion as to the behaviour of G within the domains having a density of electricity other than zero. If, as is customary, we regard electricity as a moving density of charge, by setting

$$J^\sigma = \frac{\Im^\sigma}{\sqrt{-g}} = \rho \frac{dx_\sigma}{ds}, \qquad . \quad . \quad . \qquad (7)$$

we obtain from (4a) by inner multiplication by $J^\sigma$, on account of the antisymmetry of $\phi_{\mu\nu}$, the relation

$$\frac{\partial G}{\partial x_\sigma} \frac{dx_\sigma}{ds} = 0 \qquad . \quad . \quad . \quad . \quad (8)$$

Thus the scalar of curvature is constant on every world-line of the motion of electricity. Equation (4a) can be interpreted in a graphic manner by the statement: The scalar of curvature plays the part of a negative pressure which, outside of the electric corpuscles, has a constant value $G_0$. In the interior of every corpuscle there subsists a negative pressure (positive $G - G_0$) the fall of which maintains the electro-dynamic force in equilibrium. The minimum of pressure, or, respectively, the maximum of the scalar of curvature, does not change with time in the interior of the corpuscle.

We now write the field equations (1a) in the form

$$(G_{\mu\nu} - \tfrac{1}{2} g_{\mu\nu} G) + \tfrac{1}{4} g_{\mu\nu} G_0 = -\kappa (T_{\mu\nu} + \tfrac{1}{4\kappa} g_{\mu\nu} (G - G_0)) \qquad (9)$$

On the other hand, we transform the equations supplied with the cosmological term as already given

$$G_{\mu\nu} - \lambda g_{\mu\nu} = -\kappa (T_{\mu\nu} + \tfrac{1}{2} g_{\mu\nu} T).$$

Subtracting the scalar equation multiplied by $\tfrac{1}{2}$, we next obtain

$$(G_{\mu\nu} - \tfrac{1}{2} g_{\mu\nu} G) + g_{\mu\nu} \lambda = -\kappa T_{\mu\nu}.$$

Now in regions where only electrical and gravitational fields are present, the right-hand side of this equation vanishes. For such regions we obtain, by forming the scalar,

$$-G + 4\lambda = 0.$$

In such regions, therefore, the scalar of curvature is constant, so that $\lambda$ may be replaced by $\tfrac{1}{4} G_0$. Thus we may write the earlier field equation (1) in the form

$$G_{\mu\nu} - \tfrac{1}{2} g_{\mu\nu} G + \tfrac{1}{4} g_{\mu\nu} G_0 = -\kappa T_{\mu\nu} \qquad . \quad . \quad (10)$$

Comparing (9) with (10), we see that there is no difference between the new field equations and the earlier ones, except that instead of $T_{\mu\nu}$

as tensor of "gravitating mass" there now occurs $T_{\mu\nu} + \dfrac{1}{4\kappa} g_{\mu\nu}(G - G_0)$ which is independent of the scalar of curvature. But the new formulation has this great advantage, that the quantity $\lambda$ appears in the fundamental equations as a constant of integration, and no longer as a universal constant peculiar to the fundamental law.

## § 3. ON THE COSMOLOGICAL QUESTION

The last result already permits the surmise that with our new formulation the universe may be regarded as spatially finite, without any necessity for an additional hypothesis. As in the preceding paper I shall again show that with a uniform distribution of matter, a spherical world is compatible with the equations.

In the first place we set

$$ds^2 = -\gamma_{ik}\,dx_i dx_k + dx_4^2\ (i, k = 1, 2, 3) \tag{11}$$

Then if $P_{ik}$ and $P$ are, respectively, the curvature tensor of the second rank and the curvature scalar in three-dimensional space, we have

$$G_{ik} = P_{ik}\,(i, k = 1, 2, 3)$$
$$G_4^i = G_{4i} = G_{44} = 0$$
$$G = -P$$
$$-g = \gamma.$$

It therefore follows for our case that

$$G_{ik} - \tfrac{1}{2}g_{ik}G = P_{ik} - \tfrac{1}{2}\gamma_{ik}P\,(i, k = 1, 2, 3)$$
$$G_{44} - \tfrac{1}{2}g_{44}G = \tfrac{1}{2}P.$$

We pursue our reflexions, from this point on, in two ways. Firstly, with the support of equation (1a). Here $T_{\mu\nu}$ denotes the energy-tensor of the electro-magnetic field, arising from the electrical particles constituting matter. For this field we have everywhere

$$\mathfrak{T}_1^1 + \mathfrak{T}_2^2 + \mathfrak{T}_3^3 + \mathfrak{T}_4^4 = 0.$$

The individual $\mathfrak{T}_\mu^\nu$ are quantities which vary rapidly with position; but for our purpose we no doubt may replace them by their mean values.

We therefore have to choose

$$\left.\begin{array}{l}\mathfrak{T}_1^1 = \mathfrak{T}_2^2 = \mathfrak{T}_3^3 = -\tfrac{1}{3}\mathfrak{T}_4^4 = \text{const.} \\ \mathfrak{T}_\mu^\nu = 0 \text{ (for } \mu \neq \nu), \end{array}\right\} \qquad \cdot \quad \cdot \quad (12)$$

and therefore

$$T_{ik} = \tfrac{1}{3}\frac{\mathfrak{T}_4^4}{\sqrt{\gamma}}\gamma_{ik}, \ T_{44} = \frac{\mathfrak{T}_4^4}{\sqrt{\gamma}}.$$

In consideration of what has been shown hitherto, we obtain in place of (1a)

$$P_{ik} - \tfrac{1}{4}\gamma_{ik}P = -\tfrac{1}{3}\gamma_{ik}\frac{\kappa\mathfrak{T}_4^4}{\sqrt{\gamma}} \qquad \cdot \quad \cdot \quad \cdot \quad (13)$$

$$\tfrac{1}{4}P = -\frac{\kappa\mathfrak{T}_4^4}{\sqrt{\gamma}} \qquad \cdot \quad \cdot \quad \cdot \quad \cdot \quad (14)$$

The scalar of equation (13) agrees with (14). It is on this account that our fundamental equations permit the idea of a spherical univers. For from (13) and (14) follows

$$P_{ik} + \tfrac{4}{3}\frac{\kappa\mathfrak{T}_4^4}{\sqrt{\gamma}}\gamma_{ik} = 0 \qquad \cdot \quad \cdot \quad \cdot \quad (15)$$

and it is known* that this system is satisfied by a (three-dimensional) spherical universe.

But we may also base our reflexions on the equations (9). On the right-hand side of (9) stand those terms which, from the phenomenological point of view, are to be replaced by the energy-tensor of matter; that is, they are to be replaced by

$$\begin{array}{cccc} 0 & 0 & 0 & 0 \\ 0 & 0 & 0 & 0 \\ 0 & 0 & 0 & 0 \\ 0 & 0 & 0 & \rho \end{array}$$

where $\rho$ denotes the mean density of matter assumed to be at rest. We thus obtain the equations

$$P_{ik} - \tfrac{1}{2}\gamma_{ik}P - \tfrac{1}{4}\gamma_{ik}G_0 = 0 \qquad \cdot \quad \cdot \quad \cdot \quad (16)$$

$$\tfrac{1}{2}P + \tfrac{1}{4}G_0 = -\kappa\rho \qquad \cdot \quad \cdot \quad \cdot \quad \cdot \quad (17)$$

*Cf. H. Weyl, "Raum, Zoit, Matorie," § 33.

From the scalar of equation (16) and from (17) we obtain

$$G_0 = -\tfrac{2}{3}P = 2\kappa\rho, \qquad . \quad . \quad . \quad . \quad (18)$$

and consequently from (16)

$$P_{ik} - \kappa\rho\gamma_{ik} = 0 \qquad . \quad . \quad . \quad . \quad (19)$$

which equation, with the exception of the expression for the co-efficient, agrees with (15). By comparison we obtain

$$\mathfrak{T}_4^4 = \tfrac{3}{4}\rho\sqrt{\gamma}. \qquad . \quad . \quad . \quad . \quad (20)$$

This equation signifies that of the energy constituting matter three-quarters is to be ascribed to the electromagnetic field, and one-quarter to the gravitational field.

## § 4. CONCLUDING REMARKS

The above reflexions show the possibility of a theoretical construction of matter out of gravitational field and electromagnetic field alone, without the introduction of hypothetical supplementary terms on the lines of Mie's theory. This possibility appears particularly promising in that it frees us from the necessity of introducing a special constant $\lambda$ for the solution of the cosmological problem. On the other hand, there is a peculiar difficulty. For, if we specialize (1) for the spherically symmetrical static case we obtain one equation too few for defining the $g_{\mu\nu}$ and $\phi_{\mu\nu}$, with the result that *any spherically symmetrical distribution* of electricity appears capable of remaining in equilibrium. Thus the problem of the constitution of the elementary quanta cannot yet be solved on the immediate basis of the given field equations.

# Relativity–The Special and General Theory

The earth is a slightly squashed sphere, and yet from the ground, it appears flat, and was thought to be flat for several thousand years. Likewise, our universe appears to us to be "flat" in the sense that Euclid's axioms seem obviously true; chief among these that two straight lines or beams of light could intersect at most just once. This "flat" picture of space is the simplest one, and the picture accepted by all physicists prior to Einstein.

Einstein did not immediately overturn the flat model of the universe, but simply added another dimension to height, width, and breadth: time. In "Relativity—The Special and General Theory," Einstein described physics in flat space, the domain of special relativity. His postulates were quite simple: first, the laws of physics are the same for all observers moving at constant velocity, and second, all such observers will measure the speed of light. Sir Isaac Newton would certainly have conceded the first point, but the second he would have deemed impossible. Einstein achieved this effect by noting that the laws of physics were unchanged not only under rotations between directions in space, but also under "rotations" between space and time.

Einstein recognized that the theory did not include gravity, and thus was necessarily incomplete. To remedy this, as discussed in Part II, he argued that the universe may be curved as well. The curvature of space and time has a number of profound implications: Light does not travel in straight lines, but rather is curved around massive bodies. Clocks sitting near massive bodies run slower than clocks far away. In other words, Einstein noted that not only is space curved, but time is as well. With a simple set of "field equations," Einstein derived not only both the laws of motion and of gravity put forth by Newton, but also paved the way for explanations of a number of hitherto inexplicable phenomena.

Almost immediately after Einstein published his theory of general relativity in 1915, Karl Schwarzschild showed that Einstein's field equations could be solved in the case of a single massive body. Although it was not realized at the time, and never admitted by Einstein, this solution describes compact objects from which not even light can escape: what we now call "black holes." We now believe that some stars end their lives as black holes, and at the center of most, if not all, galaxies there lie supermassive black holes. In our own Milky Way Galaxy, recent evidence suggests that there is a black hole approximately 3 million times the mass of the sun.

Since light is bent around massive objects, the images of distant galaxies can be distorted or even multiplied on their way to observers here on earth. This effect, termed "gravitational lensing," is not unlike that of a curved piece of glass. One of the first observational confirmations of general relativity was a lensing effect seen by Sir Arthur Eddington during a solar eclipse in 1919. Eddington noted that the position of a star seemed to shift in the sky relative to its normal position. The shift was consistent with the result predicted by Einstein given the mass of the sun. The bending of space is not necessarily local. Much of modern astrophysics is concerned with the question of what the overall "shape" of the universe is, and whether it is "flat," "closed" like a sphere (and thus finite), or "open" like a saddle (and thus infinite). Recent measurements from the Wilkinson Microwave Anisotropy Probe (WMAP) satellite suggest that the universe is flat, or so large that it cannot yet be distinguished from perfect flatness.

When Einstein first proposed general relativity, he recognized that his theory predicted that the universe as a whole could not be static as had always been assumed: the attraction of gravity meant that the universe had to be either expanding or contracting. Therefore, he added a "cosmological constant" to balance the attraction of gravity and keep the universe static. In 1922, the astronomer Edwin Hubble measured the expansion of the universe through observation, an expansion wholly consistent with Einstein's original theory, but not with his value of the cosmological constant. In Appendix 4 of this work, Einstein responds to the recent findings, and elsewhere notes that his *ad hoc* introduction of a cosmological constant was his "greatest blunder." As an interesting epilogue, however measurements of distant supernova explosions during the mid-1990s indicated that, though not the value proposed by Einstein, there may be a cosmological constant after all.

# PREFACE

The present book is intended, as far as possible, to give an exact insight into the theory of Relativity to those readers who, from a general scientific and philosophical point of view, are interested in the theory, but who are not conversant with the mathematical apparatus of theoretical physics. The work presumes a standard of education corresponding to that of a university matriculation examination, and, despite the shortness of the book, a fair amount of patience and force of will on the part of the reader. The author has spared himself no pains in his endeavour to present the main ideas in the simplest and most intelligible form, and on the whole, in the sequence and connection in which they actually originated. In the interest of clearness, it appeared to me inevitable that I should repeat myself frequently, without paying the slightest attention to the elegance of the presentation. I adhered scrupulously to the precept of that brilliant theoretical physicist L. Boltzmann, according to whom matters of elegance ought to be left to the tailor and to the cobbler. I make no pretence of having withheld from the reader difficulties which are inherent to the subject. On the other hand, I have purposely treated the empirical physical foundations of the theory in a "step-motherly" fashion, so that readers unfamiliar with physics may not feel like the wanderer who was unable to see the forest for trees. May the book bring some one a few happy hours of suggestive thought!

A. EINSTEIN

*December 1916*

# PART I: THE SPECIAL THEORY OF RELATIVITY

## O N E

## PHYSICAL MEANING OF GEOMETRICAL PROPOSITIONS

In your schooldays most of you who read this book made acquaintance with the noble building of Euclid's geometry, and you remember—perhaps with more respect than love—the magnificent structure, on the lofty staircase of which you were chased about for uncounted hours by conscientious teachers. By reason of your past experience, you would certainly regard everyone with disdain who should pronounce even the most out-of-the-way proposition of this science to be untrue. But perhaps this feeling of proud certainty would leave you immediately if some one were to ask you: "What, then, do you mean by the assertion that these propositions are true?" Let us proceed to give this question a little consideration.

Geometry sets out from certain conceptions such as "plane," "point," and "straight line," with which we are able to associate more or less definite ideas, and from certain simple propositions (axioms) which, in virtue of these ideas, we are inclined to accept as "true." Then, on the basis of a logical process, the justification of which we feel ourselves compelled to admit, all remaining propositions are shown to follow from those axioms, i.e. they are proven. A proposition is then correct ("true") when it has been derived in the recognised manner from the axioms. The question of the "truth" of the individual geometrical propositions is thus reduced to one of the "truth" of the axioms. Now it has long been known that the last question is not only unanswerable by the methods of geometry, but that it is in itself entirely without meaning. We cannot ask whether it is true that only one straight line goes through two points. We can only say that Euclidean geometry deals with things called "straight lines," to each of which is ascribed the property of being uniquely determined by two points situated on it. The concept "true" does not tally with the assertions of pure geometry, because by the word "true" we are eventually in the habit of designating always the correspondence with a "real" object; geometry, however, is not concerned with the relation of the ideas involved in it to objects of experience, but only with the logical connection of these ideas among themselves.

It is not difficult to understand why, in spite of this, we feel constrained to call the propositions of geometry "true." Geometrical ideas correspond to more or less exact objects in nature, and these last are undoubtedly the exclusive cause of the genesis of those ideas. Geometry ought to refrain from such a course, in order to give to its structure the largest possible logical unity. The practice, for example, of seeing in a "distance" two marked positions on a practically rigid body is something which is lodged deeply in our habit of thought. We are accustomed further to regard three points as being situated on a straight line, if their apparent positions can be made to coincide for observation with one eye, under suitable choice of our place of observation.

If, in pursuance of our habit of thought, we now supplement the propositions of Euclidean geometry by the single proposition that two points on a practically rigid body always correspond to the same distance (line-interval), independently of any changes in position to which we may subject the body, the propositions of Euclidean geometry then resolve themselves into propositions on the possible relative position of practically rigid bodies.[1] Geometry which has been supplemented in this way is then to be treated as a branch of physics. We can now legitimately ask as to the "truth" of geometrical propositions

---

[1] It follows that a natural object is associated also with a straight line. Three points $A$, $B$ and $C$ on a rigid body thus lie in a straight line when, the points $A$ and $C$ being given, $B$ is chosen such that the sum of the distances $A B$ and $B C$ is as short as possible. This incomplete suggestion will suffice for our present purpose.

interpreted in this way, since we are justified in asking whether these propositions are satisfied for those real things we have associated with the geometrical ideas. In less exact terms we can express this by saying that by the "truth" of a geometrical proposition in this sense we understand its validity for a construction with ruler and compasses.

Of course the conviction of the "truth" of geometrical propositions in this sense is founded exclusively on rather incomplete experience. For the present we shall assume the "truth" of the geometrical propositions, then at a later stage (in the general theory of relativity) we shall see that this "truth" is limited, and we shall consider the extent of its limitation.

# TWO

# THE SYSTEM OF CO-ORDINATES

On the basis of the physical interpretation of distance which has been indicated, we are also in a position to establish the distance between two points on a rigid body by means of measurements. For this purpose we require a "distance" (rod $S$) which is to be used once and for all, and which we employ as a standard measure. If, now, $A$ and $B$ are two points on a rigid body, we can construct the line joining them according to the rules of geometry; then, starting from $A$, we can mark off the distance $S$ time after time until we reach $B$. The number of these operations required is the numerical measure of the distance $A$ $B$. This is the basis of all measurement of length.[1]

Every description of the scene of an event or of the position of an object in space is based on the specification of the point on a rigid body (body of reference) with which that event or object coincides. This applies not only to scientific description, but also to everyday life. If I analyse the place specification "Trafalgar Square, London,"[2] I arrive at the following result. The earth is the rigid body to which the specification of place refers; "Trafalgar Square, London," is a well-defined point, to which a name has been assigned, and with which the event coincides in space.[3]

This primitive method of place specification deals only with places on the surface of rigid bodies, and is dependent on the existence of points on this surface which are distinguishable from each other. But we can free ourselves from both of these limitations without altering the nature of our specification of position. If, for instance, a cloud is hovering over Trafalgar Square, then we can determine its position relative to the surface of the earth by erecting a pole perpendicularly on

---

[1]Here we have assumed that there is nothing left over, *i.e.* that the measurement gives a whole number. This difficulty is got over by the use of divided measuring-rods, the introduction of which does not demand any fundamentally new method.

[2]I have chosen this as being more familiar to the English reader than the "Potsdamer Platz, Berlin," which is referred to in the original. (R. W. L.)

[3]It is not necessary here to investigate further the significance of the expression "coincidence in space." This conception is sufficiently obvious to ensure that differences of opinion are scarcely likely to arise as to its applicability in practice.

the Square, so that it reaches the cloud. The length of the pole measured with the standard measuring-rod, combined with the specification of the position of the foot of the pole, supplies us with a complete place specification. On the basis of this illustration, we are able to see the manner in which a refinement of the conception of position has been developed.

(*a*) We imagine the rigid body, to which the place specification is referred, supplemented in such a manner that the object whose position we require is reached by the completed rigid body.

(*b*) In locating the position of the object, we make use of a number (here the length of the pole measured with the measuring-rod) instead of designated points of reference.

(*c*) We speak of the height of the cloud even when the pole which reaches the cloud has not been erected. By means of optical observations of the cloud from different positions on the ground, and taking into account the properties of the propagation of light, we determine the length of the pole we should have required in order to reach the cloud.

From this consideration we see that it will be advantageous if, in the description of position, it should be possible by means of numerical measures to make ourselves independent of the existence of marked positions (possessing names) on the rigid body of reference. In the physics of measurement this is attained by the application of the Cartesian system of co-ordinates.

This consists of three plane surfaces perpendicular to each other and rigidly attached to a rigid body. Referred to a system of co-ordinates, the scene of any event will be determined (for the main part) by the specification of the lengths of the three perpendiculars or co-ordinates ($x$, $y$, $z$) which can be dropped from the scene of the event to those three plane surfaces. The lengths of these three perpendiculars can be determined by a series of manipulations with rigid measuring-rods performed according to the rules and methods laid down by Euclidean geometry.

In practice, the rigid surfaces which constitute the system of co-ordinates are generally not available; furthermore, the magnitudes of the co-ordinates are not actually determined by constructions with rigid rods, but by indirect means. If the results of physics and astronomy are to maintain their clearness, the physical meaning of specifications of position must always be sought in accordance with the above considerations.[1]

We thus obtain the following result: Every description of events in space involves the use of a rigid body to which such events have to be referred. The resulting relationship takes for granted that the laws of Euclidean geometry hold for "distances," the "distance" being represented physically by means of the convention of two marks on a rigid body.

---

[1] A refinement and modification of these views does not become necessary until we come to deal with the general theory of relativity, treated in the second part of this book.

# THREE

# SPACE AND TIME IN CLASSICAL MECHANICS

The purpose of mechanics is to describe how bodies change their position in space with "time." I should load my conscience with grave sins against the sacred spirit of lucidity were I to formulate the aims of mechanics in this way, without serious reflection and detailed explanations. Let us proceed to disclose these sins.

It is not clear what is to be understood here by "position" and "space." I stand at the window of a railway carriage which is travelling uniformly, and drop a stone on the embankment, without throwing it. Then, disregarding the influence of the air resistance, I see the stone descend in a straight line. A pedestrian who observes the misdeed from the footpath notices that the stone falls to earth in a parabolic curve. I now ask: Do the "positions" traversed by the stone lie "in reality" on a straight line or on a parabola? Moreover, what is meant here by motion "in space"? From the considerations of the previous section the answer is self-evident. In the first place we entirely shun the vague word "space," of which, we must honestly acknowledge, we cannot form the slightest conception, and we replace it by "motion relative to a practically rigid body of reference." The positions relative to the body of reference (railway carriage or embankment) have already been defined in detail in the preceding section. If instead of "body of reference" we insert "system of co-ordinates," which is a useful idea for mathematical description, we are in a position to say: The stone traverses a straight line relative to a system of co-ordinates rigidly attached to the carriage, but relative to a system of co-ordinates rigidly attached to the ground (embankment) it describes a parabola. With the aid of this example it is clearly seen that there is no such thing as an independently existing trajectory (lit. "path-curve"[1]), but only a trajectory relative to a particular body of reference.

---

[1]That is, a curve along which the body moves.

In order to have a *complete* description of the motion, we must specify how the body alters its position *with time*; i.e. for every point on the trajectory it must be stated at what time the body is situated there. These data must be supplemented by such a definition of time that, in virtue of this definition, these time-values can be regarded essentially as magnitudes (results of measurements) capable of observation. If we take our stand on the ground of classical mechanics, we can satisfy this requirement for our illustration in the following manner. We imagine two clocks of identical construction; the man at the railway-carriage window is holding one of them, and the man on the footpath the other. Each of the observers determines the position on his own reference-body occupied by the stone at each tick of the clock he is holding in his hand. In this connection we have not taken account of the inaccuracy involved by the finiteness of the velocity of propagation of light. With this and with a second difficulty prevailing here we shall have to deal in detail later.

# FOUR

# THE GALILEIAN SYSTEM
# OF CO-ORDINATES

As is well known, the fundamental law of the mechanics of Galilei-Newton, which is known as the *law of inertia*, can be stated thus: A body removed sufficiently far from other bodies continues in a state of rest or of uniform motion in a straight line. This law not only says something about the motion of the bodies, but it also indicates the reference-bodies or systems of co-ordinates, permissible in mechanics, which can be used in mechanical description. The visible fixed stars are bodies for which the law of inertia certainly holds to a high degree of approximation. Now if we use a system of co-ordinates which is rigidly attached to the earth, then, relative to this system, every fixed star describes a circle of immense radius in the course of an astronomical day, a result which is opposed to the statement of the law of inertia. So that if we adhere to this law we must refer these motions only to systems of co-ordinates relative to which the fixed stars do not move in a circle. A system of co-ordinates of which the state of motion is such that the law of inertia holds relative to it is called a "Galileian system of co-ordinates." The laws of the mechanics of Galilei-Newton can be regarded as valid only for a Galileian system of co-ordinates.

# FIVE

# THE PRINCIPLE OF RELATIVITY (IN THE RESTRICTED SENSE)

In order to attain the greatest possible clearness, let us return to our example of the railway carriage supposed to be travelling uniformly. We call its motion a uniform translation ("uniform" because it is of constant velocity and direction, "translation" because although the carriage changes its position relative to the embankment yet it does not rotate in so doing). Let us imagine a raven flying through the air in such a manner that its motion, as observed from the embankment, is uniform and in a straight line. If we were to observe the flying raven from the moving railway carriage, we should find that the motion of the raven would be one of different velocity and direction, but that it would still be uniform and in a straight line. Expressed in an abstract manner we may say: If a mass $m$ is moving uniformly in a straight line with respect to a co-ordinate system $K$, then it will also be moving uniformly and in a straight line relative to a second co-ordinate system $K'$, provided that the latter is executing a uniform translatory motion with respect to $K$. In accordance with the discussion contained in the preceding section, it follows that:

If $K$ is a Galileian co-ordinate system, then every other co-ordinate system $K'$ is a Galileian one, when, in relation to $K$, it is in a condition of uniform motion of translation. Relative to $K'$ the mechanical laws of Galilei-Newton hold good exactly as they do with respect to $K$.

We advance a step farther in our generalisation when we express the tenet thus: If, relative to $K$, $K'$ is a uniformly moving co-ordinate system devoid of rotation, then natural phenomena run their course with respect to $K'$ according to exactly the same general laws as with respect to $K$. This statement is called the *principle of relativity* (in the restricted sense).

As long as one was convinced that all natural phenomena were capable of representation with the help of classical mechanics, there was no need to doubt the validity of this principle of relativity. But in view of the more recent development of electrodynamics and optics it became more and more evident that classical mechanics affords an insufficient foundation for the physical description of all natural phenomena. At this juncture the question of the validity of the principle of relativity became ripe for discussion, and it did not appear impossible that the answer to this question might be in the negative.

Nevertheless, there are two general facts which at the outset speak very much in favour of the validity of the principle of relativity. Even though classical mechanics does not supply us with a sufficiently broad basis for the theoretical presentation of all physical phenomena, still we must grant it a considerable measure of "truth," since it supplies us with the actual motions of the heavenly bodies with a delicacy of detail little short of wonderful. The principle of relativity must therefore apply with great accuracy in the domain of *mechanics*. But that a principle of such broad generality should hold with such exactness in one domain of phenomena, and yet should be invalid for another, is *a priori* not very probable.

We now proceed to the second argument, to which, moreover, we shall return later. If the principle of relativity (in the restricted sense) does not hold, then the Galileian co-ordinate systems $K$, $K'$, $K''$, etc., which are moving uniformly relative to each other, will not be *equivalent* for the description of natural phenomena. In this case we should be constrained to believe that natural laws are capable of being formulated in a particularly simple manner, and of course only on condition that, from amongst all possible Galileian co-ordinate systems, we should have chosen *one* ($K_0$) of a particular state of motion as our body of reference. We should then be justified (because of its merits for the description of natural phenomena) in calling this system "absolutely at rest," and all other Galileian systems $K$ "in motion." If, for instance, our embankment were the system $K_0$, then our railway carriage would be a system $K$, relative to which less simple laws would hold than with respect to $K_0$. This diminished simplicity would be due to the fact that the carriage $K$ would be in motion (i.e. "really") with respect to $K_0$. In the general laws of nature which have been formulated with reference to $K$, the magnitude and direction of the velocity of the carriage would necessarily play a part. We should expect, for instance, that the note emitted by an organ-pipe placed with its axis parallel to the direction of travel would be different from that emitted if the axis of the pipe were placed perpendicular to this direction. Now in virtue of its motion in an orbit round the sun, our earth is comparable with a railway carriage travelling with a velocity of about 30 kilometres per second. If the principle of relativity were not valid we should therefore expect that the direction of motion of the earth at any moment would enter into the laws of nature, and also that physical systems in their behaviour would be dependent on the orientation in space with respect to the earth. For owing to the alteration in direction of the velocity of revolution of the earth in the course of a year, the earth cannot be at rest relative to the hypothetical system $K_0$ throughout the whole year. However, the most careful observations have never revealed such anisotropic properties in terrestrial physical space, i.e. a physical non-equivalence of different directions. This is very powerful argument in favour of the principle of relativity.

# SIX
# THE THEOREM OF THE ADDITION OF VELOCITIES EMPLOYED IN CLASSICAL MECHANICS

Let us suppose our old friend the railway carriage to be travelling along the rails with a constant velocity $v$, and that a man traverses the length of the carriage in the direction of travel with a velocity $w$. How quickly or, in other words, with what velocity $W$ does the man advance relative to the embankment during the process? The only possible answer seems to result from the following consideration: If the man were to stand still for a second, he would advance relative to the embankment through a distance $v$ equal numerically to the velocity of the carriage. As a consequence of his walking, however, he traverses an additional distance $w$ relative to the carriage, and hence also relative to the embankment, in this second, the distance $w$ being numerically equal to the velocity with which he is walking. Thus in total he covers the distance $W = v + w$ relative to the embankment in the second considered. We shall see later that this result, which expresses the theorem of the addition of velocities employed in classical mechanics, cannot be maintained; in other words, the law that we have just written down does not hold in reality. For the time being, however, we shall assume its correctness.

# SEVEN

# THE APPARENT INCOMPATIBILITY OF THE LAW OF PROPAGATION OF LIGHT WITH THE PRINCIPLE OF RELATIVITY

There is hardly a simpler law in physics than that according to which light is propagated in empty space. Every child at school knows, or believes he knows, that this propagation takes place in straight lines with a velocity $c = 300,000$ km./sec. At all events we know with great exactness that this velocity is the same for all colours, because if this were not the case, the minimum of emission would not be observed simultaneously for different colours during the eclipse of a fixed star by its dark neighbour. By means of similar considerations based on observations of double stars, the Dutch astronomer De Sitter was also able to show that the velocity of propagation of light cannot depend on the velocity of motion of the body emitting the light. The assumption that this velocity of propagation is dependent on the direction "in space" is in itself improbable.

In short, let us assume that the simple law of the constancy of the velocity of light $c$ (in vacuum) is justifiably believed by the child at school. Who would imagine that this simple law has plunged the conscientiously thoughtful physicist into the greatest intellectual difficulties? Let us consider how these difficulties arise.

Of course we must refer the process of the propagation of light (and indeed every other process) to a rigid reference-body (co-ordinate system). As such a system let us again choose our embankment. We shall imagine the air above it to have been removed. If a ray of light be sent along the embankment, we see from the above that the tip of the ray will be transmitted with the velocity $c$ relative to the embankment. Now let us suppose that our railway carriage is again travelling along the railway lines with the velocity $v$, and that its direction is the same as that of the ray of light, but its velocity of course much less. Let us inquire about the velocity of propagation of the ray of light relative to the carriage. It is obvious that we can here apply the consideration of the previous section, since the ray of light plays the part of the man walking along relatively to the carriage. The velocity $W$ of the man relative to the embankment is here replaced by the velocity of light relative to the embankment. $w$ is the required velocity of light with respect to the carriage, and we have

$$w = c - v.$$

The velocity of propagation of a ray of light relative to the carriage thus comes out smaller than $c$.

But this result comes into conflict with the principle of relativity set forth in Section 5. For, like every other general law of nature, the law of the transmission of light *in vacuo* must, according to the principle of relativity, be the same for the railway carriage as reference-body as when the rails are the body of reference. But, from our above consideration, this would appear to be impossible. If every ray of light is propagated relative to the embankment with the velocity $c$, then for this reason it would appear that another law of propagation of light must necessarily hold with respect to the carriage—a result contradictory to the principle of relativity.

In view of this dilemma there appears to be nothing else for it than to abandon either the principle of relativity or the simple law of the propagation of light *in vacuo*. Those of you who have carefully followed the preceding discussion are almost sure to expect that we should retain the principle of relativity, which appeals so convincingly to the intellect because it is so natural and simple. The law of the propagation of light *in vacuo* would then have to be replaced by a more complicated law comfortable to the principle of relativity. The development of theoretical physics shows, however, that we cannot pursue this course. The epoch-making theoretical investigations of

H. A. Lorentz on the electrodynamical and optical phenomena connected with moving bodies show that experience in this domain leads conclusively to a theory of electromagnetic phenomena, of which the law of the constancy of the velocity of light *in vacuo* is a necessary consequence. Prominent theoretical physicists were therefore more inclined to reject the principle of relativity, in spite of the fact that no empirical data had been found which were contradictory to this principle.

At this juncture the theory of relativity entered the arena. As a result of an analysis of the physical conceptions of time and space, it became evident that *in reality there is not the least incompatibility between the principle of relativity and the law of propagation of light*, and that by systematically holding fast to both these laws a logically rigid theory could be arrived at. This theory has been called the *special theory of relativity* to distinguish it from the extended theory, with which we shall deal later. In the following pages we shall present the fundamental ideas of the special theory of relativity.

# EIGHT

# ON THE IDEA OF TIME
# IN PHYSICS

Lightning has struck the rails on our railway embankment at two places $A$ and $B$ far distant from each other. I make the additional assertion that these two lightning flashes occurred simultaneously. If I ask you whether there is sense in this statement, you will answer my question with a decided "Yes." But if I now approach you with the request to explain to me the sense of the statement more precisely, you find after some consideration that the answer to this question is not so easy as it appears at first sight.

After some time perhaps the following answer would occur to you: "The significance of the statement is clear in itself and needs no further explanation; of course it would require some consideration if I were to be commissioned to determine by observations whether in the actual case the two events took place simultaneously or not." I cannot be satisfied with this answer for the following reason. Supposing that as a result of ingenious consideration an able meteorologist were to discover that the lightning must always strike the places $A$ and $B$ simultaneously, then we should be faced with the task of testing whether or not this theoretical result is in accordance with the reality. We encounter the same difficulty with all physical statements in which the conception "simultaneous" plays a part. The concept does not exist for the physicist until he has the possibility of discovering whether or not it is fulfilled in an actual case. We thus require a definition of simultaneity such that this definition supplies us with the method by means of which, in the present case, he can decide by experiment whether or not both the lightning strokes occurred simultaneously. As long as this requirement is not satisfied, I allow myself to be deceived as a physicist (and of course the same applies if I am not a physicist), when I imagine that I am able to attach a meaning to the statement

145

of simultaneity. (I would ask the reader not to proceed farther until he is fully convinced on this point.)

After thinking the matter over for some time you then offer the following suggestion with which to test simultaneity. By measuring along the rails, the connecting line $AB$ should be measured up and an observer placed at the mid-point $M$ of the distance $AB$. This observer should be supplied with an arrangement (e.g. two mirrors inclined at 90°) which allows him visually to observe both places $A$ and $B$ at the same time. If the observer perceives the two flashes of lightning at the same time, then they are simultaneous.

I am very pleased with this suggestion, but for all that I cannot regard the matter as quite settled, because I feel constrained to raise the following objection: "Your definition would certainly be right, if only I knew that the light by means of which the observer at $M$ perceives the lightning flashes travels along the length $A \rightarrow M$ with the same velocity as along the length $B \rightarrow M$. But an examination of this supposition would only be possible if we already had at our disposal the means of measuring time. It would thus appear as though we were moving here in a logical circle."

After further consideration you cast a somewhat disdainful glance at me—and rightly so—and you declare: "I maintain my previous definition nevertheless, because in reality it assumes absolutely nothing about light. There is only *one* demand to be made of the definition of simultaneity, namely, that in every real case it must supply us with an empirical decision as to whether or not the conception that has to be defined is fulfilled. That my definition satisfies this demand is indisputable. That light requires the same time to traverse the path $A \rightarrow M$ as for the path $B \rightarrow M$ is in reality neither a *supposition nor a hypothesis* about the physical nature of light, but a *stipulation* which I can make of my own freewill in order to arrive at a definition of simultaneity."

It is clear that this definition can be used to give an exact meaning not only to *two* events, but to as many events as we care to choose, and independently of the positions of the scenes of the events with respect to the body of reference[1] (here the railway embankment). We are thus led also to a definition of "time" in physics. For this purpose we suppose that clocks of identical construction are placed at the points $A$, $B$ and $C$ of the railway line (co-ordinate system), and that they are set in such a manner that the positions of their pointers are simultaneously (in the above sense) the same. Under these conditions we understand by the "time" of an event the reading (position of the hands) of that one of these clocks which is in the immediate vicinity (in space) of the event. In this manner a time-value is associated with every event which is essentially capable of observation.

This stipulation contains a further physical hypothesis, the validity of which will hardly be doubted without empirical evidence to the contrary. It has been assumed that all these clocks go *at the same rate* if they are of identical construction. Stated more exactly: When two clocks arranged at rest in different places of a reference-body are set in such a manner that a *particular* position of the pointers of the one clock is *simultaneous* (in the above sense) with the *same* position of the pointers of the other clock, then identical "settings" are always simultaneous (in the sense of the above definition).

---

[1]We suppose further, that, when three events $A$, $B$ and $C$ occur in different places in such a manner that $A$ is simultaneous with $B$, and $B$ is simultaneous with $C$ (simultaneous in the sense of the above definition), then the criterion for the simultaneity of the pair of events $A$, $C$ is also satisfied. This assumption is a physical hypothesis about the law of propagation of light; it must certainly be fulfilled if we are to maintain the law of the constancy of the velocity of light *in vacuo*.

# NINE

# THE RELATIVITY OF SIMULTANEITY

Up to now our considerations have been referred to a particular body of reference, which we have styled a "railway embankment." We suppose a very long train travelling along the rails with the constant velocity $v$ and in the direction indicated in Fig. 1. People travelling in this

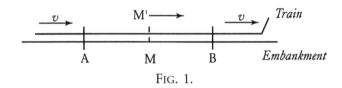

FIG. 1.

train will with advantage use the train as a rigid reference-body (co-ordinate system); they regard all events in reference to the train. Then every event which takes place along the line also takes place at a particular point of the train. Also the definition of simultaneity can be given relative to the train in exactly the same way as with respect to the embankment. As a natural consequence, however, the following question arises:

Are two events (e.g. the two strokes of lightning $A$ and $B$) which are simultaneous *with reference to the railway embankment* also simultaneous *relatively to the train?* We shall show directly that the answer must be in the negative.

When we say that the lightning strokes $A$ and $B$ are simultaneous with respect to the embankment, we mean: the rays of light emitted at the places $A$ and $B$, where the lightning occurs, meet each other at the mid-point $M$ of the length $A \rightarrow B$ of the embankment. But the events $A$ and $B$ also correspond to positions $A$ and $B$ on the train. Let $M'$ be the mid-point of the distance $A \rightarrow B$ on the travelling train. Just when the flashes[1] of lightning occur, this point $M'$ naturally coincides

---

[1] As judged from the embankment.

148

with the point $M$, but it moves towards the right in the diagram with the velocity $v$ of the train. If an observer sitting in the position $M'$ in the train did not possess this velocity, then he would remain permanently at $M$, and the light rays emitted by the flashes of lightning $A$ and $B$ would reach him simultaneously, *i.e.* they would meet just where he is situated. Now in reality (considered with reference to the railway embankment) he is hastening towards the beam of light coming from $B$, whilst he is riding on ahead of the beam of light coming from $A$. Hence the observer will see the beam of light emitted from $B$ earlier than he will see that emitted from $A$. Observers who take the railway train as their reference-body must therefore come to the conclusion that the lightning flash $B$ took place earlier than the lightning flash $A$. We thus arrive at the important result:

Events which are simultaneous with reference to the embankment are not simultaneous with respect to the train, and *vice versa* (relativity of simultaneity). Every reference-body (co-ordinate system) has its own particular time; unless we are told the reference-body to which the statement of time refers, there is no meaning in a statement of the time of an event.

Now before the advent of the theory of relativity it had always tacitly been assumed in physics that the statement of time had an absolute significance, *i.e.* that it is independent of the state of motion of the body of reference. But we have just seen that this assumption is incompatible with the most natural definition of simultaneity; if we discard this assumption, then the conflict between the law of the propagation of light *in vacuo* and the principle of relativity (developed in Section 7) disappears.

We were led to that conflict by the considerations of Section 6, which are now no longer tenable. In that section we concluded that the man in the carriage, who traverses the distance *w per second* relative to the carriage, traverses the same distance also with respect to the embankment *in each second* of time. But, according to the foregoing considerations, the time required by a particular occurrence with respect to the carriage must not be considered equal to the duration of the same occurrence as judged from the embankment (as reference-body). Hence it cannot be contended that the man in walking travels the distance $w$ relative to the railway line in a time which is equal to one second as judged from the embankment.

Moreover, the considerations of Section 6 are based on yet a second assumption, which, in the light of a strict consideration, appears to be arbitrary, although it was always tacitly made even before the introduction of the theory of relativity.

# TEN

# ON THE RELATIVITY OF THE CONCEPTION OF DISTANCE

Let us consider two particular points on the train[1] travelling along the embankment with the velocity $v$, and inquire as to their distance apart. We already know that it is necessary to have a body of reference for the measurement of a distance, with respect to which body the distance can be measured up. It is the simplest plan to use the train itself as reference-body (co-ordinate system). An observer in the train measures the interval by marking off his measuring-rod in a straight line (e.g. along the floor of the carriage) as many times as is necessary to take him from the one marked point to the other. Then the number which tells us how often the rod has to be laid down is the required distance.

It is a different matter when the distance has to be judged from the railway line. Here the following method suggests itself. If we call $A'$ and $B'$ the two points on the train whose distance apart is required, then both of these points are moving with the velocity $v$ along the embankment. In the first place we require to determine the points $A$ and $B$ of the embankment which are just being passed by the two points $A'$ and $B'$ at a particular time $t$—judged from the embankment. These points $A$ and $B$ of the embankment can be determined by applying the definition of time given in Section 8. The distance between these points $A$ and $B$ is then measured by repeated application of the measuring-rod along the embankment.

*A priori* it is by no means certain that this last measurement will supply us with the same result as the first. Thus the length of the train as measured from the embankment may be different from that obtained by measuring in the train itself. This circumstance leads us

---

[1] E.g. the middle of the first and of the twentieth carriage.

to a second objection which must be raised against the apparently obvious consideration of Section 6. Namely, if the man in the carriage covers the distance $w$ in a unit of time—*measured from the train*—then this distance—*as measured from the embankment*—is not necessarily also equal to $w$.

# ELEVEN
# THE LORENTZ
# TRANSFORMATION

The results of the last three sections show that the apparent incompatibility of the law of propagation of light with the principle of relativity (Section 7) has been derived by means of a consideration which borrowed two unjustifiable hypotheses from classical mechanics; these are as follows:

(1) The time-interval (time) between two events is independent of the condition of motion of the body of reference.

(2) The space-interval (distance) between two points of a rigid body is independent of the condition of motion of the body of reference.

If we drop these hypotheses, then the dilemma of Section 7 disappears, because the theorem of the addition of velocities derived in Section 6 becomes invalid. The possibility presents itself that the law of the propagation of light *in vacuo* may be compatible with the principle of relativity, and the question arises: How have we to modify the considerations of Section 6 in order to remove the apparent disagreement between these two fundamental results of experience? This question leads to a general one. In the discussion of Section 6 we have to do with places and times relative both to the train and to the embankment. How are we to find the place and time of an event in relation to the train, when we know the place and time of the event with respect to the railway embankment? Is there a thinkable answer to this question of such a nature that the law of transmission of light *in vacuo* does not contradict the principle of relativity? In other words: Can we conceive of a relation between place and time of the individual events relative to both reference-bodies, such that every ray of light possesses the velocity of transmission *c* relative to the embankment and relative to the train? This question leads to a quite definite positive answer, and to a perfectly definite transformation law for the space-time magnitudes of an event when changing over from one body of reference to another.

Before we deal with this, we shall introduce the following incidental consideration. Up to the present we have only considered events taking place along the embankment, which had mathematically to assume the function of a straight line. In the manner indicated in Section 2 we can imagine this reference-body supplemented laterally and in a vertical direction by means of a framework of rods, so that an event which takes place anywhere can be localised with reference to this framework. Similarly, we can imagine the train travelling with the velocity $v$ to be continued across the whole of space, so that every event, no matter how far off it may be, could also be localised with respect to the second framework. Without committing any fundamental error, we can disregard the fact that in reality these frameworks would continually interfere with each other, owing to the impenetrability of solid bodies. In every such framework we imagine three surfaces perpendicular to each other marked out, and designated as "co-ordinate planes" ("co-ordinate system"). A co-ordinate system $K$ then corresponds to the embankment, and a co-ordinate system $K'$ to the train. An event, wherever it may have taken place, would be fixed in space with respect to $K$ by the three perpendiculars $x$, $y$, $z$ on the co-ordinate planes, and with regard to time by a time-value $t$. Relative to $K'$, *the same event* would be fixed in respect of space and time by corresponding values $x'$, $y'$, $z'$, $t'$, which of course are not identical with $x$, $y$, $z$, $t$. It has already been set forth in detail how these magnitudes are to be regarded as results of physical measurements.

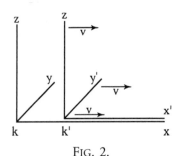

FIG. 2.

Obviously our problem can be exactly formulated in the following manner. What are the values $x'$, $y'$, $z'$, $t'$, of an event with respect to $K'$, when the magnitudes $x$, $y$, $z$, $t$, of the same event with respect to $K$ are given? The relations must be so chosen that the law of the transmission of light *in vacuo* is satisfied for one and the same ray of light (and of course for every ray) with respect to $K$ and $K'$. For the relative orientation in space of the co-ordinate systems indicated in the diagram (Fig. 2), this problem is solved by means of the equations:

$$x' = \frac{x - vt}{\sqrt{1 - \frac{v^2}{c^2}}}$$

$$y' = y$$

$$z' = z$$

$$t' = \frac{t - \frac{v}{c^2} \cdot x}{\sqrt{1 - \frac{v^2}{c^2}}}$$

This system of equations is known as the "Lorentz transformation."[1]

If in place of the law of transmission of light we had taken as our basis the tacit assumptions of the older mechanics as to the absolute character of times and lengths, then instead of the above we should have obtained the following equations:

$$x' = x - vt$$

$$y' = y$$

$$z' = z$$

$$t' = t.$$

This system of equations is often termed the "Galilei transformation." The Galilei transformation can be obtained from the Lorentz transformation by substituting an infinitely large value for the velocity of light $c$ in the latter transformation.

Aided by the following illustration, we can readily see that, in accordance with the Lorentz transformation, the law of the transmission of

[1] A simple derivation of the Lorentz transformation is given in Appendix 1.

light *in vacuo* is satisfied both for the reference-body $K$ and for the reference-body $K'$. A light-signal is sent along the positive $x$-axis, and this light-stimulus advances in accordance with the equation.

$$x = ct,$$

i.e. with the velocity $c$. According to the equations of the Lorentz transformation, this simple relation between $x$ and $t$ involves a relation between $x'$ and $t'$. In point of fact, if we substitute for $x$ the value $ct$ in the first and fourth equations of the Lorentz transformation, we obtain:

$$x' = \frac{(c - v)t}{\sqrt{1 - \frac{v^2}{c^2}}}$$

$$t' = \frac{\left(1 - \frac{v}{c}\right)t}{\sqrt{1 - \frac{v^2}{c^2}}},$$

from which, by division, the expression

$$x' = ct'$$

immediately follows. If referred to the system $K'$, the propagation of light takes place according to this equation. We thus see that the velocity of transmission relative to the reference-body $K'$ is also equal to $c$. The same result is obtained for rays of light advancing in any other direction whatsoever. Of course this is not surprising, since the equations of the Lorentz transformation were derived conformably to this point of view.

# TWELVE

# THE BEHAVIOUR OF MEASURING-RODS AND CLOCKS IN MOTION

I place a metre-rod in the $x'$-axis of $K'$ in such a manner that one end (the beginning) coincides with the point $x' = 0$, whilst the other end (the end of the rod) coincides with the point $x' = 1$. What is the length of the metre-rod relatively to the system $K$? In order to learn this, we need only ask where the beginning of the rod and the end of the rod lie with respect to $K$ at a particular time $t$ of the system $K$. By means of the first equation of the Lorentz transformation the values of these two points at the time $t = 0$ can be shown to be

$$x_{(\text{beginning of rod})} = 0\sqrt{1 - \frac{v^2}{c^2}}$$

$$x_{(\text{end of rod})} = 1 \cdot \sqrt{1 - \frac{v^2}{c^2}},$$

the distance between the points being $\sqrt{1 - \frac{v^2}{c^2}}$. But the metre-rod is moving with the velocity $v$ relative to $K$. It therefore follows that the length of a rigid metre-rod moving in the direction of its length with a velocity $v$ is $\sqrt{1 - v^2/c^2}$ of a metre. The rigid rod is thus shorter when in motion than when at rest, and the more quickly it is moving, the shorter is the rod. For the velocity $v = c$ we should have $\sqrt{1 - v^2/c^2} = 0$, and for still greater velocities the square-root becomes imaginary. From this we conclude that in the theory of relativity the velocity $c$ plays the part of a limiting velocity, which can neither be reached nor exceeded by any real body.

Of course this feature of the velocity $c$ as a limiting velocity also clearly follows from the equations of the Lorentz transformation, for these become meaningless if we choose values of $v$ greater than $c$.

If, on the contrary, we had considered a metre-rod at rest in the x-axis with respect to $K$, then we should have found that the length of the rod as judged from $K'$ would have been $\sqrt{1 - v^2/c^2}$; this is quite in accordance with the principle of relativity which forms the basis of our considerations.

*A priori* it is quite clear that we must be able to learn something about the physical behaviour of measuring-rods and clocks from the equations of transformation, for the magnitudes $x$, $y$, $z$, $t$, are nothing more nor less than the results of measurements obtainable by means of measuring-rods and clocks. If we had based our considerations on the Galileian transformation we should not have obtained a contraction of the rod as a consequence of its motion.

Let us now consider a seconds-clock which is permanently situated at the origin $(x' = 0)$ of $K'$ . $t' = 0$ and $t' = 1$ are two successive ticks of this clock. The first and fourth equations of the Lorentz transformation give for these two ticks:

$$t = 0$$

and

$$t = \frac{1}{\sqrt{1 - \dfrac{v^2}{c^2}}}$$

As judged from $K$, the clock is moving with the velocity $v$; as judged from this reference-body, the time which elapses between two strokes of the clock is not one second, but $\dfrac{1}{\sqrt{1 - \dfrac{v^2}{c^2}}}$ seconds, i.e. a somewhat larger time. As a consequence of its motion the clock goes more slowly than when at rest. Here also the velocity $c$ plays the part of an unattainable limiting velocity.

# THIRTEEN

# THEOREM OF THE ADDITION OF THE VELOCITIES. THE EXPERIMENT OF FIZEAU

Now in practice we can move clocks and measuring-rods only with velocities that are small compared with the velocity of light; hence we shall hardly be able to compare the results of the previous section directly with the reality. But, on the other hand, these results must strike you as being very singular, and for that reason I shall now draw another conclusion from the theory, one which can easily be derived from the foregoing considerations, and which has been most elegantly confirmed by experiment.

In Section 6 we derived the theorem of the addition of velocities in one direction in the form which also results from the hypotheses of classical mechanics. This theorem can also be deduced readily from the Galilei transformation (Section 11). In place of the man walking inside the carriage, we introduce a point moving relatively to the co-ordinate system $K'$ in accordance with the equation

$$x = wt'.$$

By means of the first and fourth equations of the Galilei transformation we can express $x'$ and $t'$ in terms of $x$ and $t$, and we then obtain

$$x = (v + w)t.$$

This equation expresses nothing else than the law of motion of the point with reference to the system $K$ (of the man with reference to the embankment). We denote this velocity by the symbol $W$, and we then obtain, as in Section 6,

$$W = v + w \qquad . \qquad . \qquad . \quad \text{(A)}.$$

But we can carry out this consideration just as well on the basis of the theory of relativity. In the equation

$$x' = wt'$$

we must then express $x'$ and $t'$ in terms of $x$ and $t$, making use of the first and fourth equations of the *Lorentz transformation*. Instead of the equation (A) we then obtain the equation

$$W = \frac{v + w}{1 + \dfrac{vw}{c^2}} \qquad \cdot \qquad \cdot \qquad \cdot \qquad (B),$$

which corresponds to the theorem of addition for velocities in one direction according to the theory of relativity. The question now arises as to which of these two theorems is the better in accord with experience. On this point we are enlightened by a most important experiment which the brilliant physicist Fizeau performed more than half a century ago, and which has been repeated since then by some of the best experimental physicists, so that there can be no doubt about its result. The experiment is concerned with the following question. Light travels in a motionless liquid with a particular velocity $w$. How quickly does it travel in the direction of the arrow in the tube $T$ (see the accompanying diagram, Fig. 3) when the liquid above mentioned is flowing through the tube with a velocity $v$?

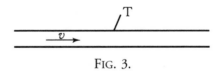

FIG. 3.

In accordance with the principle of relativity we shall certainly have to take for granted that the propagation of light always takes place with the same velocity $w$ *with respect to the liquid*, whether the latter is in motion with reference to other bodies or not. The velocity of light relative to the liquid and the velocity of the latter relative to the tube are thus known, and we require the velocity of light relative to the tube.

It is clear that we have the problem of Section 6 again before us. The tube plays the part of the railway embankment or of the co-ordinate system $K$, the liquid plays the part of the carriage or of the co-ordinate system $K'$, and finally, the light plays the part of the man walking along the carriage, or of the moving point in the present section. If we denote the velocity of the light relative to the tube by $W$, then this is given by the equation (A) or (B), according as the Galilei transformation or the Lorentz transformation corresponds to the facts. Experiment[1] decides in favour of equation (B) derived from the theory of relativity, and the agreement is, indeed, very exact. According to recent and most excellent measurements by Zeeman, the influence of the velocity of flow $v$ on the propagation of light is represented by formula (B) to within one per cent.

Nevertheless we must now draw attention to the fact that a theory of this phenomenon was given by H. A. Lorentz long before the statement of the theory of relativity. This theory was of a purely electrodynamical nature, and was obtained by the use of particular hypotheses as to the electromagnetic structure of matter. This circumstance, however, does not in the least diminish the conclusiveness of the experiment as a crucial test in favour of the theory of relativity, for the electrodynamics of Maxwell-Lorentz, on which the original theory was based, in no way opposes the theory of relativity. Rather has the latter been developed from electrodynamics as an astoundingly simple combination of generalisation of the hypotheses, formerly independent of each other, on which electrodynamics was built.

[1] Fizeau found $W = w + v\left(1 - \frac{1}{n^2}\right)$, where $n = \frac{c}{w}$ is the index of refraction of the liquid. On the other hand, owing to the smallness of $\frac{vw}{c^2}$ as compared with 1, we can replace (B) in the first place by $W = w + v\left(1 - \frac{vw}{c^2}\right)$, or to the same order of approximation by $w + v\left(1 - \frac{1}{n^2}\right)$, which agrees with Fizeau's result.

# FOURTEEN

# THE HEURISTIC VALUE OF THE THEORY OF RELATIVITY

Our train of thought in the foregoing pages can be epitomised in the following manner. Experience has led to the conviction that, on the one hand, the principle of relativity holds true and that on the other hand the velocity of transmission of light *in vacuo* has to be considered equal to a constant $c$. By uniting these two postulates we obtained the law of transformation for the rectangular co-ordinates $x$, $y$, $z$ and the time $t$ of the events which constitute the processes of nature. In this connection we did not obtain the Galilei transformation, but, differing from classical mechanics, the *Lorentz transformation*.

The law of transmission of light, the acceptance of which is justified by our actual knowledge, played an important part in this process of thought. Once in possession of the Lorentz transformation, however, we can combine this with the principle of relativity, and sum up the theory thus:

Every general law of nature must be so constituted that it is transformed into a law of exactly the same form when, instead of the space-time variables $x$, $y$, $z$, $t$ of the original co-ordinate system $K$, we introduce new space-time variables $x'$, $y'$, $z'$, $t'$ of a co-ordinate system $K'$. In this connection the relation between the ordinary and the accented magnitudes is given by the Lorentz transformation. Or in brief: General laws of nature are co-variant with respect to Lorentz transformations.

This is a definite mathematical condition that the theory of relativity demands of a natural law, and in virtue of this, the theory becomes a valuable heuristic aid in the search for general laws of nature. If a general law of nature were to be found which did not satisfy this condition, then at least one of the two fundamental assumptions of the theory would have been disproved. Let us now examine what general results the latter theory has hitherto evinced.

# FIFTEEN

# GENERAL RESULTS OF THE THEORY

It is clear from our previous considerations that the (special) theory of relativity has grown out of electrodynamics and optics. In these fields it has not appreciably altered the predictions of theory, but it has considerably simplified the theoretical structure, *i.e.* the derivation of laws, and—what is incomparably more important—it has considerably reduced the number of independent hypotheses forming the basis of theory. The special theory of relativity has rendered the Maxwell-Lorentz theory so plausible, that the latter would have been generally accepted by physicists even if experiment had decided less unequivocally in its favour.

Classical mechanics required to be modified before it could come into line with the demands of the special theory of relativity. For the main part, however, this modification affects only the laws for rapid motions, in which the velocities of matter $v$ are not very small as compared with the velocity of light. We have experience of such rapid motions only in the case of electrons and ions; for other motions the variations from the laws of classical mechanics are too small to make themselves evident in practice. We shall not consider the motion of stars until we come to speak of the general theory of relativity. In accordance with the theory of relativity the kinetic energy of a material point of mass $m$ is no longer given by the well-known expression

$$m \frac{v^2}{2},$$

but by the expression

$$\frac{mc^2}{\sqrt{1 - \frac{v^2}{c^2}}}.$$

This expression approaches infinity as the velocity $v$ approaches the velocity of light $c$. The velocity must therefore always remain less than

$c$, however great may be the energies used to produce the acceleration. If we develop the expression for the kinetic energy in the form of a series, we obtain

$$mc^2 + m\frac{v^2}{2} - \frac{3}{8}\,m\frac{v^4}{c^2} + \cdots$$

When $\frac{v^2}{c^2}$ is small compared with unity, the third of these terms is always small in comparison with the second, which last is alone considered in classical mechanics. The first term $mc^2$ does not contain the velocity, and requires no consideration if we are only dealing with the question as to how the energy of a point-mass depends on the velocity. We shall speak of its essential significance later.

The most important result of a general character to which the special theory of relativity has led is concerned with the conception of mass. Before the advent of relativity, physics recognised two conservation laws of fundamental importance, namely, the law of the conservation of energy and the law of the conservation of mass; these two fundamental laws appeared to be quite independent of each other. By means of the theory of relativity they have been united into one law. We shall now briefly consider how this unification came about, and what meaning is to be attached to it.

The principle of relativity requires that the law of the conservation of energy should hold not only with reference to a co-ordinate system $K$, but also with respect to every co-ordinate system $K'$ which is in a state of uniform motion of translation relative to $K$, or, briefly, relative to every "Galileian" system of co-ordinates. In contrast to classical mechanics, the Lorentz transformation is the deciding factor in the transition from one such system to another.

By means of comparatively simple considerations we are led to draw the following conclusion from these premises, in conjunction with the fundamental equations of the electrodynamics of Maxwell: A body moving with the velocity $v$, which absorbs[1] an amount of energy

---

[1] $E_0$ is the energy taken up, as judged from a co-ordinate system moving with the body.

$E_0$ in the form of radiation without suffering an alteration in velocity in the process, has, as a consequence, its energy increased by an amount

$$\frac{E_0}{\sqrt{1 - \dfrac{v^2}{c^2}}}.$$

In consideration of the expression given above for the kinetic energy of the body, the required energy of the body comes out to be

$$\frac{\left(m + \dfrac{E_0}{c^2}\right) c^2}{\sqrt{1 - \dfrac{v^2}{c^2}}}.$$

Thus the body has the same energy as a body of mass $\left(m + \dfrac{E_0}{c^2}\right)$ moving with the velocity $v$. Hence we can say: If a body takes up an amount of energy $E_0$, then its inertial mass increases by an amount $\dfrac{E_0}{c^2}$; the inertial mass of a body is not a constant, but varies according to the change in the energy of the body. The inertial mass of a system of bodies can even be regarded as a measure of its energy. The law of the conservation of the mass of a system becomes identical with the law of the conservation of energy, and is only valid provided that the system neither takes up nor sends out energy. Writing the expression for the energy in the form

$$\frac{mc^2 + E_0}{\sqrt{1 - \dfrac{v^2}{c^2}}},$$

we see that the term $mc^2$, which has hitherto attracted our attention, is nothing else than the energy possessed by the body[1] before it absorbed the energy $E_0$.

A direct comparison of this relation with experiment is not possible at the present time (1920; see Note, p. 165), owing to the fact

---

[1] As judged from a co-ordinate system moving with the body.

that the changes in energy $E_0$ to which we can subject a system are not large enough to make themselves perceptible as a change in the inertial mass of the system. $\dfrac{E_0}{c^2}$ is too small in comparison with the mass $m$, which was present before the alteration of the energy. It is owing to this circumstance that classical mechanics was able to establish successfully the conservation of mass as a law of independent validity.

Let me add a final remark of a fundamental nature. The success of the Faraday-Maxwell interpretation of electromagnetic action at a distance resulted in physicists becoming convinced that there are no such things as instantaneous actions at a distance (not involving an intermediary medium) of the type of Newton's law of gravitation. According to the theory of relativity, action at a distance with the velocity of light always takes the place of instantaneous action at a distance or of action at a distance with an infinite velocity of transmission. This is connected with the fact that the velocity $c$ plays a fundamental rôle in this theory. In Part II we shall see in what way this result becomes modified in the general theory of relativity.

NOTE.—With the advent of nuclear transformation processes, which result from the bombardment of elements by $\alpha$-particles, protons, deuterons, neutrons or $\gamma$-rays, the equivalence of mass and energy expressed by the relation $E = mc^2$ has been amply confirmed. The sum of the reacting masses, together with the mass equivalent of the kinetic energy of the bombarding particle (or photon), is always greater than the sum of the resulting masses. The difference is the equivalent mass of the kinetic energy of the particles generated, or of the released electromagnetic energy ($\gamma$-photons). In the same way, the mass of a spontaneously disintegrating radioactive atom is always greater than the sum of the masses of the resulting atoms by the mass equivalent of the kinetic energy of the particles generated (or of the photonic energy). Measurements of the energy of the rays emitted in nuclear reactions, in combination with the equations of such reactions, render it possible to evaluate atomic weights to a high degree of accuracy.

R. W. L.

# SIXTEEN

# EXPERIENCE AND THE SPECIAL THEORY OF RELATIVITY

To what extent is the special theory of relativity supported by experience? This question is not easily answered for the reason already mentioned in connection with the fundamental experiment of Fizeau. The special theory of relativity has crystallised out from the Maxwell-Lorentz theory of electromagnetic phenomena. Thus all facts of experience which support the electromagnetic theory also support the theory of relativity. As being of particular importance, I mention here the fact that the theory of relativity enables us to predict the effects produced on the light reaching us from the fixed stars. These results are obtained in an exceedingly simple manner, and the effects indicated, which are due to the relative motion of the earth with reference to those fixed stars, are found to be in accord with experience. We refer to the yearly movement of the apparent position of the fixed stars resulting from the motion of the earth round the sun (aberration), and to the influence of the radial components of the relative motions of the fixed stars with respect to the earth on the colour of the light reaching us from them. The latter effect manifests itself in a slight displacement of the spectral lines of the light transmitted to us from a fixed star, as compared with the position of the same spectral lines when they are produced by a terrestrial source of light (Doppler principle). The experimental arguments in favour of the Maxwell-Lorentz theory, which are at the same time arguments in favour of the theory of relativity, are too numerous to be set forth here. In reality they limit the theoretical possibilities to such an extent, that no other theory than that of Maxwell and Lorentz has been able to hold its own when tested by experience.

But there are two classes of experimental facts hitherto obtained which can be represented in the Maxwell-Lorentz theory only by the introduction of an auxiliary hypothesis, which in itself—i.e. without making use of the theory of relativity—appears extraneous.

It is known that cathode rays and the so-called β-rays emitted by radioactive substances consist of negatively electrified particles (electrons) of very small inertia and large velocity. By examining the deflection of these rays under the influence of electric and magnetic fields, we can study the law of motion of these particles very exactly.

In the theoretical treatment of these electrons, we are faced with the difficulty that electrodynamic theory of itself is unable to give an account of their nature. For since electrical masses of one sign repel each other, the negative electrical masses constituting the electron would necessarily be scattered under the influence of their mutual repulsions, unless there are forces of another kind operating between them, the nature of which has hitherto remained obscure to us.[1] If we now assume that the relative distances between the electrical masses constituting the electron remain unchanged during the motion of the electron (rigid connection in the sense of classical mechanics), we arrive at a law of motion of the electron which does not agree with experience. Guided by purely formal points of view, H. A. Lorentz was the first to introduce the hypothesis that the form of the electron experiences a contraction in the direction of motion in consequence of that motion, the contracted length being proportional to the expression

$\sqrt{1 - \dfrac{v^2}{c^2}}$. This hypothesis, which is not justifiable by any electrodynamical facts, supplies us then with that particular law of motion which has been confirmed with great precision in recent years.

The theory of relativity leads to the same law of motion, without requiring any special hypothesis whatsoever as to the structure and the behaviour of the electron. We arrived at a similar conclusion of Section 13 in connection with the experiment of Fizeau, the result of which is foretold by the theory of relativity without the necessity of drawing on hypotheses as to the physical nature of the liquid.

---

[1] The general theory of relativity renders it likely that the electrical masses of an electron are held together by gravitational forces.

The second class of facts to which we have alluded has reference to the question whether or not the motion of the earth in space can be made perceptible in terrestrial experiments. We have already remarked in Section 5 that all attempts of this nature led to a negative result. Before the theory of relativity was put forward, it was difficult to become reconciled to this negative result, for reasons now to be discussed. The inherited prejudices about time and space did not allow any doubt to arise as to the prime importance of the Galileian transformation for changing over from one body of reference to another. Now assuming that the Maxwell-Lorentz equations hold for a reference-body $K$, we then find that they do not hold for a reference-body $K'$ moving uniformly with respect to $K$, if we assume that the relations of the Galileian transformation exist between the co-ordinates of $K$ and $K'$. It thus appears that, of all Galileian co-ordinate systems, one $(K)$ corresponding to a particular state of motion is physically unique. This result was interpreted physically by regarding $K$ as at rest with respect to a hypothetical æther of space. On the other hand, all co-ordinate systems $K'$ moving relatively to $K$ were to be regarded as in motion with respect to the æther. To this motion of $K'$ against the æther ("æther-drift" relative to $K'$) were attributed the more complicated laws which were supposed to hold relative to $K'$. Strictly speaking, such an æther-drift ought also to be assumed relative to the earth, and for a long time the efforts of physicists were devoted to attempts to detect the existence of an æther-drift at the earth's surface.

In one of the most notable of these attempts Michelson devised a method which appears as though it must be decisive. Imagine two mirrors so arranged on a rigid body that the reflecting surfaces face each other. A ray of light requires a perfectly definite time $T$ to pass from one mirror to the other and back again, if the whole system be at rest with respect to the æther. It is found by calculation, however, that a slightly different time $T'$ is required for this process, if the body, together with the mirrors, be moving relatively to the æther. And yet another point: it is shown by calculation that for a given velocity $v$ with reference to the æther, this time $T'$ is different when the body is moving perpendicularly to the planes of the mirrors from that resulting when the motion is parallel to these planes. Although the estimated difference between these two times is exceedingly small, Michelson and Morley performed an experiment involving interference in which this difference should have been clearly detectable. But the experiment gave a negative result—a fact very perplexing to physicists. Lorentz and FitzGerald rescued the theory from this difficulty by assuming that the motion of the body relative to the æther produces a contraction of the body in the direction of motion, the amount of contraction being just sufficient to compensate for the difference in time mentioned above. Comparison with the discussion in Section 12 shows that also from the standpoint of the theory of relativity this solution of the difficulty was the right one. But on the basis of the theory of relativity the method of interpretation is incomparably more satisfactory. According to this theory there is no such things as a "specially favoured" (unique) co-ordinate system to occasion the introduction of the æther-idea, and hence there can be no æther-drift, nor any experiment with which to demonstrate it. Here the contraction of moving bodies follows from the two fundamental principles of the theory, without the introduction of particular hypotheses; and as the prime factor involved in this contraction we find, not the motion in itself, to which we cannot attach any meaning, but the motion with respect to the body of reference chosen in the particular case in point. Thus for a co-ordinate system moving with the earth the mirror system of Michelson and Morley is not shortened, but it *is* shortened for a co-ordinate system which is at rest relatively to the sun.

# SEVENTEEN

# MINKOWSKI'S FOUR-DIMENSIONAL SPACE

The non-mathematician is seized by a mysterious shuddering when he hears of "four-dimensional" things, by a feeling not unlike that awakened by thoughts of the occult. And yet there is no more common-place statement than that the world in which we live is a four-dimensional space-time continuum.

Space is a three-dimensional continuum. By this we mean that it is possible to describe the position of a point (at rest) by means of three numbers (co-ordinates) $x$, $y$, $z$, and that there is an indefinite number of points in the neighbourhood of this one, the position of which can be described by co-ordinates such as $x_1$, $y_1$, $z_1$, which may be as near as we choose to the respective values of the co-ordinates $x$, $y$, $z$ of the first point. In virtue of the latter property we speak of a "continuum," and owing to the fact that there are three co-ordinates we speak of it as being "three-dimensional."

Similarly, the world of physical phenomena which was briefly called "world" by Minkowski is naturally four-dimensional in the space-time sense. For it is composed of individual events, each of which is described by four numbers, namely, three space co-ordinates $x$, $y$, $z$ and a time co-ordinate, the time-value $t$. The "world" is in this sense also a continuum; for to every event there are as many "neighbouring" events (realised or at least thinkable) as we care to choose, the coordinates $x_1$, $y_1$, $z_1$, $t_1$ of which differ by an indefinitely small amount from those of the event $x$, $y$, $z$, $t$ originally considered. That we have not been accustomed to regard the world in this sense as a four-dimensional continuum is due to the fact that in physics, before the advent of the theory of relativity, time played a different and more independent rôle, as compared with the space co-ordinates. It is for this reason that we have been in the habit of treating time as an independent continuum. As a matter of fact, according to classical mechanics, time is absolute, *i.e.* it is independent of the position and the condition of motion of the system of co-ordinates. We see this expressed in the last equation of the Galileian transformation ($t' = t$).

The four-dimensional mode of consideration of the "world" is natural on the theory of relativity, since according to this theory time is robbed of its independence. This is shown by the fourth equation of the Lorentz transformation:

$$t' = \frac{t - \frac{v}{c^2}x}{\sqrt{1 - \frac{v^2}{c^2}}}.$$

Moreover, according to this equation the time difference $\Delta t'$ of two events with respect to $K'$ does not in general vanish, even when the time difference $\Delta t$ of the same events with reference to $K$ vanishes. Pure "space-distance" of two events with respect to $K$ results in "time-distance" of the same events with respect to $K'$. But the discovery of Minkowski, which was of importance for the formal development of the theory of relativity, does not lie here. It is to be found rather in the fact of his recognition that the four-dimensional space-time continuum of the theory of relativity, in its most essential formal properties, shows a pronounced relationship to the three-dimensional continuum of Euclidean geometrical space.[1] In order to give due prominence to this relationship, however, we must replace the usual time co-ordinate $t$ by an imaginary magnitude $\sqrt{-1}$. $ct$ proportional to it. Under these conditions, the natural laws satisfying the demands of the (special) theory of relativity assume mathematical forms, in which the time co-ordinate plays exactly the same rôle as the three space co-ordinates. Formally, these four co-ordinates correspond exactly to the three space co-ordinates in Euclidean geometry. It must be clear even to the non-mathematician that, as a consequence of this purely formal addition

---

[1] Cf. the somewhat more detailed discussion in Appendix 2.

to our knowledge, the theory perforce gained clearness in no mean measure.

These inadequate remarks can give the reader only a vague notion of the important idea contributed by Minkowski. Without it the general theory of relativity, of which the fundamental ideas are developed in the following pages, would perhaps have got no farther than its long clothes. Minkowski's work is doubtless difficult of access to anyone inexperienced in mathematics, but since it is not necessary to have a very exact grasp of this work in order to understand the fundamental ideas of either the special or the general theory of relativity, I shall leave it here at present, and revert to it only towards the end of Part II.

# PART II: THE GENERAL THEORY OF RELATIVITY

## E I G H T E E N

# SPECIAL AND GENERAL PRINCIPLE OF RELATIVITY

The basal principle, which was the pivot of all our previous consider-ations, was the *special* principle of relativity, i.e. the principle of the physical relativity of all *uniform* motion. Let us once more analyse its meaning carefully.

It was at all times clear that, from the point of view of the idea it conveys to us, every motion must be considered only as a relative motion. Returning to the illustration we have frequently used of the embankment and the railway carriage, we can express the fact of the motion here taking place in the following two forms, both of which are equally justifiable:

(*a*) The carriage is in motion relative to the embankment.

(*b*) The embankment is in motion relative to the carriage.

In (*a*) the embankment, in (*b*) the carriage, serves as the body of reference in our statement of the motion taking place. If it is sim-ply a question of detecting or of describing the motion involved, it is in principle immaterial to what reference-body we refer the motion. As already mentioned, this is self-evident, but it must not be confused with the much more comprehensive statement called "the principle of relativity," which we have taken as the basis of our investigations.

The principle we have made use of not only maintains that we may equally well choose the carriage or the embankment as our reference-body for the description of any event (for this, too, is self-evident). Our principle rather asserts what follows: If we formulate the general laws of nature as they are obtained from experience, by making use of

(*a*) the embankment as reference-body,

(*b*) the railway carriage as reference-body,

then these general laws of nature (*e.g.* the laws of mechanics or the law of the propagation of light *in vacuo*) have exactly the same form in both cases. This can also be expressed as follows: For the *physical* description of natural processes, neither of the reference-bodies $K$, $K'$ is unique (lit. "specially marked out") as compared with the other. Unlike the first, this latter statement need not of necessity hold *a priori*; it is not contained in the conceptions of "motion" and "reference-body" and derivable from them; only *experience* can decide as to its correctness or incorrectness.

Up to the present, however, we have by no means maintained the equivalence of *all* bodies of reference $K$ in connection with the formulation of natural laws. Our course was more on the following lines. In the first place, we started out from the assumption that there exists a reference-body $K$, whose condition of motion is such that the Galileian law holds with respect to it: A particle left to itself and sufficiently far removed from all other particles moves uniformly in a straight line. With reference to $K$ (Galileian reference-body) the laws of nature were to be as simple as possible. But in addition to $K$, all bodies of reference $K'$ should be given preference in this sense, and they should be exactly equivalent to $K$ for the formulation of natural laws, provided that they are in a state of *uniform rectilinear and nonrotary motion* with respect to $K$; all these bodies of reference are to be regarded as Galileian reference-bodies. The validity of the principle of relativity was assumed only for these reference-bodies, but not for others (*e.g.* those possessing motion of a different kind). In this sense we speak of the *special* principle of relativity, or special theory of relativity.

In contrast to this we wish to understand by the "general principle of relativity" the following statement: All bodies of reference $K$, $K'$, etc., are equivalent for the description of natural phenomena (formulation of the general laws of nature), whatever may be their state of motion. But before proceeding farther, it ought to be pointed out that this formulation must be replaced later by a more abstract one, for reasons which will become evident at a later stage.

Since the introduction of the special principle of relativity has been justified, every intellect which strives after generalisation must feel the temptation to venture the step towards the general principle of relativity. But a simple and apparently quite reliable consideration seems to suggest that, for the present at any rate, there is little hope of success in such an attempt. Let us imagine ourselves transferred to our old friend the railway carriage, which is travelling at a uniform rate. As long as it is moving uniformly, the occupant of the carriage is not sensible of its motion, and it is for this reason that he can without reluctance interpret the facts of the case as indicating that the carriage is at rest, but the embankment in motion. Moreover, according to the special principle of relativity, this interpretation is quite justified also from a physical point of view.

If the motion of the carriage is now changed into a nonuniform motion, as for instance by a powerful application of the brakes, then the occupant of the carriage experiences a correspondingly powerful jerk forwards. The retarded motion is manifested in the mechanical behaviour of bodies relative to the person in the railway carriage. The mechanical behaviour is different from that of the case previously considered, and for this reason it would appear to be impossible that the same mechanical laws hold relatively to the non-uniformly moving carriage, as hold with reference to the carriage when at rest or in uniform motion. At all events it is clear that the Galileian law does not hold with respect to the non-uniformly moving carriage. Because of this, we feel compelled at the present juncture to grant a kind of absolute physical reality to non-uniform motion, in opposition to the general principle of relativity. But in what follows we shall soon see that this conclusion cannot be maintained.

# NINETEEN

# THE GRAVITATIONAL FIELD

If we pick up a stone and then let it go, why does it fall to the ground? The usual answer to this question is: "Because it is attracted by the earth." Modern physics formulates the answer rather differently for the following reason. As a result of the more careful study of electromagnetic phenomena, we have come to regard action at a distance as a process impossible without the intervention of some intermediary medium. If, for instance, a magnet attracts a piece of iron, we cannot be content to regard this as meaning that the magnet acts directly on the iron through the intermediate empty space, but we are constrained to imagine—after the manner of Faraday—that the magnet always calls into being something physically real in the space around it, that something being what we call a "magnetic field." In its turn this magnetic field operates on the piece of iron, so that the latter strives to move towards the magnet. We shall not discuss here the justification for this incidental conception, which is indeed a somewhat arbitrary one. We shall only mention that with its aid electromagnetic phenomena can be theoretically represented much more satisfactorily than without it, and this applies particularly to the transmission of electromagnetic waves. The effects of gravitation also are regarded in an analogous manner.

The action of the earth on the stone takes place indirectly. The earth produces in its surroundings a gravitational field, which acts on the stone and produces its motion of fall. As we know from experience, the intensity of the action on a body diminishes according to a quite definite law, as we proceed farther and farther away from the earth. From our point of view this means: The law governing the properties of the gravitational field in space must be a perfectly definite one, in order correctly to represent the diminution of gravitational action with the distance from operative bodies. It is something like this: The body (*e.g.* the earth) produces a field in its immediate neighbourhood directly; the intensity and direction of the field at points farther removed from the body are thence determined by the law which governs the properties in space of the gravitational fields themselves.

In contrast to electric and magnetic fields, the gravitational field exhibits a most remarkable property, which is of fundamental importance for what follows. Bodies which are moving under the sole influence of a gravitational field receive an acceleration, *which does not in the least depend either on the material or on the physical state of the body.* For instance, a piece of lead and a piece of wood fall in exactly the same manner in a gravitational field (*in vacuo*), when they start off from rest or with the same initial velocity. This law, which holds most accurately, can be expressed in a different form in the light of the following consideration.

According to Newton's law of motion, we have

$$(\text{Force}) = (\text{inertial mass}) \times (\text{acceleration}),$$

where the "inertial mass" is a characteristic constant of the accelerated body. If now gravitation is the cause of the acceleration, we then have

$$(\text{Force}) = (\text{gravitational mass}) \times (\text{intensity of the gravitational field}),$$

where the "gravitational mass" is likewise a characteristic constant for the body. From these two relations follows:

$$(\text{acceleration}) = \frac{(\text{gravitational mass})}{(\text{inertial mass})} (\text{intensity of the gravitational field}).$$

If now, as we find from experience, the acceleration is to be independent of the nature and the condition of the body and always the same for a given gravitational field, then the ratio of the gravitational to the inertial mass must likewise be the same for all bodies. By a suitable choice of units we can thus make this ratio equal to unity. We then have the following law: The *gravitational* mass of a body is equal to its *inertial* mass.

It is true that this important law had hitherto been recorded in mechanics, but it had not been *interpreted.* A satisfactory interpretation can be obtained only if we recognise the following fact: *The same* quality of a body manifests itself according to circumstances as "inertia" or as "weight" (lit. "heaviness"). In the following section we shall show to what extent this is actually the case, and how this question is connected with the general postulate of relativity.

# TWENTY

# THE EQUALITY OF INERTIAL AND GRAVITATIONAL MASS AS AN ARGUMENT FOR THE GENERAL POSTULATE OF RELATIVITY

We imagine a large portion of empty space, so far removed from stars and other appreciable masses, that we have before us approximately the conditions required by the fundamental law of Galilei. It is then possible to choose a Galileian reference-body for this part of space (world), relative to which points at rest remain at rest and points in motion continue permanently in uniform rectilinear motion. As reference-body let us imagine a spacious chest resembling a room with an observer inside who is equipped with apparatus. Gravitation naturally does not exist for this observer. He must fasten himself with strings to the floor, otherwise the slightest impact against the floor will cause him to rise slowly towards the ceiling of the room.

To the middle of the lid of the chest is fixed externally a hook with rope attached, and now a "being" (what kind of a being is immaterial to us) begins pulling at this with a constant force. The chest together with the observer then begins to move "upwards" with a uniformly accelerated motion. In course of time their velocity will reach unheard-of values—provided that we are viewing all this from another reference-body which is not being pulled with a rope.

But how does the man in the chest regard the process? The acceleration of the chest will be transmitted to him by the reaction of the floor of the chest. He must therefore take up this pressure by means of his legs if he does not wish to be laid out full length on the floor. He is then standing in the chest in exactly the same way as anyone stands in a room of a house on our earth. If he releases a body which he previously had in his hand, the acceleration of the chest will no longer be transmitted to this body, and for this reason the body will approach the floor of the chest with an accelerated relative motion. The observer will further convince himself *that the acceleration of the body towards the floor of the chest is always of the same magnitude, whatever kind of body he may happen to use for the experiment.*

Relying on his knowledge of the gravitational field (as it was discussed in the preceding section), the man in the chest will thus come to the conclusion that he and the chest are in a gravitational field which is constant with regard to time. Of course he will be puzzled for a moment as to why the chest does not fall in this gravitational field. Just then, however, he discovers the hook in the middle of the lid of the chest and the rope which is attached to it, and he consequently comes to the conclusion that the chest is suspended at rest in the gravitational field.

Ought we to smile at the man and say that he errs in his conclusion? I do not believe we ought to if we wish to remain consistent; we must rather admit that his mode of grasping the situation violates neither reason nor known mechanical laws. Even though it is being accelerated with respect to the "Galileian space" first considered, we can nevertheless regard the chest as being at rest. We have thus good grounds for extending the principle of relativity to include bodies of reference which are accelerated with respect to each other, and as a result we have gained a powerful argument for a generalised postulate of relativity.

We must note carefully that the possibility of this mode of interpretation rests on the fundamental property of the gravitational field of giving all bodies the same acceleration, or, what comes to the same thing, on the law of the equality of inertial and gravitational mass. If this natural law did not exist, the man in the accelerated chest would not be able to interpret the behaviour of the bodies around him on the supposition of a gravitational field, and he would not be justified on the grounds of experience in supposing his reference-body to be "at rest."

Suppose that the man in the chest fixes a rope to the inner side of the lid, and that he attaches a body to the free end of the rope. The result of this will be to stretch the rope so that it will hang "vertically" downwards. If we ask for an opinion of the cause of tension in the rope, the man in the chest will say: "The suspended body experiences a downward force in the gravitational field, and this is neutralised by the tension of the rope; what determines the magnitude of the tension of the rope is the *gravitational mass* of the suspended body." On the other hand, an observer who is poised freely in space will interpret the condition of things thus: "The rope must perforce take part in the accelerated motion of the chest, and it transmits this motion to the body attached to it. The tension of the rope is just large enough to effect the acceleration of the body. That which determines the magnitude of the tension of the rope is the *inertial mass* of the body." Guided by this example, we see that our extension of the principle of relativity implies the *necessity* of the law of the equality of inertial and gravitational mass. Thus we have obtained a physical interpretation of this law.

From our consideration of the accelerated chest we see that a general theory of relativity must yield important results on the laws of gravitation. In point of fact, the systematic pursuit of the general idea of relativity has supplied the laws satisfied by the gravitational field. Before proceeding farther, however, I must warn the reader against a misconception suggested by these considerations. A gravitational field exists for the man in the chest, despite the fact that there was no such field for the co-ordinate system first chosen. Now we might easily suppose that the existence of a gravitational field is always only an *apparent* one. We might also think that, regardless of the kind of gravitational field which may be present, we could always choose another reference-body such that *no* gravitational field exists with reference to it. This is by no means true for all gravitational fields, but only for those of quite special form. It is, for instance, impossible to choose a body of reference such that, as judged from it, the gravitational field of the earth (in its entirety) vanishes.

We can now appreciate why that argument is not convincing, which we brought forward against the general principle of relativity at the end of Section 18. It is certainly true that the observer in the railway carriage experiences a jerk forwards as a result of the application of the brake, and that he recognises in this the non-uniformity of motion (retardation) of the carriage. But he is compelled by nobody to refer this jerk to a "real" acceleration (retardation) of the carriage. He might also interpret his experience, thus: "My body of reference (the carriage) remains permanently at rest. With reference to it, however, there exists (during the period of application of the brakes) a gravitational field which is directed forwards and which is variable with respect to time. Under the influence of this field, the embankment together with the earth moves nonuniformly in such a manner that their original velocity in the backwards direction is continuously reduced."

# TWENTY-ONE

# IN WHAT RESPECTS ARE THE FOUNDATIONS OF CLASSICAL MECHANICS AND OF THE SPECIAL THEORY OF RELATIVITY UNSATISFACTORY?

We have already stated several times that classical mechanics starts out from the following law: Material particles sufficiently far removed from other material particles continue to move uniformly in a straight line or continue in a state of rest. We have also repeatedly emphasised that this fundamental law can only be valid for bodies of reference $K$ which possess certain unique states of motion, and which are in uniform translational motion relative to each other. Relative to other reference-bodies $K$ the law is not valid. Both in classical mechanics and in the special theory of relativity we therefore differentiate between reference-bodies $K$ relative to which the recognised "laws of nature" can be said to hold, and reference-bodies $K$ relative to which these laws do not hold.

But no person whose mode of thought is logical can rest satisfied with this condition of things. He asks: "How does it come that certain reference-bodies (or their states of motion) are given priority over other reference-bodies (or their states of motion)?" *What is the reason for this preference*? In order to show clearly what I mean by this question, I shall make use of a comparison.

I am standing in front of a gas range. Standing alongside of each other on the range are two pans so much alike that one may be mistaken for the other. Both are half full of water. I notice that steam is being emitted continuously from the one pan, but not from the other. I am surprised at this, even if I have never seen either a gas range or a pan before. But if I now notice a luminous something of bluish colour under the first pan but not under the other, I cease to be astonished, even if I have never before seen a gas flame. For I can only say that this bluish something will cause the emission of the stream, or at least *possibly* it may do so. If, however, I notice the bluish something in neither case, and if I observe that the one continuously emits steam whilst the other does not, then I shall remain astonished and dissatisfied until I have discovered some circumstance to which I can attribute the different behaviour of the two pans.

Analogously, I seek in vain for a real something in classical mechanics (or in the special theory of relativity) to which I can attribute the different behaviour of bodies considered with respect to the reference-systems $K$ and $K'$.[1] Newton saw this objection and attempted to invalidate it, but without success. But E. Mach recognised it most clearly of all, and because of this objection he claimed that mechanics must be placed on a new basis. It can only be got rid of by means of a physics which is comfortable to the general principle of relativity, since the equations of such a theory hold for every body of reference, whatever may be its state of motion.

---

[1] The objection is of importance more especially when the state of motion of the reference-body is of such a nature that it does not require any external agency for its maintenance, *e.g.* in the case when the reference-body is rotating uniformly.

# TWENTY-TWO
# A Few Inferences from the General Principle of Relativity

The considerations of Section 20 show that the general principle of relativity puts us in a position to derive properties of the gravitational field in a purely theoretical manner. Let us suppose, for instance, that we know the space-time "course" for any natural process whatsoever, as regards the manner in which it takes place in the Galileian domain relative to a Galileian body of reference $K$. By means of purely theoretical operations (*i.e.* simply by calculation) we are then able to find how this known natural process appears, as seen from a reference-body $K'$ which is accelerated relatively to $K$. But since a gravitational field exists with respect to this new body of reference $K'$ our consideration also teaches us how the gravitational field influences the process studied.

For example, we learn that a body which is in a state of uniform rectilinear motion with respect to $K$ (in accordance with the law of Gelilei) is executing an accelerated and in general curvilinear motion with respect to the accelerated reference-body $K'$ (chest). This acceleration or curvature corresponds to the influence on the moving body of the gravitational field prevailing relatively to $K'$. It is known that a gravitational field influences the movement of bodies in this way, so that our consideration supplies us with nothing essentially new.

However, we obtain a new result of fundamental importance when we carry out the analogous consideration for a ray of light. With respect to the Galileian reference-body $K$, such a ray of light is transmitted rectilinearly with the velocity $c$. It can easily be shown that the path of the same ray of light is no longer a straight line when we consider it with reference to the accelerated chest (reference-body $K'$). From this we conclude, *that, in general, rays of light are propagated curvilinearly in gravitational fields.* In two respects this result is of great importance.

In the first place, it can be compared with the reality. Although a detailed examination of the question shows that the curvature of light rays required by the general theory of relativity is only exceedingly small for the gravitational fields at our disposal in practice, its estimated magnitude for light rays passing the sun at grazing incidence is nevertheless 1.7 seconds of arc. This ought to manifest itself in the following way. As seen from the earth, certain fixed stars appear to be in the neighbourhood of the sun, and are thus capable of observation during a total eclipse of the sun. At such times, these stars ought to appear to be displaced outwards from the sun by an amount indicated above, as compared with their apparent position in the sky when the sun is situated at another part of the heavens. The examination of the correctness or otherwise of this deduction is a problem of the greatest importance, the early solution of which is to be expected of astronomers.[1]

In the second place our result shows that, according to the general theory of relativity, the law of the constancy of the velocity of light *in vacuo*, which constitutes one of the two fundamental assumptions in the special theory of relativity and to which we have already frequently referred, cannot claim any unlimited validity. A curvature of rays of light can only take place when the velocity of propagation of light varies with position. Now we might think that as a consequence of this, the special theory of relativity and with it the whole theory of relativity would be laid in the dust. But in reality this is not the case. We can only conclude that the special theory of relativity cannot claim an unlimited domain of validity; its results hold only so long as we are able to disregard the influences of gravitational fields on the phenomena (*e.g.* of light).

Since it has often been contended by opponents of the theory of relativity that the special theory of relativity is overthrown by the general

---

[1] By means of the star photographs of two expeditions equipped by a Joint Committee of the Royal and Royal Astronomical Societies, the existence of the deflection of light demanded by theory was first confirmed during the solar eclipse of 29th May, 1919. (Cr. Appendix 3.)

theory of relativity, it is perhaps advisable to make the facts of the case clearer by means of an appropriate comparison. Before the development of electrodynamics the laws of electrostatics were looked upon as the laws of electricity. At the present time we know that electric fields can be derived correctly from electrostatic considerations only for the case, which is never strictly realised, in which the electrical masses are quite at rest relatively to each other, and to the co-ordinate system. Should we be justified in saying that for this reason electrostatics is overthrown by the field-equations of Maxwell in electrodynamics? Not in the least. Electrostatics is contained in electrodynamics as a limiting case; the laws of the latter lead directly to those of the former for the case in which the fields are invariable with regard to time. No fairer destiny could be allotted to any physical theory, than that it should of itself point out the way to the introduction of a more comprehensive theory, in which it lives on as a limiting case.

In the example of the transmission of light just dealt with, we have seen that the general theory of relativity enables us to derive theoretically the influence of a gravitational field on the course of natural processes, the laws of which are already known when a gravitational field is absent. But the most attractive problem, to the solution of which the general theory of relativity supplies the key, concerns the investigation of the laws satisfied by the gravitational field itself. Let us consider this for a moment.

We are acquainted with space-time domains which behave (approximately) in a "Galileian" fashion under suitable choice of reference-body, *i.e.* domains in which gravitational fields are absent. If we now refer such a domain to a reference-body $K'$ possessing any kind of motion, then relative to $K'$ there exists a gravitational field which is variable with respect to space and time.[1] The character of this field will of course depend on the motion chosen for $K'$. According to the general theory of relativity, the general law of the gravitational field must be satisfied for all gravitational fields obtainable in this way.

[1] This follows from a generalisation of the discussion in Section 20.

Even though by no means all gravitational fields can be produced in this way, yet we may entertain the hope that the general law of gravitation will be derivable from such gravitational fields of a special kind. This hope has been realised in the most beautiful manner. But between the clear vision of this goal and its actual realisation it was necessary to surmount a serious difficulty, and as this lies deep at the root of things, I dare not withhold it from the reader. We require to extend our ideas of the space-time continuum still farther.

# TWENTY-THREE
# BEHAVIOUR OF CLOCKS AND MEASURING-RODS ON A ROTATING BODY OF REFERENCE

Hitherto I have purposely refrained from speaking about the physical interpretation of space- and time-data in the case of the general theory of relativity. As a consequence, I am guilty of a certain slovenliness of treatment, which, as we know from the special theory of relativity, is far from being unimportant and pardonable. It is now high time that we remedy this defect; but I would mention at the outset, that this matter lays no small claims on the patience and on the power of abstraction of the reader.

We start off again from quite special cases, which we have frequently used before. Let us consider a space-time domain in which no gravitational field exists relative to a reference-body $K$ whose state of motion has been suitably chosen. $K$ is then a Galileian reference-body as regards the domain considered, and the results of the special theory of relativity hold relative to $K$. Let us suppose the same domain referred to a second body of reference $K'$, which is rotating uniformly with respect to $K$. In order to fix our ideas, we shall imagine $K'$ to be in the form of a plane circular disc, which rotates uniformly in its own plane about its centre. An observer who is sitting eccentrically on the disc $K'$ is sensible of a force which acts outwards in a radial direction, and which would be interpreted as an effect of inertia (centrifugal force) by an observer who was at rest with respect to the original reference-body $K$. But the observer on the disc may regard his disc as a reference-body which is "at rest"; on the basis of the general principle of relativity he is justified in doing this. The force acting on himself, and in fact on all other bodies which are at rest relative to the disc, he regards as the effect of a gravitational field. Nevertheless, the space-distribution of this gravitational field is of a kind that would not be possible on Newton's theory of gravitation.[1] But since the observer believes in the general theory of relativity, this does not disturb him; he is quite in the right when he believes that a general law of gravitation can be formulated—a law which not only explains the motion of the stars correctly, but also the field of force experienced by himself.

The observer performs experiments on his circular disc with clocks and measuring-rods. In doing so, it is his intention to arrive at exact definitions for the signification of time- and space-data with reference to the circular disc $K'$, these definitions being based on his observations. What will be his experience in this enterprise?

To start with, he places one of two identically constructed clocks at the centre of the circular disc, and the other on the edge of the disc, so that they are at rest relative to it. We now ask ourselves whether both clocks go at the same rate from the standpoint of the non-rotating Galilean reference-body $K$. As judged from this body, the clock at the centre of the disc has no velocity, whereas the clock at the edge of the disc is in motion relative to $K$ in consequence of the rotation. According to a result obtained in Section 12, it follows that the latter clock goes at a rate permanently slower than that of the clock at the centre of the circular disc, i.e. as observed from $K$. It is obvious that the same effect would be noted by an observer whom we will imagine sitting alongside his clock at the centre of the circular disc. Thus on our circular disc, or, to make the case more general, in every gravitational field, a clock will go more quickly or less quickly, according to the position in which the clock is situated (at rest). For this reason it is not possible to obtain a reasonable definition of time with the aid of clocks which are arranged at rest with respect to the body of reference. A similar difficulty presents itself when we attempt to apply our earlier definition of simultaneity in such a case, but I do not wish to go any farther into this question.

[1] The field disappears at the centre of the disc and increases proportionally to the distance from the centre as we proceed outwards.

Moreover, at this stage the definition of the space coordinates also presents insurmountable difficulties. If the observer applies his standard measuring-rod (a rod which is short as compared with the radius of the disc) tangentially to the edge of the disc, then, as judged from the Galileian system, the length of this rod will be less than 1, since, according to Section 12, moving bodies suffer a shortening in the direction of the motion. On the other hand, the measuring-rod will not experience a shortening in length, as judged from $K$, if it is applied to the disc in the direction of the radius. If, then, the observer first measures the circumference of the disc with his measuring-rod and then the diameter of the disc, on dividing the one by the other, he will not obtain as quotient the familiar number $\pi = 3.14 \ldots$, but a larger number,[1] whereas of course, for a disc which is at rest with respect to $K$, this operation would yield $\pi$ exactly. This proves that the propositions of Euclidean geometry cannot hold exactly on the rotating disc, nor in general in a gravitational field, at least if we attribute the length 1 to the rod in all positions and in every orientation. Hence the idea of a straight line also loses its meaning. We are therefore not in a position to define exactly the co-ordinates $x$, $y$, $z$ relative to the disc by means of the method used in discussing the special theory, and as long as the co-ordinates and times of events have not been defined, we cannot assign an exact meaning to the natural laws in which these occur.

Thus all our previous conclusions based on general relativity would appear to be called in question. In reality we must make a subtle detour in order to be able to apply the postulate of general relativity exactly. I shall prepare the reader for this in the following paragraphs.

---

[1]Throughout this consideration we have to use the Galileian (non-rotating) system $K$ as reference-body, since we may only assume the validity of the results of the special theory of relativity relative to $K$ (relative to $K'$ a gravitational field prevails).

# TWENTY-FOUR

# EUCLIDEAN AND NON-EUCLIDEAN CONTINUUM

The surface of a marble table is spread out in front of me. I can get from any one point on this table to any other point by passing continuously from one point to a "neighbouring" one, and repeating this process a (large) number of times, or, in other words, by going from point to point without executing "jumps." I am sure the reader will appreciate with sufficient clearness what I mean here by "neighbouring" and by "jumps" (if he is not too pedantic). We express this property of the surface by describing the latter as a continuum.

Let us now imagine that a large number of little rods of equal length have been made, their lengths being small compared with the dimensions of the marble slab. When I say they are of equal length, I mean that one can be laid on any other without the ends overlapping. We next lay four of these little rods on the marble slab so that they constitute a quadrilateral figure (a square), the diagonals of which are equally long. To ensure the equality of the diagonals, we make use of a little testing-rod. To this square we add similar ones, each of which has one rod in common with the first. We proceed in like manner with each of these squares until finally the whole marble slab is laid out with squares. The arrangement is such, that each side of a square belongs to two squares and each corner to four squares.

It is a veritable wonder that we can carry out this business without getting into the greatest difficulties. We only need to think of the following. If at any moment three squares meet at a corner, then two sides of the fourth square are already laid, and, as a consequence, the arrangement of the remaining two sides of the square is already completely determined. But I am now no longer able to adjust the quadrilateral so that its diagonals may be equal. If they are equal of their own accord, then this is an especial favour of the marble slab and of the little rods, about which I can only be thankfully surprised. We must needs experience many such surprises if the construction is to be successful.

If everything has really gone smoothly, then I say that the points of the marble slab constitute a Euclidean continuum with respect to the little rod, which has been used as a "distance" (line-interval). By choosing one corner of a square as "origin," I can characterize every other corner of a square with reference to this origin by means of two numbers. I only need state how many rods I must pass over when, starting from the origin, I proceed towards the "right" and then "upwards," in order to arrive at the corner of the square under consideration. These two numbers are then the "Cartesian co-ordinates" of this corner with reference to the "Cartesian co-ordinate system" which is determined by the arrangement of little rods.

By making use of the following modification of this abstract experiment, we recognise that there must also be cases in which the experiment would be unsuccessful. We shall suppose that the rods "expand" by an amount proportional to the increase of temperature. We heat the central part of the marble slab, but not the periphery, in which case two of our little rods can still be brought into coincidence at every position on the table. But our construction of squares must necessarily come into disorder during the heating, because the little rods on the central region of the table expand, whereas those on the outer part do not.

With reference to our little rods—defined as unit lengths—the marble slab is no longer a Euclidean continuum, and we are also no longer in the position of defining Cartesian coordinates directly with their aid, since the above construction can no longer be carried out. But since there are other things which are not influenced in a similar manner to the little rods (or perhaps not at all) by the temperature of the table, it is possible quite naturally to maintain the point of view that the marble slab is a "Euclidean continuum." This can be done in a satisfactory manner by making a more subtle stipulation about the measurement or the comparison of lengths.

But if rods of every kind (*i.e.* of every material) were to behave *in the same way* as regards the influence of temperature when they are on the variably heated marble slab, and if we had no other means of detecting the effect of temperature than the geometrical behaviour of our rods in experiments analogous to the one described above, then our best plan would be to assign the distance *one* to two points on the slab, provided that the ends of one of our rods could be made to coincide with these two points; for how else should we define the distance without our proceeding being in the highest measure grossly arbitrary? The method of Cartesian co-ordinates must then be discarded, and replaced by another which does not assume the validity of Euclidean geometry for rigid bodies.[1] The reader will notice that the situation depicted here corresponds to the one brought about by the general postulate of relativity (Section 23).

---

[1] Mathematicians have been confronted with our problem in the following form. If we are given a surface (*e.g.* an ellipsoid) in Euclidean three-dimensional space, then there exists for this surface a two-dimensional geometry, just as much as for a plane surface. Gauss undertook the task of treating this two-dimensional geometry from first principles, without making use of the fact that the surface belongs to a Euclidean continuum of three dimensions. If we imagine constructions to be made with rigid rods *in the surface* (similar to that above with the marble slab), we should find that different laws hold for these from those resulting on the basis of Euclidean plane geometry. The surface is not a Euclidean continuum with respect to the rods, and we cannot define Cartesian co-ordinates *in the surface*. Gauss indicated the principles according to which we can treat the geometrical relationships in the surface, and thus pointed out the way to the method of Ricmann of treating multi-dimensional, non-Euclidean *continua*. Thus it is that mathematicians long ago solved the formal problems to which we are led by the general postulate of relativity.

# TWENTY-FIVE

# GAUSSIAN CO-ORDINATES

According to Gauss, this combined analytical and geometrical mode of handling the problem can be arrived at in the following way. We imagine a system of arbitrary curves (see Fig. 4) drawn on the surface of the table.

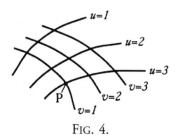

FIG. 4.

These we designate as *u*-curves, and we indicate each of them by means of a number. The curves *u* = 1, *u* = 2 and *u* = 3 are drawn in the diagram. Between the curves *u* = 1 and *u* = 2 we must imagine an infinitely large number to be drawn, all of which correspond to real numbers lying between 1 and 2. We have then a system of *u*-curves, and this "infinitely dense" system covers the whole surface of the table. These *u*-curves must not intersect each other, and through each point of the surface one and only one curve must pass. Thus a perfectly definite value of *u* belongs to every point on the surface of the marble slab. In like manner we imagine a system of *v*-curves drawn on the surface. These satisfy the same conditions as the *u*-curves, they are provided with numbers in a corresponding manner, and they may likewise be of arbitrary shape. It follows that a value of *u* and a value of *v* belong to every point on the surface of the table. We call these two numbers the co-ordinates of the surface of the table (Gaussian co-ordinates). For example, the point *P* in the diagram has the Gaussian co-ordinates *u* = 3, *v* = 1. Two neighbouring points *P* and *P'* on the surface then correspond to the co-ordinates

$$P: \qquad u, v$$
$$P': \qquad u + du, v + dv,$$

where $du$ and $dv$ signify very small numbers. In a similar manner we may indicate the distance (line-interval) between $P$ and $P'$, as measured with a little rod, by means of the very small number $ds$. Then according to Gauss we have

$$ds^2 = g_{11} du^2 + 2g_{12} du dv + g_{22} dv^2,$$

where $g_{11}$, $g_{12}$, $g_{22}$, are magnitudes which depend in a perfectly definite way to $u$ and $v$. The magnitudes $g_{11}$, $g_{12}$ and $g_{22}$ determine the behaviour of the rods relative to the $u$-curves and $v$-curves, and thus also relative to the surface of the table. For the case in which the points of the surface considered form a Euclidean continuum with reference to the measuring-rods, but only in this case, it is possible to draw the $u$-curves and $v$-curves and to attach numbers to them, in such a manner, that we simply have:

$$ds^2 = du^2 + dv^2.$$

Under these conditions, the $u$-curves and $v$-curves are straight lines in the sense of Euclidean geometry, and they are perpendicular to each other. Here the Gaussian co-ordinates are simply Cartesian ones. It is clear that Gauss co-ordinates are nothing more than an association of two sets of numbers with the points of the surface considered, of such a nature that numerical values differing very slightly from each other are associated with neighbouring points "in space."

So far, these considerations hold for a continuum of two dimensions. But the Gaussian method can be applied also to a continuum of three, four or more dimensions. If, for instance, a continuum of four dimensions be supposed available, we may represent it in the following way. With every point of the continuum we associate arbitrarily four numbers, $x_1$, $x_2$, $x_3$, $x_4$, which are known as "co-ordinates." Adjacent points correspond to adjacent values of the co-ordinates. If a distance $ds$ is associated with the adjacent points $P$ and $P'$, this distance being measurable and well-defined from a physical point of view, then the following formula holds:

$$ds^2 = g_{11} dx_1{}^2 + 2g_{12} dx_1 dx_2 \ldots + g_{44} dx_4{}^2,$$

where the magnitudes $g_{11}$, etc., have values which vary with the position in the continuum. Only when the continuum is a Euclidean one is it possible to associate the co-ordinates $x_1 \ldots x_4$ with the points of the continuum so that we have simply

$$ds^2 = dx_1^2 + dx_2^2 + dx_3^2 + dx_4^2.$$

In this case relations hold in the four-dimensional continuum which are analogous to those holding in our three-dimensional measurements.

However, the Gauss treatment for $ds^2$ which we have given above is not always possible. It is only possible when sufficiently small regions of the continuum under consideration may be regarded as Euclidean continua. For example, this obviously holds in the case of the marble slab of the table and local variation of temperature. The temperature is practically constant for a small part of the slab, and thus the geometrical behaviour of the rods is *almost* as it ought to be according to the rules of Euclidean geometry. Hence the imperfections of the construction of squares in the previous section do not show themselves clearly until this construction is extended over a considerable portion of the surface of the table.

We can sum this up as follows: Gauss invented a method for the mathematical treatment of continua in general, in which "size-relations" ("distances" between neighbouring points) are defined. To every point of a continuum are assigned as many numbers (Gaussian co-ordinates) as the continuum has dimensions. This is done in such a way, that only one meaning can be attached to the assignment, and that numbers (Gaussian co-ordinates) which differ by an indefinitely small amount are assigned to adjacent points. The Gaussian co-ordinate system is a logical generalisation of the Cartesian co-ordinate system. It is also applicable to non-Euclidean continua, but only when, with respect to the defined "size" or "distance," small parts of the continuum under consideration behave more nearly like a Euclidean system, the smaller the part of the continuum under our notice.

TWENTY-SIX

# THE SPACE-TIME CONTINUUM OF THE SPECIAL THEORY OF RELATIVITY CONSIDERED AS A EUCLIDEAN CONTINUUM

We are now in a position to formulate more exactly the the idea of Minkowski, which was only vaguely indicated in Section 17. In accordance with the special theory of relativity, certain co-ordinate systems are given preference for the description of the four-dimensional, space-time continuum. We called these "Galileian co-ordinate systems." For these systems, the four co-ordinates $x$, $y$, $z$, $t$, which determine an event or— in other words—a point of the four-dimensional continuum, are defined physically in a simple manner, as set forth in detail in the first part of this book. For the transition from one Galileian system to another, which is moving uniformly with reference to the first, the equations of the Lorentz transformation are valid. These last form the basis for the derivation of deductions from the special theory of relativity, and in themselves they are nothing more than the expression of the universal validity of the law of transmission of light for all Galileian systems of reference.

Minkowski found that the Lorentz transformations satisfy the following simple conditions. Let us consider two neighbouring events, the relative position of which in the four-dimensional continuum is given with respect to a Galileian reference-body $K$ by the space co-ordinate differences $dx$, $dy$, $dz$ and the time-difference $dt$. With reference to a second Galileian system we shall suppose that the corresponding differences for these two events are $dx'$, $dy'$, $dz'$, $dt'$. Then these magnitudes always fulfil the condition[1]

$$dx^2 + dy^2 + dz^2 - c^2dt^2 = dx'^2 + dy'^2 + dz'^2 - c^2dt'^2.$$

---

[1]Cf. Appendices 1 and 2. The relations which are derived there for the co-ordinates themselves are valid also for co-ordinate *differences*, and thus also for co-ordinate differentials (indefinitely small differences).

The validity of the Lorentz transformation follows from this condition. We can express this as follows: The magnitude

$$ds^2 = dx^2 + dy^2 + dz^2 - c^2dt^2,$$

which belongs to two adjacent points of the four-dimensional spacetime continuum, has the same value for all selected (Galileian) reference-bodies. If we replace $x, y, z \sqrt{-1} \, ct$, by $x_1, x_2, x_3, x_4$, we also obtain the result that

$$ds^2 = dx_1^2 + dx_2^2 + dx_3^2 + dx_4^2$$

is independent of the choice of the body of reference. We call the magnitude $ds$ the "distance" apart of the two events or four-dimensional points.

Thus, if we choose as time-variable the imaginary variable $\sqrt{-1} \, ct$ instead of the real quantity $t$, we can regard the spacetime continuum—in accordance with the special theory of relativity—as a "Euclidean" four-dimensional continuum, a result which follows from the considerations of the preceding section.

# TWENTY-SEVEN

# THE SPACE-TIME CONTINUUM OF THE GENERAL THEORY OF RELATIVITY IS NOT A EUCLIDEAN CONTINUUM

In the first part of this book we were able to make use of space-time co-ordinates which allowed of a simple and direct physical interpretation, and which, according to Section 26, can be regarded as four-dimensional Cartesian co-ordinates. This was possible on the basis of the law of the constancy of the velocity of light. But according to Section 21, the general theory of relativity cannot retain this law. On the contrary, we arrived at the result that according to this latter theory the velocity of light must always depend on the coordinates when a gravitational field is present. In connection with a specific illustration in Section 23, we found that the presence of a gravitational field invalidates the definition of the co-ordinates and the time, which led us to our objective in the special theory of relativity.

In view of the results of these considerations we are led to the conviction that, according to the general principle of relativity, the space-time continuum cannot be regarded as a Euclidean one, but that here we have the general case, corresponding to the marble slab with local variations of temperature, and with which we made acquaintance as an example of a two-dimensional continuum. Just as it was there impossible to construct a Cartesian co-ordinate system from equal rods, so here it is impossible to build up a system (reference-body) from rigid bodies and clocks, which shall be of such a nature that measuring-rods and clocks, arranged rigidly with respect to one another, shall indicate position and time directly. Such was the essence of the difficulty with which we were confronted in Section 23.

But the considerations of Sections 25 and 26 show us the way to surmount this difficulty. We refer the four-dimensional space-time continuum in an arbitrary manner to Gauss co-ordinates. We assign to every point of the continuum (event) four numbers, $x_1$, $x_2$, $x_3$, $x_4$ (co-ordinates), which have not the least direct physical significance, but only serve the purpose of numbering the points of the continuum in a definite but arbitrary manner. This arrangement does not even need to be of such a kind that we must regard $x_1$, $x_2$, $x_3$ as "space" co-ordinates and $x_4$ as a "time" co-ordinate.

The reader may think that such a description of the world would be quite inadequate. What does it mean to assign to an event the particular co-ordinates $x_1$, $x_2$, $x_3$, $x_4$, if in themselves these co-ordinates have no significance? More careful consideration shows, however, that this anxiety is unfounded. Let us consider, for instance, a material point with any kind of motion. If this point had only a momentary existence without duration, then it would be described in space-time by a single system of values $x_1$, $x_2$, $x_3$, $x_4$. Thus its permanent existence must be characterised by an infinitely large number of such systems of values, the co-ordinate values of which are so close together as to give continuity; corresponding to the material point, we thus have a (uni-dimensional) line in the four-dimensional continuum. In the same way, any such lines in our continuum correspond to many points in motion. The only statements having regard to these points which can claim a physical existence are in reality the statements about their encounters. In our mathematical treatment, such an encounter is expressed in the fact that the two lines which represent the motions of the points in question have a particular system of co-ordinate values, $x_1$, $x_2$, $x_3$, $x_4$, in common. After mature consideration the reader will doubtless admit that in reality such encounters constitute the only actual evidence of a time-space nature with which we meet in physical statements.

When we were describing the motion of a material point relative to a body of reference, we stated nothing more than the encounters of this point with particular points of the reference-body. We can also determine the corresponding values of the time by the observation of encounters of the body with clocks, in conjunction with the observation of the encounter of the hands of clocks with particular points on the dials. It is just the same in the case of space-measurements by means of measuring-rods, as a little consideration will show.

The following statements hold generally: Every physical description resolves itself into a number of statements, each of which refers to the space-time coincidence of two events $A$ and $B$. In terms of Gaussian co-ordinates, every such statement is expressed by the agreement of their four co-ordinates $x_1$, $x_2$, $x_3$, $x_4$. Thus in reality, the description of the time-space continuum by means of Gauss co-ordinates completely replaces the description with the aid of a body of reference, without suffering from the defects of the latter mode of description; it is not tied down to the Euclidean character of the continuum which has to be represented.

# TWENTY-EIGHT

# EXACT FORMULATION OF THE GENERAL PRINCIPLE OF RELATIVITY

We are now in a position to replace the provisional formulation of the general principle of relativity given in Section 18 by an exact formulation. The form there used, "All bodies of reference $K$, $K'$, etc., are equivalent for the description of natural phenomena (formulation of the general laws of nature), whatever may be their state of motion," cannot be maintained, because the use of rigid reference-bodies, in the sense of the method followed in the special theory of relativity, is in general not possible in space-time description. The Gauss co-ordinate system has to take the place of the body of reference. The following statement corresponds to the fundamental idea of the general principle of relativity: *"All Gaussian co-ordinate systems are essentially equivalent for the formulation of the general laws of nature."*

We can state this general principle of relativity in still another form, which renders it yet more clearly intelligible than it is when in the form of the natural extension of the special principle of relativity. According to the special theory of relativity, the equations which express the general laws of nature pass over into equations of the same form when, by making use of the Lorentz transformation, we replace the space-time variables $x$, $y$, $z$, $t$, of a (Galileian) reference-body $K$ by the space-time variables $x'$, $y'$, $z'$, $t'$, of a new reference-body $K'$. According to the general theory of relativity, on the other hand, by application of *arbitrary substitutions* of the Gauss variables $x_1$, $x_2$, $x_3$, $x_4$, the equations must pass over into equations of the same form; for every transformation (not only the Lorentz transformation) corresponds to the transition of one Gauss co-ordinate system into another.

If we desire to adhere to our "old-time" three-dimensional view of things, then we can characterise the development which is being undergone by the fundamental idea of the general theory of relativity as follows: The special theory of relativity has reference to Galileian domains, *i.e.* to those in which no gravitational field exists. In this connection a Galileian reference-body serves as body of reference, *i.e.* a rigid body the state of motion of which is so chosen that the Galileian law of the uniform rectilinear motion of "isolated" material points holds relatively to it.

Certain considerations suggest that we should refer the same Galileian domains to *non-Galileian* reference-bodies also. A gravitational field of a special kind is then present with respect to these bodies (cf. Sections 20 and 23).

In gravitational fields there are no such things as rigid bodies with Euclidean properties; thus the fictitious rigid body of reference is of no avail in the general theory of relativity. The motion of clocks is also influenced by gravitational fields, and in such a way that a physical definition of time which is made directly with the aid of clocks has by no means the same degree of plausibility as in the special theory of relativity.

For this reason non-rigid reference-bodies are used, which are as a whole not only moving in any way whatsoever, but which also suffer alterations in form *ad lib.* during their motion. Clocks, for which the law of motion is of any kind, however irregular, serve for the definition of time. We have to imagine each of these clocks fixed at a point on the non-rigid reference-body. These clocks satisfy only the one condition, that the "readings" which are observed simultaneously on adjacent clocks (in space) differ from each other by an indefinitely small amount. This non-rigid reference-body, which might appropriately be termed a "reference-mollusc," is in the main equivalent to a Gaussian four-dimensional co-ordinate system chosen arbitrarily. That which gives the "mollusc" a certain comprehensibility as compared with the Gauss co-ordinate system is the (really unjustified) formal retention of the separate existence of the space co-ordinates as opposed to the time co-ordinate. Every point on the mollusc is treated as a space-point, and every material point which is at rest relatively to it as at rest, so long as the mollusc is considered as reference-body. The general principle of relativity requires that all these molluscs can be used as reference-bodies with equal right and equal success in the formulation of the general laws of nature; the laws themselves must be quite independent of the choice of mollusc.

The great power possessed by the general principle of relativity lies in the comprehensive limitation which is imposed on the laws of nature in consequence of what we have seen above.

# TWENTY-NINE

# THE SOLUTION OF THE PROBLEM OF GRAVITATION ON THE BASIS OF THE GENERAL PRINCIPLE OF RELATIVITY

If the reader has followed all our previous considerations, he will have no further difficulty in understanding the methods leading to the solution of the problem of gravitation.

We start off from a consideration of a Galileian domain, i.e. a domain in which there is no gravitational field relative to the Galileian reference-body $K$. The behaviour of measuring-rods and clocks with reference to $K$ is known from the special theory of relativity, likewise the behaviour of "isolated" material points; the latter move uniformly and in straight lines.

Now let us refer this domain to a random Gauss co-ordinate system or to a "mollusc" as reference-body $K'$. Then with respect to $K'$ there is a gravitational field $G$ (of a particular kind). We learn the behaviour of measuring-rods and clocks and also of freely-moving material points with reference to $K'$ simply by mathematical transformation. We interpret this behaviour as the behaviour of measuring-rods, clocks and material points under the influence of the gravitational field $G$. Hereupon we introduce a hypothesis: that the influence of the gravitational field on measuring-rods, clocks and freely-moving material points continues to take place according to the same laws, even in the case where the prevailing gravitational field is *not* derivable from the Galileian special case, simply by means of a transformation of co-ordinates.

The next step is to investigate the space-time behaviour of the gravitational field $G$, which was derived from the Galileian special case simply by transformation of the co-ordinates. This behaviour is formulated in a law, which is always valid, no matter how the reference-body (mollusc) used in the description may be chosen.

This law is not yet the *general* law of the gravitational field, since the gravitational field under consideration is of a special kind. In order to find out the general law-of-field of gravitation we still require to obtain a generalisation of the law as found above. This can be obtained without caprice, however, by taking into consideration the following demands:

(*a*) The required generalisation must likewise satisfy the general postulate of relativity.

(*b*) If there is any matter in the domain under consideration, only its inertial mass, and thus according to Section 15 only its energy is of importance for its effect in exciting a field.

(*c*) Gravitational field and matter together must satisfy the law of the conservation of energy (and of impulse).

Finally, the general principle of relativity permits us to determine the influence of the gravitational field on the course of all those processes which take place according to known laws when a gravitational field is absent, *i.e.* which have already been fitted into the frame of the special theory of relativity. In this connection we proceed in principle according to the method which has already been explained for measuring-rods, clocks and freely-moving material points.

The theory of gravitation derived in this way from the general postulate of relativity excels not only in its beauty; nor in removing the defect attaching to classical mechanics which was brought to light in Section 21; nor in interpreting the empirical law of the equality of inertial and gravitational mass; but it has also already explained a result of observation in astronomy, against which classical mechanics is powerless.

If we confine the application of the theory to the case where the gravitational fields can be regarded as being weak, and in which all masses move with respect to the co-ordinate system with velocities which are small compared with the velocity of light, we then obtain as a first approximation the Newtonian theory. Thus the latter theory is obtained here without any particular assumption, whereas Newton had to introduce the hypothesis that the force of attraction between mutually attracting material points is inversely proportional to the square of the distance between them. If we increase the accuracy of the calculation, deviations from the theory of Newton make their appearance, practically all of which must nevertheless escape the test of observation owing to their smallness.

We must draw attention here to one of these deviations. According to Newton's theory, a planet moves round the sun in an ellipse, which would permanently maintain its position with respect to the fixed stars, if we could disregard the motion of the fixed stars themselves and the action of the other planets under consideration. Thus, if we correct the observed motion of the planets for these two influences, and if Newton's theory be strictly correct, we ought to obtain for the orbit of the planet an ellipse, which is fixed with reference to the fixed stars. This deduction, which can be tested with great accuracy, has been confirmed for all the planets save one, with the precision that is capable of being obtained by the delicacy of observation attainable at the present time. The sole exception is Mercury, the planet which lies nearest the sun. Since the time of Leverrier, it has been known that the ellipse corresponding to the orbit of Mercury, after it has been corrected for the influences mentioned above, is not stationary with respect to the fixed stars, but that it rotates exceedingly slowly in the plane of the orbit and in the sense of the orbital motion. The value obtained for this rotary movement of the orbital ellipse was 43 seconds of arc per century, an amount ensured to be correct to within a few seconds of arc. This effect can be explained by means of classical mechanics only on the assumption of hypotheses which have little probability, and which were devised solely for this purpose.

On the basis of the general theory of relativity, it is found that the ellipse of every planet round the sun must necessarily rotate in the manner indicated above; that for all the planets, with the exception of Mercury, this rotation is too small to be detected with the delicacy of observation possible at the present time; but that in the case of Mercury it must amount to 43 seconds of arc per century, a result which is strictly in agreement with observation.

Apart from this one, it has hitherto been possible to make only two deductions from the theory which admit of being tested by observation, to wit, the curvature of light rays by the gravitational field of the sun,[1] and a displacement of the spectral lines of light reaching us from large stars, as compared with the corresponding lines for light produced in an analogous manner terrestrially (i.e. by the same kind of atom).[2] These two deductions from the theory have both been confirmed.

[1]First observed by Eddington and others in 1919. (Cf. Appendix 3, pp. 227–228).
[2]Established by Adams in 1924. (Cf. pp. 229–231).

# PART III: CONSIDERATIONS ON THE UNIVERSE AS A WHOLE

## THIRTY

# COSMOLOGICAL DIFFICULTIES OF NEWTON'S THEORY

Apart from the difficulty discussed in Section 21, there is a second fundamental difficulty attending classical celestial mechanics, which, to the best of my knowledge, was first discussed in detail by the astronomer Seeliger. If we ponder over the questions as to how the universe, considered as a whole, is to be regarded, the first answer that suggests itself to us is surely this: As regards space (and time) the universe is infinite. There are stars everywhere, so that the density of matter, although very variable in detail, is nevertheless on the average everywhere the same. In other words: However far we might travel through space, we should find everywhere an attenuated swarm of fixed stars of approximately the same kind and density.

This view is not in harmony with the theory of Newton. The latter theory rather requires that the universe should have a kind of centre in which the density of the stars is a maximum, and that as we proceed outwards from this centre the group-density of the stars should diminish, until finally, at great distances, it is succeeded by an infinite region of emptiness. The stellar universe ought to be a finite island in the infinite ocean of space.[1]

---

[1] *Proof*—According to the theory of Newton, the number of "lines of force" which come from infinity and terminate in a mass $m$ is proportional to the mass $m$. If, on the average, the mass density $\rho_0$ is constant throughout the universe, then a sphere of volume $V$ will enclose the average mass $\rho_0 V$. Thus the number of lines of force passing through the surface $F$ of the sphere into its interior is proportional to $\rho_0 V$. For unit area of the surface of the sphere the number of lines of force which enters the sphere is thus proportional to $\rho_0 \dfrac{V}{F}$ or to $\rho_0 R$. Hence the intensity of the field at the surface would ultimately become infinite with increasing radius $R$ of the sphere, which is impossible.

This conception is in itself not very satisfactory. It is still less satisfactory because it leads to the result that the light emitted by the stars and also individual stars of the stellar system are perpetually passing out into infinite space, never to return, and without ever again coming into interaction with other objects of nature. Such a finite material universe would be destined to become gradually but systematically impoverished.

In order to escape this dilemma, Seeliger suggested a modification of Newton's law, in which he assumes that for great distances the force of attraction between two masses diminishes more rapidly than would result from the inverse square law. In this way it is possible for the mean density of matter to be constant everywhere, even to infinity, without infinitely large gravitational fields being produced. We thus free ourselves from the distasteful conception that the material universe ought to possess something of the nature of a centre. Of course we purchase our emancipation from the fundamental difficulties mentioned, at the cost of a modification and complication of Newton's law which has neither empirical nor theoretical foundation. We can imagine innumerable laws which would serve the same purpose, without our being able to state a reason why one of them is to be preferred to the others; for any one of these laws would be founded just as little on more general theoretical principles as is the law of Newton.

# THIRTY-ONE

# THE POSSIBILITY OF A "FINITE" AND YET "UNBOUNDED" UNIVERSE

But speculations on the structure of the universe also move in quite another direction. The development of non-Euclidean geometry led to the recognition of the fact, that we can cast doubt on the *infiniteness* of our space without coming into conflict with the laws of thought or with experience (Riemann, Helmholtz). These questions have already been treated in detail and with unsurpassable lucidity by Helmholtz and Poincaré, whereas I can only touch on them briefly here.

In the first place, we imagine an existence in two-dimensional space. Flat beings with flat implements, and in particular flat rigid measuring-rods, are free to move in a *plane*. For them nothing exists outside of this plane: that which they observe to happen to themselves and to their flat "things" is the all-inclusive reality of their plane. In particular, the constructions of plane Euclidean geometry can be carried out by means of the rods, e.g. the lattice construction, considered in Section 24. In contrast to ours, the universe of these beings is two-dimensional; but, like ours, it extends to infinity. In their universe there is room for an infinite number of identical squares made up of rods, i.e. its volume (surface) is infinite. If these beings say their universe is "plane," there is sense in the statement, because they mean that they can perform the constructions of plane Euclidean geometry with their rods. In this connection the individual rods always represent the same distance, independently of their position.

Let us consider now a second two-dimensional existence, but this time on a spherical surface instead of on a plane. The flat beings with their measuring-rods and other objects fit exactly on this surface and they are unable to leave it. Their whole universe of observation extends exclusively over the surface of the sphere. Are these beings able to regard the geometry of their universe as being plane geometry and their rods withal as the realisation of "distance"? They cannot do this. For if they attempt to realize a straight line, they will obtain a curve, which we "three-dimensional beings" designate as a great circle, i.e. a self-contained line of definite finite length, which can be measured up by means of a measuring-rod. Similarly, this universe has a finite area that can be compared with the area of a square constructed with rods. The great charm resulting from this consideration lies in the recognition of the fact that *the universe of these beings is finite and yet has no limits.*

But the spherical-surface beings do not need to go on a world-tour in order to perceive that they are not living in a Euclidean universe. They can convince themselves of this on every part of their "world," provided they do not use too small a piece of it. Starting from a point, they draw "straight lines" (arcs of circles as judged in three-dimensional space) of equal length in all directions. They will call the line joining the free ends of these lines a "circle." For a plane surface, the ratio of the circumference of a circle to its diameter, both lengths being measured with the same rod, is, according to Euclidean geometry of the plane, equal to a constant value $\pi$, which is independent of the diameter of the circle. On their spherical surface our flat beings would find for this ratio the value

$$\pi = \frac{\sin\left(\dfrac{r}{R}\right)}{\left(\dfrac{v}{R}\right)},$$

*i.e.* a smaller value than $\pi$, the difference being the more considerable, the greater is the radius of the circle in comparison with the radius $R$ of the "world-sphere." By means of this relation the spherical beings can determine the radius of their universe ("world"), even when only a relatively small part of their world-sphere is available for their measurements. But if this part is very small indeed, they will no longer be able to demonstrate that they are on a spherical "world" and not on a Euclidean plane, for a small part of a spherical surface differs only slightly from a piece of a plane of the same size.

Thus if the spherical-surface beings are living on a planet of which the solar system occupies only a negligibly small part of the spherical universe, they have no means of determining whether they are living in a finite or in an infinite universe, because the "piece of universe" to which they have access is in both cases practically plane, or Euclidean. It follows directly from this discussion, that for our sphere-beings the circumference of a circle first increases with the radius until the "circumference of the universe" is reached, and that it thenceforward gradually decreases to zero for still further increasing values of the radius. During this process the area of the circle continues to increase more and more, until finally it becomes equal to the total area of the whole "world-sphere."

Perhaps the reader will wonder why we have placed our "beings" on a sphere rather than on another closed surface. But this choice has its justification in the fact that, of all closed surfaces, the sphere is unique in possessing the property that all points on it are equivalent. I admit that the ratio of the circumference $c$ of circle to its radius $r$ depends on $r$, but for a given value of $r$ it is the same for all points of the "world-sphere"; in other words, the "world-sphere" is a "surface of constant curvature."

To this two-dimensional sphere-universe there is a three-dimensional analogy, namely, the three-dimensional spherical space which was discovered by Riemann. Its points are likewise all equivalent. It possesses a finite volume, which is determined by its "radius" $(2\pi^2 R^3)$. Is it possible to imagine a spherical space? To imagine a space means nothing else than that we imagine an epitome of our "space" experience, $i.e.$ of experience that we can have in the movement of "rigid" bodies. In this sense we $can$ imagine a spherical space.

Suppose we draw lines or stretch strings in all directions from a point, and mark off from each of these the distance $\gamma$ with a measuring-rod. All the free end-points of these lengths lie on a spherical surface. We can specially measure up the area $(F)$ of this surface by means of a square made up of measuring-rods. If the universe is Euclidean, then $F = 4\pi\gamma^2$; if it is spherical, then $F$ is always less than $4\pi\gamma^2$. With increasing values of $\gamma$, $F$ increases from zero up to a maximum value which is determined by the "world-radius," but for still further increasing values of $\gamma$, the area gradually diminishes to zero. At first, the straight lines which radiate from the starting point diverge farther and farther from one another, but later they approach each other, and finally they run together again at a "counter-point" to the starting point. Under such conditions they have traversed the whole spherical space. It is easily seen that the three-dimensional spherical space is quite analogous to the two-dimensional spherical surface. It is finite (*i.e.* of finite volume), and has no bounds.

It may be mentioned that there is yet another kind of curved space: "elliptical space." It can be regarded as a curved space in which the two "counter-points" are identical (indistinguishable from each other). An elliptical universe can thus be considered to some extent as a curved universe possessing central symmetry.

It follows from what has been said, that closed spaces without limits are conceivable. From amongst these, the spherical space (and the elliptical) excels in its simplicity, since all points on it are equivalent. As a result of this discussion, a most interesting question arises for astronomers and physicists, and that is whether the universe in which we live is infinite, or whether it is finite in the manner of the spherical universe. Our experience is far from being sufficient to enable us to answer this question. But the general theory of relativity permits of our answering it with a moderate degree of certainty, and in this connection the difficulty mentioned in Section 30 finds its solution.

# THIRTY-TWO

# THE STRUCTURE OF SPACE ACCORDING TO THE GENERAL THEORY OF RELATIVITY

According to the general theory of relativity, the geometrical properties of space are not independent, but they are determined by matter. Thus we can draw conclusions about the geometrical structure of the universe only if we base our considerations on the state of the matter as being something that is known. We know from experience that, for a suitably chosen co-ordinate system, the velocities of the stars are small as compared with the velocity of transmission of light. We can thus as a rough approximation arrive at a conclusion as to the nature of the universe as a whole, if we treat the matter as being at rest.

We already know from our previous discussion that the behaviour of measuring-rods and clocks is influenced by gravitational fields, *i.e.* by the distribution of matter. This in itself is sufficient to exclude the possibility of the exact validity of Euclidean geometry in our universe. But it is conceivable that our universe differs only slightly from a Euclidean one, and this notion seems all the more probable, since calculations show that the metrics of surrounding space is influenced only to an exceedingly small extend by masses even of the magnitude of our sun. We might imagine that, as regards geometry, our universe behaves analogously to a surface which is irregularly curved in its individual parts, but which nowhere departs appreciably from a plane: Something like the rippled surface of a lake. Such a universe might fittingly be called a quasi-Euclidean universe. As regards its space it would be infinite. But calculation shows that in a quasi-Euclidean universe the average density of matter would necessarily be *nil*. Thus such a universe could not be inhabited by matter everywhere; it would present to us that unsatisfactory picture which we portrayed in Section 30.

If we are to have in the universe an average density of matter which differs from zero, however small may be that difference, then the universe cannot be quasi-Euclidean. On the contrary, the results of calculation indicate that if matter be distributed uniformly, the universe would necessarily be spherical (or elliptical). Since in reality the detailed distribution of matter is not uniform, the real universe will deviate in individual parts from the spherical, *i.e.* the universe will be quasi-spherical. But it will be necessarily finite. In fact, the theory supplies us with a simple connection[1] between the space-expanse of the universe and the average density of matter in it.

---

[1]For the "radius" $R$ of the universe we obtain the equation

$$R^2 = \frac{2}{\kappa\rho}.$$

The use of the C.G.S. system is in this equation gives $\frac{2}{\kappa} = 108.10^{37}$; $\rho$ is the average density of the matter and $x$ is a constant connected with the Newtonian constant of gravitational.

# APPENDIX ONE

# SIMPLE DERIVATION OF THE LORENTZ TRANSFORMATION

## [SUPPLEMENTARY TO SECTION 11]

For the relative orientation of the co-ordinate systems indicated in Fig. 2, the $x$-axes of both systems permanently coincide. In the present case we can divide the problem into parts by considering first only events which are localised on the $x$-axis. Any such event is represented with respect to the co-ordinate system $K$ by the abscissa $x$ and the time $t$, and with respect to the system $K'$ by the abscissa $x'$ and the time $t'$. We require to find $x'$ and $t'$ when $x$ and $t$ are given.

A light-signal, which is proceeding along the positive axis of $x$, is transmitted according to the equation

$$x = ct$$

or

$$x - ct = 0 \qquad . \qquad . \qquad . \qquad . \qquad (1).$$

Since the same light-signal has to be transmitted relative to $K'$ with the velocity $c$, the propagation relative to the system $K'$ will be represented by the analogous formula

$$x' - ct' = 0 \qquad . \qquad . \qquad . \qquad . \qquad (2).$$

Those space-time points (events) which satisfy (1) must also satisfy (2). Obviously this will be the case when the relation

$$(x' - ct') = \lambda(x - ct) \qquad . \qquad . \qquad . \qquad (3),$$

is fulfilled in general, where $\lambda$ indicates a constant; for, according to (3), the disappearance of $(x - ct)$ involves the disappearance of $(x' - ct')$.

If we apply quite similar considerations to light rays which are being transmitted along the negative $x$-axis, we obtain the condition

$$(x' + ct') = \mu(x + ct) \qquad . \qquad . \qquad . \qquad . \qquad (4).$$

By adding (or subtracting) equations (3) and (4), and introducing for convenience the constants $a$ and $b$ in place of the constants $\lambda$, and $\mu$, where

$$a = \frac{\lambda + \mu}{2}$$

and

$$b = \frac{\lambda - \mu}{2},$$

we obtain the equations

$$\left.\begin{array}{l} x' = ax - bct \\ ct' = act - bx \end{array}\right\} \quad \cdot \quad \cdot \quad \cdot \quad \cdot \quad \cdot \quad (5).$$

We should thus have the solution of our problem, if the constants $a$ and $b$ were known. These result from the following discussion.

For the origin of $K'$ we have permanently $x' = 0$, and hence according to the first of the equations (5)

$$x = \frac{bc}{a}t.$$

If we call $v$ the velocity with which the origin of $K'$ is moving relative to $K$, we then have

$$v = \frac{bc}{a} \quad \cdot \quad \cdot \quad \cdot \quad \cdot \quad \cdot \quad (6).$$

The same value $v$ can be obtained from equations (5), if we calculate the velocity of another point of $K'$ relative to $K$, or the velocity (directed towards the negative x-axis) of a point of $K$ with respect to $K'$. In short, we can designate $v$ as the relative velocity of the two systems.

Furthermore, the principle of relativity teaches us that, as judged from $K$, the length of a unit measuring-rod which is at rest with reference to $K'$ must be exactly the same as the length, as judged from $K'$, of a unit measuring-rod which is at rest relative to $K$. In order to see how the points of the x'-axis appear as viewed from $K$, we only require to take a "snapshot" of $K'$ from $K$; this means that we have to insert a particular value of $t$ (time of $K$), e.g. $t = 0$. For this value of $t$ we then obtain from the first of the equations (5)

$$x' = ax.$$

Two points of the $x'$-axis which are separated by the distance $\Delta x' = 1$ when measured in the $K'$ system are thus separated in our instantaneous photograph by the distance

$$\Delta x = \frac{1}{a} \qquad . \quad . \quad . \quad . \quad (7).$$

But if the snapshot be taken from $K'(t' = 0)$, and if we eliminate $t$ from the equations (5), taking into account the expression (6), we obtain

$$x' = a\left(1 - \frac{v^2}{c^2}\right)x.$$

From this we conclude that two points of the $x$-axis separated by the distance 1 (relative to $K$) will be represented on our snapshot by the distance

$$\Delta x' = a\left(1 - \frac{v^2}{c^2}\right) \qquad . \quad . \quad . \quad (7a).$$

But from what has been said, the two snapshots must be identical; hence $\Delta x$ in (7) must be equal to $\Delta x'$ in (7a), so that we obtain

$$a^2 = \frac{1}{1 - \dfrac{v^2}{c^2}} \qquad . \quad . \quad . \quad (7b).$$

The equations (6) and (7b) determine the constants $a$ and $b$. By inserting the values of these constants in (5), we obtain the first and the fourth of the equations given in Section XI.

$$\left.\begin{array}{l} x' = \dfrac{x - vt}{\sqrt{1 - \dfrac{v^2}{c^2}}} \\[3em] t' = \dfrac{t - \dfrac{v}{c^2}x}{\sqrt{1 - \dfrac{v^2}{c^2}}} \end{array}\right\} \qquad . \quad . \quad . \quad (8).$$

Thus we have obtained the Lorentz transformation for events on the $x$-axis. It satisfies the condition

$$x'^2 - c^2 t'^2 = x^2 - c^2 t^2 \qquad . \quad . \quad . \quad (8a).$$

220

The extension of this result, to include events which take place outside the x-axis, is obtained by retaining equations (8) and supplementing them by the relations

$$\left. \begin{array}{c} y' = y \\ z' = z \end{array} \right\} \qquad . \quad . \quad . \quad . \quad . \quad (9).$$

In this way we satisfy the postulate of the constancy of the velocity of light *in vacuo* for rays of light of arbitrary directions, both for the system $K$ and for the system $K'$. This may be shown in the following manner.

We suppose a light-signal sent out from the origin of $K$ at the time $t = 0$. It will be propagated according to the equation

$$r = \sqrt{x^2 + y^2 + z^2} = ct,$$

or, if we square this equation, according to the equation

$$x^2 + y^2 + z^2 - c^2t^2 = 0 \qquad . \qquad . \qquad . \qquad (10).$$

It is required by the law of propagation of light, in conjunction with the postulate of relativity, that the transmission of the signal in question should take place—as judged from $K'$—in accordance with the corresponding formula

$$r' = ct',$$

or,

$$x'^2 + y'^2 + z'^2 - c^2t'^2 = 0 \qquad . \qquad . \qquad . \qquad (10a).$$

In order that equation (10a) may be a consequence of equation (10), we must have

$$x' + y'^2 + z'^2 - c^2t'^2 = \sigma(x^2 + y^2 + z^2 - c^2t^2) \qquad (11).$$

Since equation (8a) must hold for points on the x-axis, we thus have $\sigma = 1$. It is easily seen that the Lorentz transformation really satisfies equation (11) for $\sigma = 1$; for (11) is a consequence of (8a) and (9), and hence also of (8) and (9). We have thus derived the Lorentz transformation.

The Lorentz transformation represented by (8) and (9) still requires to be generalised. Obviously it is immaterial whether the axes of $K'$ be chosen so that they are spatially parallel to those of $K$. It is also not essential that the velocity of translation of $K'$ with respect to $K$ should be in the direction of the x-axis. A simple consideration shows that we are able to construct the Lorentz transformation in this general sense from two kinds of transformations, viz. from Lorentz transformations in the special sense and from purely spatial transformations, which corresponds to the replacement of the rectangular co-ordinate system by a new system with its axes pointing in other directions.

Mathematically, we can characterize the generalized Lorentz transformation thus:

It expresses $x'$, $y'$, $z'$, $t'$, in terms of linear homogeneous functions of $x$, $y$, $z$, $t$, of such a kind that the relation

$$x'^2 + y'^2 + z'^2 - c^2 t'^2 = x^2 + y^2 + z^2 - c^2 t^2 \qquad (11a)$$

is satisfied identically. That is to say: If we substitute their expressions in $x$, $y$, $z$, $t$ in place of $x'$, $y'$, $z'$, $t'$, on the left-hand side, then the left-hand side of (11$a$) agrees with the right-hand side.

# APPENDIX TWO
# MINKOWSKI'S
# FOUR-DIMENSIONAL SPACE
# ("WORLD")

[SUPPLEMENTARY TO SECTION 17]

We can characterise the Lorentz transformation still more simply if we introduce the imaginary $\sqrt{-1}$. $ct$ in place of $t$, as time-variable, If, in accordance with this, we insert

$$x_1 = x$$
$$x_2 = y$$
$$x_3 = z$$
$$x_4 = \sqrt{-1}.\ ct,$$

and similarly for the accented system $K'$, then the condition which is identically satisfied by the transformation can be expressed thus:

$$x_1'^2 + x_2'^2 + x_3'^2 + x_4'^2 = x_1^2 + x_2^2 + x_3^2 + x_4^2 \qquad (12).$$

That is, by the afore-mentioned choice of "co-ordinates," $(11a)$ is transformed into this equation.

We see from (12) that the imaginary time co-ordinate $x_4$ enters into the condition of transformation in exactly the same way as the space co-ordinates $x_1$, $x_2$, $x_3$. It is due to this fact that, according to the theory of relativity, the "time" $x_4$ enters into natural laws in the same form as the space co-ordinates $x_1$, $x_2$, $x_3$.

A four-dimensional continuum described by the "co-ordinates" $x_1$, $x_2$, $x_3$, $x_4$, was called "world" by Minkowski, who also termed a point-event a "world-point." From a "happening" in three-dimensional space, physics becomes, as it were, an "existence" in the four-dimensional "world."

This four-dimensional "world" bears a close similarity to the three-dimensional "space" of (Euclidean) analytical geometry. If we introduce into the latter a new Cartesian co-ordinate system $(x_1', x_2', x_3')$ with the same origin, then $x'_1, x'_2, x'_3$, are linear homogeneous functions of $x_1, x_2, x_3$, which identically satisfy the equation

$$x_1'^2 + x_2'^2 + x_3'^2 = x_1^2 + x_2^2 + x_3^2.$$

The analogy with (12) is a complete one. We can regard Minkowski's "world" in a formal manner as a four-dimensional Euclidean space (with imaginary time co-ordinate); the Lorentz transformation corresponds to a "rotation" of the co-ordinate system in the four-dimensional "world."

# APPENDIX THREE

# THE EXPERIMENTAL CONFIRMATION OF THE GENERAL THEORY OF RELATIVITY

From a systematic theoretical point of view, we may imagine the process of evolution of an empirical science to be a continuous process of induction. Theories are evolved and are expressed in short compass as statements of a large number of individual observations in the form of empirical laws, from which the general laws can be ascertained by comparison. Regarded in this way, the development of a science bears some resemblance to the compilation of a classified catalogue. It is, as it were, a purely empirical enterprise.

But this point of view by no means embraces the whole of the actual process; for it slurs over the important part played by intuition and deductive thought in the development of an exact science. As soon as a science has emerged from its initial stages, theoretical advances are no longer achieved merely by a process of arrangement. Guided by empirical data, the investigator rather develops a system of thought which, in general, is built up logically from a small number of fundamental assumptions, the so-called axioms. We call such a system of thought a *theory*. The theory finds the justification for its existence in the fact that it correlates a large number of single observations, and it is just here that the "truth" of the theory lies.

Corresponding to the same complex of empirical data, there may be several theories, which differ from one another to a considerable extent. But as regards the deductions from the theories which are capable of being tested, the agreement between the theories may be so complete, that it becomes difficult to find any deductions in which the two theories differ from each other. As an example, a case of general interest is available in the province of biology, in the Darwinian theory of the development of species by selection in the struggle for existence, and in the theory of development which is based on the hypothesis of the hereditary transmission of acquired characters.

We have another instance of far-reaching agreement between the deductions from two theories in Newtonian mechanics on the one hand, and the general theory of relativity on the other. This agreement goes so far, that up to the present we have been able to find only a few deductions from the general theory of relativity which are capable of investigation, and to which the physics of pre-relativity days does not also lead, and this despite the profound difference in the fundamental assumptions of the two theories. In what follows, we shall again consider these important deductions, and we shall also discuss the empirical evidence appertaining to them which has hitherto been obtained.

# (A) MOTION OF THE PERIHELION
# OF MERCURY

According to Newtonian mechanics and Newton's law of gravitation, a planet which is revolving round the sun would describe an ellipse round the latter, or, more correctly, round the common centre of gravity of the sun and the planet. In such a system, the sun, or the common centre of gravity, lies in one of the foci of the orbital ellipse in such a manner that, in the course of a planet-year, the distance sun-planet grows from a minimum to a maximum, and then decreases again to a minimum. If instead of Newton's law we insert a somewhat different law of attraction into the calculation, we find that, according to this new law, the motion would still take place in such a manner that the distance sun-planet exhibits periodic variations; but in this case the angle described by the line joining sun and planet during such a period (from perihelion—closest proximity to the sun—to perihelion) would differ from 360°. The line of the orbit would not then be a closed one but in the course of time it would fill up an annular part of the orbital plane, viz. between the circle of least and the circle of greatest distance of the planet from the sun.

According also to the general theory of relativity, which differs of course from the theory of Newton, a small variation from the Newton-Kepler motion of a planet in its orbit should take place, and in such a way, that the angle described by the radius sun-planet between one perihelion and the next should exceed that corresponding to one complete revolution by an amount given by

$$+\frac{24\pi^3 a^2}{T^2 c^2 (1 - e^2)}.$$

(*N.B.*—One complete revolution corresponds to the angle $2\pi$ in the absolute angular measure customary in physics, and the above expression gives the amount by which the radius sun-planet exceeds this angle during the interval between one perihelion and the next.) In this expression $a$ represents the major semi-axis of the ellipse, $e$ its eccentricity, $c$ the velocity of light, and $T$ the period of revolution of the planet. Our result may also be stated as follows: According to the general theory of relativity, the major axis of the ellipse rotates round the sun in the same sense as the orbital motion of the planet. Theory requires that this rotation should amount to 43 seconds of arc per century for the planet Mercury, but for the other planets of our solar system its magnitude should be so small that it would necessarily escape detection.[1]

In point of fact, astronomers have found that the theory of Newton does not suffice to calculate the observed motion of Mercury with an exactness corresponding to that of the delicacy of observation attainable at the present time. After taking account of all the disturbing influences exerted on Mercury by the remaining planets, it was found (Leverrier—1859—and Newcomb—1895) that an unexplained perihelial movement of the orbit of Mercury remained over, the amount of which does not differ sensibly from the above-mentioned +43 seconds of arc per century. The uncertainty of the empirical result amounts to a few seconds only.

---

[1] Especially since the next planet Venus has an orbit that is almost an exact circle, which makes it more difficult to locate the perihelion with precision.

# (B) DEFLECTION OF LIGHT BY A GRAVITATIONAL FIELD

In Section 12 it has been already mentioned that according to the general theory of relativity, a ray of light will experience a curvature of its path when passing through a gravitational field, this curvature being similar to that experienced by the path of a body which is projected through a gravitational field. As a result of this theory, we should expect that a ray of light which is passing close to a heavenly body would be deviated towards the latter. For a ray of light which passes the sun at a distance of $\Delta$ sun-radii from its centre, the angle of deflection ($a$) should amount to

$$a = \frac{1.7 \text{ seconds of arc}}{\Delta}.$$

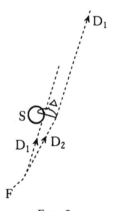

It may be added that, according to the theory, half of this deflection is produced by the Newtonian field of attraction of the sun, and the other half by the geometrical modification ("curvature") of space caused by the sun.

This result admits of an experimental test by means of the photographic registration of stars during a total eclipse of the sun. The only reason why we must wait for a total eclipse is because at every other time the atmosphere is so strongly illuminated by the light from the sun that the

FIG. 5.

stars situated near the sun's disc are invisible. The predicted effect can be seen clearly from the accompanying diagram. If the sun ($S$) were not present, a star which is practically infinitely distant would be seen in the direction $D_1$, as observed from the earth. But as a consequence of the deflection of light from the star by the sun, the star will be seen in the direction $D_2$, *i.e.* at a somewhat greater distance from the centre of the sun that corresponds to its real position.

In practice, the question is tested in the following way. The stars in the neighbourhood of the sun are photographed during a solar eclipse. In addition, a second photograph of the same stars is taken when the sun is situated at another position in the sky, i.e. a few months earlier or later. As compared with the standard photograph, the positions of the stars on the eclipse-photograph ought to appear displaced radially outwards (away from the centre of the sun) by an amount corresponding to the angle $a$.

We are indebted to the Royal Society and to the Royal Astronomical Society for the investigation of this important deduction. Undaunted by the war and by difficulties of both a material and a psychological nature aroused by the war, these societies equipped two expeditions—to Sobral (Brazil), and to the island of Principe (West Africa)—and sent several of Britain's most celebrated astronomers (Eddington, Cottingham, Crommelin, Davidson), in order to obtain photographs of the solar eclipse of 29th May, 1919. The relative discrepancies to be expected between the stellar photographs obtained during the eclipse and the comparison photographs amounted to a few hundredths of a millimetre only. Thus great accuracy was necessary in making the adjustments required for the taking of the photographs, and in their subsequent measurement.

The results of the measurements confirmed the theory in a thoroughly satisfactory manner. The rectangular components of the observed and of the calculated deviations of the stars (in seconds of arc) are set forth in the following table of results:

| Number of the Star. | First Co-ordinate. | | Second Co-ordinate. | |
|---|---|---|---|---|
| | Observed. | Calculated. | Observed. | Calculated. |
| 11 | −0.19 | −0.22 | +0.16 | +0.02 |
| 5 | +0.29 | +0.31 | −0.46 | −0.43 |
| 4 | +0.11 | +0.10 | +0.83 | +0.74 |
| 3 | +0.20 | +0.12 | +1.00 | +0.87 |
| 6 | +0.10 | +0.04 | +0.57 | +0.40 |
| 10 | −0.08 | +0.09 | +0.35 | +0.32 |
| 2 | +0.95 | +0.85 | −0.27 | −0.09 |

# (C) DISPLACEMENT OF SPECTRAL LINES TOWARDS THE RED

In Section 23 it has been shown that in a system $K'$ which is in rotation with regard to a Galileian system $K$, clocks of identical construction, and which are considered at rest with respect to the rotating reference-body, go at rates which are dependent on the positions of the clocks. We shall now examine this dependence quantitatively. A clock, which is situated at a distance $\gamma$ from the centre of the disc, has a velocity relative to $K$ which is given by

$$v = \omega\gamma$$

where $\omega$ represents the angular velocity of rotation of the disc $K'$ with respect to $K$. If $v_0$ represents the number of ticks of the clock per unit time ("rate" of the clock) relative to $K$ when the clock is at rest, then the "rate" of the clock ($v$) when it is moving relative to $K$ with a velocity $v$, but at rest with respect to the disc, will, in accordance with Section 12, be given by

$$v = v_0\sqrt{1 - \frac{v^2}{c^2}},$$

or with sufficient accuracy by

$$v = v_0\left(1 - \frac{1}{2}\frac{v^2}{c^2}\right).$$

This expression may also be stated in the following form:

$$v = v_0\left(1 - \frac{1}{c^2}\frac{\omega^2\gamma^2}{2}\right).$$

If we represent the difference of potential of the centrifugal force between the position of the clock and the centre of the disc by $\phi$, i.e. the work, considered negatively, which must be performed on the unit of mass against the centrifugal force in order to transport it from the position of the clock on the rotating disc to the centre of the disc, then we have

$$\phi = -\frac{\omega^2\gamma^2}{2}.$$

From this it follows that

$$\nu = \nu_0\left(1 + \frac{\phi}{c^2}\right).$$

In the first place, we see from this expression that two clocks of identical construction will go at different rates when situated at different distances from the centre of the disc. This result is also valid from the standpoint of an observer who is rotating with the disc.

Now, as judged from the disc, the latter is in a gravitational field of potential $\phi$, hence the result we have obtained will hold quite generally for gravitational fields. Furthermore, we can regard an atom which is emitting spectral lines as a clock, so that the following statement will hold:

*An atom absorbs or emits light of a frequency which is dependent on the potential of the gravitational field in which it is situated.*

The frequency of an atom situated on the surface of a heavenly body will be somewhat less than the frequency of an atom of the same element which is situated in free space (or on the surface of a smaller celestial body). Now $\phi = -K\dfrac{M}{\gamma}$, where $K$ is Newton's constant of gravitation, and $M$ is the mass of the heavenly body. Thus a displacement towards the red ought to take place for spectral lines produced at the surface of stars as compared with the spectral lines of the same element produced at the surface of the earth, the amount of this displacement being

$$\frac{\nu_0 - \nu}{\nu_0} = \frac{K}{c^2}\frac{M}{\gamma}.$$

For the sun, the displacement towards the red predicted by theory amounts to about two millionths of the wave-length. A trustworthy calculation is not possible in the case of the stars, because in general neither the mass $M$ nor the radius $\gamma$ are known.

It is an open question whether or not this effect exists, and at the present time (1920) astronomers are working with great zeal towards the solution. Owing to the smallness of the effect in the case of the sun, it is difficult to form an opinion as to its existence. Whereas

Grebe and Bachem (Bonn), as a result of their own measurements and those of Evershed and Schwarzschild on the cyanogen bands, have placed the existence of the effect almost beyond doubt, other investigators, particularly St. John, have been led to the opposite opinion in consequence of their measurements.

Mean displacements of lines towards the less refrangible end of the spectrum are certainly revealed by statistical investigations of the fixed stars; but up to the present the examination of the available data does not allow of any definite decision being arrived at, as to whether or not these displacements are to be referred in reality to the effect of gravitation. The results of observation have been collected together, and discussed in detail from the standpoint of the question which has been engaging our attention here, in a paper by E. Freundlich entitled "Zur Prüfung der aligemeinen Relativitäts-Theorie" (*Die Naturwissenschaften*, 1919, No. 35, p. 520: Julius Springer, Berlin).

At all events, a definite decision will be reached during the next few years. If the displacement of spectral lines towards the red by the gravitational potential does not exist, then the general theory of relativity will be untenable. On the other hand, if the cause of the displacement of spectral lines be definitely traced to the gravitational potential, then the study of this displacement will furnish us with important information as to the mass of the heavenly bodies.

NOTE.—The displacement of spectral lines towards the red end of the spectrum was definitely established by Adams in 1924, by observations on the dense companion of Sirius, for which the effect is about thirty times greater than for the sun.

R. W. L.

# APPENDIX FOUR

# THE STRUCTURE OF SPACE ACCORDING TO THE GENERAL THEORY OF RELATIVITY

[SUPPLEMENTARY TO SECTION 32]

Since the publication of the first edition of this little book, our knowledge about the structure of space in the large ("cosmo-logical problem") has had an important development, which ought to be mentioned even in a popular presentation of the subject.

My original considerations on the subject were based on two hypotheses:

1. There exists an average density of matter in the whole of space which is everywhere the same and different from zero.

2. The magnitude ("radius") of space is independent of time.

Both these hypotheses proved to be consistent, according to the general theory of relativity, but only after a hypothetical term was added to the field equations, a term which was not required by the theory as such nor did it seem natural from a theoretical point of view ("cosmological term of the field equations").

Hypothesis (2) appeared unavoidable to me at the time, since I thought that one would get into bottomless speculations if one departed from it.

However, already in the 'twenties, the Russian mathematician Friedman showed that a different hypothesis was natural from a purely theoretical point of view. He realized that it was possible to preserve hypothesis (1) without introducing the less natural cosmological term into the field equations of gravitation, if one was ready to drop hypothesis (2). Namely, the original field equations admit a solution in which the "world-radius" depends on time (expanding space). In that sense one can say, according to Friedman, that the theory demands an expansion of space.

A few years later Hubble showed, by a special investigation of the extra-galactic nebulae ("milky ways"), that the spectral lines emitted showed a red shift which increased regularly with the distance of the nebulae. This can be interpreted in regard to our present knowledge only in the sense of Doppler's principle, as an expansive motion of the system of stars in the large—as required, according to Friedman, by the field equations of gravitation. Hubble's discovery can, therefore, be considered to some extent as a confirmation of the theory.

There does arise, however, a stranger difficulty. The interpretation of the galactic line-shift discovered by Hubble as an expansion (which can hardly be doubted from a theoretical point of view), leads to an origin of this expansion which lies "only" about $10^9$ years ago, while physical astronomy makes it appear likely that the development of individual stars and systems of stars takes considerably longer. It is in no way known how this incongruity is to be overcome.

I further want to remark that the theory of expanding space, together with the empirical data of astronomy, permit no decision to be reached about the finite or infinite character of (three-dimensional) space, while the original "static" hypothesis of space yielded the closure (finiteness) of space.

# Sidelights on Relativity

What really happens when one billiard ball strikes another? Prior to the twentieth century, it was understood that the cue ball and target ball only interacted during the brief moment when they were in contact with one another, as is dictated by common sense. This is all well and good for billiard balls, but what about the forces of gravity and electromagnetism, which appear to act at a distance? Scientists had hypothesized that those forces must propagate through some ponderable medium, known as the "luminiferous ether," much as a shock wave propagates through the air.

The ether, however, did not stand up to serious scientific scrutiny. In 1887, Albert Michelson and Edward Morley showed that, whatever the ether might be, it did not behave like normal matter. For example, a water wave traveling along a flowing river will propagate faster in the direction of the water's motion than against it. In the case of light, however, Michelson and Morley showed that the speed of propagation was the same regardless of the relative motion of the observer and the hypothetical ether.

In "Ether and the Theory of Relativity," Einstein notes that special relativity is predicated on the observational fact that light travels at a constant speed for all observers, and thus ether, whatever it is, cannot be like ordinary matter. His theory of general relativity further complicates this matter by proposing that gravity gives rise to the structure of space itself. To put this plainly, gravity is defined even in "empty" space, and thus, there must be *something*.

That "something" is the ether, or, in modern language, a field. General relativity and Maxwell's theory of electromagnetism represented the first field theories: descriptions of how the world works in terms of omnipresent fields, rather than tiny particles. In many respects, this is one of the most important contributions of relativity to physics. In the modern view, all forces arise from fields. The billiard balls described above don't really collide at all, but their electromagnetic fields repel each other on very small scales. In quantum field theory, developed in the mid-twentieth century, about forty years after the present work, not only do the forces, but the particles themselves arise from the field. Consider this work, then, as a transitional commentary between Isaac Newton's classical particle picture and the modern picture in which the universe is comprised fundamentally of fields.

# ETHER AND THE THEORY OF RELATIVITY

*An Address delivered on May 5th, 1920, in the University of Leyden*

How does it come about that alongside of the idea of ponderable matter, which is derived by abstraction from everyday life, the physicists set the idea of the existence of another kind of matter, the ether? The explanation is probably to be sought in those phenomena which have given rise to the theory of action at a distance, and in the properties of light which have led to the undulatory theory. Let us devote a little while to the consideration of these two subjects.

Outside of physics we know nothing of action at a distance. When we try to connect cause and effect in the experiences which natural objects afford us, it seems at first as if there were no other mutual actions than those of immediate contact, e.g. the communication of motion by impact, push and pull, heating or inducing combustion by means of a flame, etc. It is true that even in everyday experience weight, which is in a sense action at a distance, plays a very important part. But since in daily experience the weight of bodies meets us as something constant, something not linked to any cause which is variable in time or place, we do not in everyday life speculate as to the cause of gravity, and therefore do not become conscious of its character as action at a distance. It was Newton's theory of gravitation that first assigned a cause for gravity by interpreting it as action at a distance, proceeding from masses. Newton's theory is probably the greatest stride ever made in the effort towards the causal nexus of natural phenomena. And yet this theory evoked a lively sense of discomfort among Newton's contemporaries, because it seemed to be in conflict with the principle springing from the rest of experience, that there can be reciprocal action only through contact, and not through immediate action at a distance. It is only with reluctance that man's desire for knowledge endures a dualism of this kind. How was unity to be preserved in his comprehension of the forces of nature? Either by trying to look upon contact forces as being themselves distant forces which admittedly are observable only at a very small distance—and this was the road which Newton's followers, who were entirely under the spell of his doctrine, mostly preferred to take; or by assuming that the Newtonian action at a distance is only *apparently* immediate action at a distance, but in truth is conveyed by a medium permeating space, whether by movements or by elastic deformation of this medium. Thus the endeavour toward a unified view of the nature of forces leads to the hypothesis of an ether. This hypothesis, to be sure, did not at first bring with it any advance in the theory of gravitation or in physics generally, so that it became customary to treat Newton's law of force as an axiom not further reducible. But the ether hypothesis was bound always to play some part in physical science, even if at first only a latent part.

When in the first half of the nineteenth century the far-reaching similarity was revealed which subsists between the properties of light and those of elastic waves in ponderable bodies, the ether hypothesis found fresh support. It appeared beyond question that light must be interpreted as a vibratory process in an elastic, inert medium filling up universal space. It also seemed to be a necessary consequence of the fact that light is capable of polarisation that this medium, the ether, must be of the nature of a solid body, because transverse waves are not possible in a fluid, but only in a solid. Thus the physicists were bound to arrive at the theory of the "quasi-rigid" luminiferous ether, the parts of which can carry out no movements relatively to one another except the small movements of deformation which correspond to light-waves.

This theory—also called the theory of the stationary luminiferous ether—moreover found a strong support in an experiment which is also of fundamental importance in the special theory of relativity, the experiment of Fizeau, from which one was obliged to infer that the luminiferous ether does not take part in the movements of bodies. The phenomenon of aberration also favoured the theory of the quasi-rigid ether.

The development of the theory of electricity along the path opened up by Maxwell and Lorentz gave the development of our ideas concerning the ether quite a peculiar and unexpected turn. For Maxwell himself the ether indeed still had properties which were purely mechanical, although of a much more complicated kind than the mechanical properties of tangible solid bodies. But neither Maxwell nor his followers succeeded in elaborating a mechanical model for the ether which might furnish a satisfactory mechanical interpretation of Maxwell's laws of the electro-magnetic field. The laws were clear and simple, the mechanical interpretations clumsy and con-tradictory. Almost imperceptibly the theoretical physicists adapted themselves to a situation which, from the standpoint of their mechanical programme, was very depressing. They were particularly influenced by the electro-dynamical investigations of Heinrich Hertz. For whereas they previously had required of a conclusive theory that it should content itself with the fundamental concepts which belong exclusively to mechanics (e.g. densities, velocities, deformations, stresses) they gradually accustomed themselves to admitting electric and magnetic force as fundamental concepts side by side with those of mechanics, without requiring a mechanical interpretation for them. Thus the purely mechanical view of nature was gradually abandoned. But this change led to a fundamental dualism which in the long-run was insupportable. A way of escape was now sought in the reverse direction, by reducing the principles of mechanics to those of electricity, and this especially as confidence in the strict validity of the equations of Newton's mechanics was shaken by the experiments with beta-rays and rapid kathode rays.

This dualism still confronts us in unextenuated form in the theory of Hertz, where matter appears not only as the bearer of velocities, kinetic energy, and mechanical pressures, but also as the bearer of electromagnetic fields. Since such fields also occur *in vacuo*—i.e. in free ether—the ether also appears as bearer of electromagnetic fields. The ether appears indistinguishable in its functions from ordinary matter. Within matter it takes part in the motion of matter and in empty space it has everywhere a velocity; so that the ether has a definitely assigned velocity throughout the whole of space. There is no fundamental difference between Hertz's ether and ponderable matter (which in part subsists in the ether).

The Hertz theory suffered not only from the defect of ascribing to matter and ether, on the one hand mechanical states, and on the other hand electrical states, which do not stand in any conceivable relation to each other; it was also at variance with the result of Fizeau's important experiment on the velocity of the propagation of light in moving fluids, and with other established experimental results.

Such was the state of things when H. A. Lorentz entered upon the scene. He brought theory into harmony with experience by means of a wonderful simplification of theoretical principles. He achieved this, the most important advance in the theory of electricity since Maxwell, by taking from ether its mechanical, and from matter its electromagnetic qualities. As in empty space, so too in the interior of material bodies, the ether, and not matter viewed atomistically, was exclusively the seat of electromagnetic fields. According to Lorentz the elementary particles of matter alone are capable of carrying out movements; their electromagnetic activity is entirely confined to the carrying of electric charges. Thus Lorentz succeeded in reducing all electromagnetic happenings to Maxwell's equations for free space.

As to the mechanical nature of the Lorentzian ether, it may be said of it, in a somewhat playful spirit, that immobility is the only mechanical property of which it has not been deprived by H. A. Lorentz. It may be added that the whole change in the conception of the ether which the special theory of relativity brought about, consisted in taking away from the ether its last mechanical quality, namely, its immobility. How this is to be understood will forthwith be expounded.

The space-time theory and the kinematics of the special theory of relativity were modelled on the Maxwell-Lorentz theory of the electromagnetic field. This theory therefore satisfies the conditions of the special theory of relativity, but when viewed from the latter it acquires a novel aspect. For if K be a system of co-ordinates relatively to which the Lorentzian ether is at rest, the Maxwell-Lorentz equations are valid primarily with reference to K. But by the special theory of relativity the same equations without any change of meaning also hold in relation to any new system of co-ordinates K′ which is moving in uniform translation relatively to K. Now comes the anxious question:— Why must I in the theory distinguish the K system above all K′ systems, which are physically equivalent to it in all respects, by assuming that the ether is at rest relatively to the K system? For the theoretician such an asymmetry in the theoretical structure, with no corresponding asymmetry in the system of experience, is intolerable. If we assume the ether to be at rest relatively to K, but in motion relatively to K′, the physical equivalence of K and K′ seems to me from the logical standpoint, not indeed downright incorrect, but nevertheless inacceptable.

The next position which it was possible to take up in face of this state of things appeared to be the following. The ether does not exist at all. The electromagnetic fields are not states of a medium, and are not bound down to any bearer, but they are independent realities which are not reducible to anything else, exactly like the atoms of ponderable matter. This conception suggests itself the more readily as, according to Lorentz's theory, electromagnetic radiation, like ponderable matter, brings impulse and energy with it, and as, according to the special theory of relativity, both matter and radiation are but special forms of distributed energy, ponderable mass losing its isolation and appearing as a special form of energy.

More careful reflection teaches us, however, that the special theory of relativity does not compel us to deny ether. We may assume the existence of an ether; only we must give up ascribing a definite state of motion to it, i.e. we must by abstraction take from it the last mechanical characteristic which Lorentz had still left it. We shall see later that this point of view, the conceivability of which I shall at once endeavour to make more intelligible by a somewhat halting comparison, is justified by the results of the general theory of relativity.

Think of waves on the surface of water. Here we can describe two entirely different things. Either we may observe how the undulatory surface forming the boundary between water and air alters in the course of time; or else—with the help of small floats, for instance—we can observe how the position of the separate particles of water alters in the course of time. If the existence of such floats for tracking the motion of the particles of a fluid were a fundamental impossibility in physics—if, in fact, nothing else whatever were observable than the shape of the space occupied by the water as it varies in time, we should have no ground for the assumption that water consists of movable particles. But all the same we could characterise it as a medium.

We have something like this in the electromagnetic field. For we may picture the field to ourselves as consisting of lines of force. If we wish to interpret these lines of force to ourselves as something material in the ordinary sense, we are tempted to interpret the dynamic processes as motions of these lines of force, such that each separate line of force is tracked through the course of time. It is well known, however, that this way of regarding the electromagnetic field leads to contradictions.

Generalising we must say this:—There may be supposed to be extended physical objects to which the idea of motion cannot be applied. They may not be thought of as consisting of particles which allow themselves to be separately tracked through time. In Minkowski's idiom this is expressed as follows:—Not every extended conformation in the four-dimensional world can be regarded as composed of world-threads. The special theory of relativity forbids us to assume the ether to consist of particles observable through time, but the hypothesis of ether in itself is not in conflict with the special theory of relativity. Only we must be on our guard against ascribing a state of motion to the ether.

Certainly, from the standpoint of the special theory of relativity, the ether hypothesis appears at first to be an empty hypothesis. In the equations of the electromagnetic field there occur, in addition to the densities of the electric charge, *only* the intensities of the field. The career of electromagnetic processes *in vacuo* appears to be completely determined by these equations, uninfluenced by other physical quantities. The electromagnetic fields appear as ultimate, irreducible realities, and at first it seems superfluous to postulate a homogeneous, isotropic ether-medium, and to envisage electromagnetic fields as states of this medium.

But on the other hand there is a weighty argument to be adduced in favour of the ether hypothesis. To deny the ether is ultimately to assume that empty space has no physical qualities whatever. The fundamental facts of mechanics do not harmonize with this view. For the mechanical behaviour of a corporeal system hovering freely in empty space depends not only on relative positions (distances) and relative velocities, but also on its state of rotation, which physically may be taken as a characteristic not appertaining to the system in itself. In order to be able to look upon the rotation of the system, at least formally, as something real, Newton objectivises space.

Since he classes his absolute space together with real things, for him rotation relative to an absolute space is also something real. Newton might no less well have called his absolute space "Ether"; what is essential is merely that besides observable objects, another thing, which is not perceptible, must be looked upon as real, to enable acceleration or rotation to be looked upon as something real.

It is true that Mach tried to avoid having to accept as real something which is not observable by endeavouring to substitute in mechanics a mean acceleration with reference to the totality of the masses in the universe in place of an acceleration with reference to absolute space. But inertial resistance opposed to relative acceleration of distant masses presupposes action at a distance; and as the modern physicist does not believe that he may accept this action at a distance, he comes back once more, if he follows Mach, to the ether, which has to serve as medium for the effects of inertia. But this conception of the ether to which we are led by Mach's way of thinking differs essentially from the ether as conceived by Newton, by Fresnel, and by

Lorentz. Mach's ether not only *conditions* the behaviour of inert masses, but *is also conditioned* in its state by them.

Mach's idea finds its full development in the ether of the general theory of relativity. According to this theory the metrical qualities of the continuum of space-time differ in the environment of different points of space-time, and are partly conditioned by the matter existing outside of the territory under consideration. This space-time variability of the reciprocal relations of the standards of space and time, or, perhaps, the recognition of the fact that "empty space" in its physical relation is neither homogeneous nor isotropic, compelling us to describe its state by ten functions (the gravitation potentials g (mn)), has, I think, finally disposed of the view that space is physically empty. But therewith the conception of the ether has again acquired an intelligible content, although this content differs widely from that of the ether of the mechanical undulatory theory of light. The ether of the general theory of relativity is a medium which is itself devoid of *all* mechanical and kinematical qualities, but helps to determine mechanical (and electromagnetic) events.

What is fundamentally new in the ether of the general theory of relativity as opposed to the ether of Lorentz consists in this, that the state of the former is at every place determined by connections with the matter and the state of the ether in neighbouring places, which are amenable to law in the form of differential equations; whereas the state of the Lorentzian ether in the absence of electromagnetic fields is conditioned by nothing outside itself, and is everywhere the same. The ether of the general theory of relativity is transmuted conceptually into the ether of Lorentz if we substitute constants for the functions of space which describe the former, disregarding the causes which condition its state. Thus we may also say, I think, that the ether of the general theory of relativity is the outcome of the Lorentzian ether, through relativation.

As to the part which the new ether is to play in the physics of the future we are not yet clear. We know that it determines the metrical relations in the space-time continuum, e.g. the configurative possibilities of solid bodies as well as the gravitational fields; but we do not know whether it has an essential share in the structure of the electrical elementary particles constituting matter. Nor do we know whether it is only in the proximity of ponderable masses that its structure differs essentially from that of the Lorentzian ether; whether the geometry of spaces of cosmic extent is approximately Euclidean. But we can assert by reason of the relativistic equations of gravitation that there must be a departure from Euclidean relations, with spaces of cosmic order of magnitude, if there exists a positive mean density, no matter how small, of the matter in the universe. In this case the universe must of necessity be spatially unbounded and of finite magnitude, its magnitude being determined by the value of that mean density.

If we consider the gravitational field and the electromagnetic field from the stand-point of the ether hypothesis, we find a remarkable difference between the two. There can be no space nor any part of space without gravitational potentials; for these confer upon space its metrical qualities, without which it cannot be imagined at all. The existence of the gravitational field is inseparably bound up with the existence of space. On the other hand a part of space may very well be imagined without an electromagnetic field; thus in contrast with the gravitational field, the electromagnetic field seems to be only secondarily linked to the ether, the formal nature of the electromagnetic field being as yet in no way determined by that of gravitational ether. From the present state of theory it looks as if the electromagnetic field, as opposed to the gravitational field, rests upon an entirely new formal *motif*, as though nature might just as well have endowed the gravitational ether with fields of quite another type, for example, with fields of a scalar potential, instead of fields of the electromagnetic type.

Since according to our present conceptions the elementary particles of matter are also, in their essence, nothing else than condensations of the electromagnetic field, our present view of the universe presents two realities which are completely separated from each other conceptually, although connected causally, namely, gravitational ether and electromagnetic field, or—as they might also be called—space and matter.

Of course it would be a great advance if we could succeed in comprehending the gravitational field and the electromagnetic field together as one unified conformation. Then for the first time the epoch of theoretical physics founded by Faraday and Maxwell would reach a satisfactory conclusion. The contrast between ether and matter would fade away, and, through the general theory of relativity, the whole of physics would become a complete system of thought, like geometry, kinematics, and the theory of gravitation. An exceedingly ingenious attempt in this direction has been made by the mathematician H. Weyl; but I do not believe that his theory will hold its ground in relation to reality. Further, in contemplating the immediate future of theoretical physics we ought not unconditionally to reject the possibility that the facts comprised in the quantum theory may set bounds to the field theory beyond which it cannot pass.

Recapitulating, we may say that according to the general theory of relativity space is endowed with physical qualities; in this sense, therefore, there exists an ether. According to the general theory of relativity space without ether is unthinkable; for in such space there not only would be no propagation of light, but also no possibility of existence for standards of space and time (measuring-rods and clocks), nor therefore any space-time intervals in the physical sense. But this ether may not be thought of as endowed with the quality characteristic of ponderable media, as consisting of parts which may be tracked through time. The idea of motion may not be applied to it.

# Introduction to
# "Geometry and Experience"

Mathematics and physics are seen as two sides of the same coin. They are quite different, however. The truth of physics and other natural sciences can only be established through observation and experiment. Even then, the best we can hope to say is that a theory hasn't been proven wrong, or "falsified" in the terminology of Karl Popper, rather than being proven correct. Mathematics, on the other hand, could be developed and established even by someone with no direct experience of the physical world.

Geometry seems to lie in an intermediate position between the physical sciences and pure math. In "Geometry and Experience" (1921), Einstein notes that while propositions may be proven from stated axioms in geometry, the axioms themselves cannot be. If one is to study "practical geometry," the postulates must be based on the physical properties of the real universe.

Euclid's *Elements*, one of Einstein's first inspirations as a young thinker, laid down a set of geometric postulates that seem obvious from every experience; two straight lines, for example, can have at most one intersection, parallel lines remain parallel, and so on. Euclidean geometry seems to so perfectly describe our world on the human scale, and is in such accord with our intuition, that Newton's mechanics seem to flow straight from Euclid's geometry.

In the mid-nineteenth century, however, a number of thinkers began exploring non-Euclidean geometry, which starts from very different axioms than Euclid did and describes curved surfaces. Lines drawn on a sphere, for example, can intersect each other more than once. Consider that on a globe, lines of longitude meet at both the North and South poles. One of Einstein's greatest contributions to physics was his recognition that non-Euclidean geometry might fundamentally be the correct description of the shape of the universe.

General relativity is thus a way of describing the geometry of the universe. As John Archibald Wheeler has put it, "Matter tells space-time how to curve, and space-time tells matter how to move." What does the curvature of the universe mean? On a small scale, it describes the motions of the planets around the sun, and the gravitational pull between you and the earth.

On larger scales, there may be an overall curvature to the universe as well. This can be likened to the earth, which has both an overall sphericity and little bumps and ridges (mountain ranges) as well. The shape of the universe describes whether it is finite or infinite as well as what its ultimate fate will be.

The issues which Einstein raises in "Geometry and Experience" are still with us. Recent measurements from the Wilkinson Microwave Anistropy Probe and other experiments suggest that on the largest scales, the universe is flat, while gravitational wave experiments like the Laser Interferometer Gravitational Wave Observatory (LIGO) and the Laser Interfoerometer Space Antenna (LISA), scheduled to launch in 2015, aim to measure the bumps and wiggles in space-time on the smallest scales. On all scales, though, our intuition is of little use, and it is only by direct observation that we may make a "practical geometric" measurement of the shape of space-time.

# GEOMETRY AND EXPERIENCE

*An expanded form of an Address to the Prussian Academy of Sciences in Berlin on January 27th, 1921.*

One reason why mathematics enjoys special esteem, above all other sciences, is that its laws are absolutely certain and indisputable, while those of all other sciences are to some extent debatable and in constant danger of being overthrown by newly discovered facts. In spite of this, the investigator in another department of science would not need to envy the mathematician if the laws of mathematics referred to objects of our mere imagination, and not to objects of reality. For it cannot occasion surprise that different persons should arrive at the same logical conclusions when they have already agreed upon the fundamental laws (axioms), as well as the methods by which other laws are to be deduced therefrom. But there is another reason for the high repute of mathematics, in that it is mathematics which affords the exact natural sciences a certain measure of security, to which without out mathematics they could not attain.

At this point an enigma presents itself which in all ages has agitated inquiring minds. How can it be that mathematics, being after all a product of human thought which is independent of experience, is so admirably appropriate to the objects of reality? Is human reason, then, without experience, merely by taking thought, able to fathom the properties of real things?

In my opinion the answer to this question is, briefly, this:—As far as the laws of mathematics refer to reality, they are not certain; and as far as they are certain, they do not refer to reality. It seems to me that complete clearness as to this state of things first became common property through that new departure in mathematics which is known by the name of mathematical logic or "Axiomatics." The progress achieved by axiomatics consists in its having neatly separated the logical-formal from its objective or intuitive content; according to axiomatics the logical-formal alone forms the subject-matter of mathematics, which is not concerned with the intuitive or other content associated with the logical-formal.

Let us for a moment consider from this point of view any axiom of geometry, for instance, the following:—Through two points in space there always passes one and only one straight line. How is this axiom to be interpreted in the older sense and in the more modern sense?

The older interpretation:—Every one knows what a straight line is, and what a point is. Whether this knowledge springs from an ability of the human mind or from experience, from some collaboration of the two or from some other source, is not for the mathematician to decide. He leaves the question to the philosopher. Being based upon this knowledge, which precedes all mathematics, the axiom stated above is, like all other axioms, self-evident, that is, it is the expression of a part of this *a priori* knowledge.

The more modern interpretation:—Geometry treats of entities which are denoted by the words straight line, point, etc. These entities do not take for granted any knowledge or intuition whatever, but they presuppose only the validity of the axioms, such as the one stated above, which are to be taken in a purely formal sense, i.e. as void of all content of intuition or experience. These axioms are free creations of the human mind. All other propositions of geometry are logical inferences from the axioms (which are to be taken in the nominalistic sense only). The matter of which geometry treats is first defined by the axioms. Schlick in his book on epistemology has therefore characterised axioms very aptly as "implicit definitions."

This view of axioms, advocated by modern axiomatics, purges mathematics of all extraneous elements, and thus dispels the mystic obscurity which formerly surrounded the principles of mathematics.

But a presentation of its principles thus clarified makes it also evident that mathematics as such cannot predicate anything about perceptual objects or real objects. In axiomatic geometry the words "point," "straight line," etc., stand only for empty conceptual schemata. That which gives them substance is not relevant to mathematics.

Yet on the other hand it is certain that mathematics generally, and particularly geometry, owes its existence to the need which was felt of learning something about the relations of real things to one another. The very word geometry, which, of course, means earth-measuring, proves this. For earth-measuring has to do with the possibilities of the disposition of certain natural objects with respect to one another, namely, with parts of the earth, measuring-lines, measuring-wands, etc. It is clear that the system of concepts of axiomatic geometry alone cannot make any assertions as to the relations of real objects of this kind, which we will call practically-rigid bodies. To be able to make such assertions, geometry must be stripped of its merely logical-formal character by the co-ordination of real objects of experience with the empty conceptual frame-work of axiomatic geometry. To accomplish this, we need only add the proposition:—Solid bodies are related, with respect to their possible dispositions, as are bodies in Euclidean geometry of three dimensions. Then the propositions of Euclid contain affirmations as to the relations of practically-rigid bodies.

Geometry thus completed is evidently a natural science; we may in fact regard it as the most ancient branch of physics. Its affirmations rest essentially on induction from experience, but not on logical inferences only. We will call this completed geometry "practical geometry," and shall distinguish it in what follows from "purely axiomatic geometry." The question whether the practical geometry of the universe is Euclidean or not has a clear meaning, and its answer can only be furnished by experience. All linear measurement in physics is practical geometry in this sense, so too is geodetic and astronomical linear measurement, if we call to our help the law of experience that light is propagated in a straight line, and indeed in a straight line in the sense of practical geometry.

I attach special importance to the view of geometry which I have just set forth, because without it I should have been unable to formulate the theory of relativity. Without it the following reflection would have been impossible:—In a system of reference rotating relatively to an inert system, the laws of disposition of rigid bodies do not correspond to the rules of Euclidean geometry on account of the Lorentz contraction; thus if we admit non-inert systems we must abandon Euclidean geometry. The decisive step in the transition to general co-variant equations would certainly not have been taken if the above interpretation had not served as a stepping-stone. If we deny the relation between the body of axiomatic Euclidean geometry and the practically-rigid body of reality, we readily arrive at the following view, which was entertained by that acute and profound thinker, H. Poincaé:— Euclidean geometry is distinguished above all other imaginable axiomatic geometries by its simplicity. Now since axiomatic geometry by itself contains no assertions as to the reality which can be experienced, but can do so only in combination with physical laws, it should be possible and reasonable—whatever may be the nature of reality— to retain Euclidean geometry. For if contradictions between theory and experience manifest themselves, we should rather decide to change physical laws than to change axiomatic Euclidean geometry. If we deny the relation between the practically-rigid body and geometry, we shall indeed not easily free ourselves from the convention that Euclidean geometry is to be retained as the simplest. Why is the equivalence of the practically-rigid body and the body of geometry—which suggests itself so readily—denied by Poincaré and other investigators? Simply because under closer inspection the real solid bodies in nature are not rigid, because their geometrical behaviour, that is, their possibilities of relative disposition, depend upon temperature, external forces, etc. Thus the original, immediate relation between geometry and physical reality appears destroyed, and we feel impelled toward the following more general view, which characterizes Poincaré's standpoint. Geometry (G) predicates nothing about the relations of real things, but only geometry together with the purport (P) of physical laws can do so. Using symbols, we may say that only the sum of (G) + (P) is subject to the control of experience. Thus (G) may be chosen arbitrarily, and also parts of (P); all these laws are conventions. All that is necessary to avoid contradictions is to choose the remainder of (P) so that (G) and the whole of (P) are together in accord with experience. Envisaged in this way, axiomatic geometry and the part of natural law which has been given a conventional status appear as epistemologically equivalent.

*Sub specie aeterni* Poincaré, in my opinion, is right. The idea of the measuring-rod and the idea of the clock co-ordinated with it in the theory of relativity do not find their exact correspondence in the real world. It is also clear that the solid body and the clock do not in the conceptual edifice of physics play the part of irreducible elements, but that of composite structures, which may not play any independent part in theoretical physics. But it is my conviction that in the present stage of development of theoretical physics these ideas must still be employed as independent ideas; for we are still far from possessing such certain knowledge of theoretical principles as to be able to give exact theoretical constructions of solid bodies and clocks.

Further, as to the objection that there are no really rigid bodies in nature, and that therefore the properties predicated of rigid bodies do not apply to physical reality,—this objection is by no means so radical as might appear from a hasty examination. For it is not a difficult task to determine the physical state of a measuring-rod so accurately that its behaviour relatively to other measuring-bodies shall be sufficiently free from ambiguity to allow it to be substituted for the "rigid" body. It is to measuring-bodies of this kind that statements as to rigid bodies must be referred.

All practical geometry is based upon a principle which is accessible to experience, and which we will now try to realise. We will call that which is enclosed between two boundaries, marked upon a practically-rigid body, a tract. We imagine two practically-rigid bodies, each with a tract marked out on it. These two tracts are said to be "equal to one another" if the boundaries of the one tract can be brought to coincide permanently with the boundaries of the other. We now assume that:

If two tracts are found to be equal once and anywhere, they are equal always and everywhere.

Not only the practical geometry of Euclid, but also its nearest generalisation, the practical geometry of Riemann, and therewith the general theory of relativity, rest upon this assumption. Of the experimental reasons which warrant this assumption I will mention only one. The phenomenon of the propagation of light in empty space assigns a tract, namely, the appropriate path of light, to each interval of local time, and conversely. Thence it follows that the above assumption for tracts must also hold good for intervals of clock-time in the theory of relativity. Consequently it may be formulated as follows:—If two ideal clocks are going at the same rate at any time and at any place (being then in immediate proximity to each other), they will always go at the same rate, no matter where and when they are again compared with each other at one place.—If this law were not valid for real clocks, the proper frequencies for the separate atoms of the same chemical element would not be in such exact agreement as experience demonstrates. The existence of sharp spectral lines is a convincing experimental proof of the above-mentioned principle of practical geometry. This is the ultimate foundation in fact which enables us to speak with meaning of the mensuration, in Riemann's sense of the word, of the four-dimensional continuum of space-time.

The question whether the structure of this continuum is Euclidean, or in accordance with Riemann's general scheme, or otherwise, is, according to the view which is here being advocated, properly speaking a physical question which must be answered by experience, and not a question of a mere convention to be selected on practical grounds. Riemann's geometry will be the right thing if the laws of disposition of practically-rigid bodies are transformable into those of the bodies of Euclid's geometry with an exactitude which increases in proportion as the dimensions of the part of space-time under consideration are diminished.

It is true that this proposed physical interpretation of geometry breaks down when applied immediately to spaces of sub-molecular order of magnitude. But nevertheless, even in questions as to the constitution of elementary particles, it retains part of its importance. For even when it is a question of describing the electrical elementary particles constituting matter, the attempt may still be made to ascribe physical importance to those ideas of fields which have been physically defined for the purpose of describing the geometrical behaviour of bodies which are large as compared with the molecule. Success alone can decide as to the justification of such an attempt, which postulates physical reality for the fundamental principles of Riemann's geometry outside of the domain of their physical definitions. It might possibly turn out that this extrapolation has no better warrant than the extrapolation of the idea of temperature to parts of a body of molecular order of magnitude.

It appears less problematical to extend the ideas of practical geometry to spaces of cosmic order of magnitude. It might, of course, be objected that a construction composed of solid rods departs more and more from ideal rigidity in proportion as its spatial extent becomes greater. But it will hardly be possible, I think, to assign fundamental significance to this objection. Therefore the question whether the universe is spatially finite or not seems to me decidedly a pregnant question in the sense of practical geometry. I do not even consider it impossible that this question will be answered before long by astronomy. Let us call to mind what the general theory of relativity teaches in this respect. It offers two possibilities:—

1. The universe is spatially infinite. This can be so only if the average spatial density of the matter in universal space, concentrated in the stars, vanishes, i.e. if the ratio of the total mass of the stars to the magnitude of the space through which they are scattered approximates indefinitely to the value zero when the spaces taken into consideration are constantly greater and greater.

2. The universe is spatially finite. This must be so, if there is a mean density of the ponderable matter in universal space differing from zero. The smaller that mean density, the greater is the volume of universal space.

I must not fail to mention that a theoretical argument can be adduced in favour of the hypothesis of a finite universe. The general theory of relativity teaches that the inertia of a given body is greater as there are more ponderable masses in proximity to it; thus it seems very natural to reduce the total effect of inertia of a body to action and reaction between it and the other bodies in the universe, as indeed, ever since Newton's time, gravity has been completely reduced to action and reaction between bodies. From the equations of the general theory of relativity it can be deduced that this total reduction of inertia to reciprocal action between masses—as required by E. Mach, for example—is possible only if the universe is spatially finite.

On many physicists and astronomers this argument makes no impression. Experience alone can finally decide which of the two possibilities is realised in nature. How can experience furnish an answer? At first it might seem possible to determine the mean density of matter by observation of that part of the universe which is accessible to our perception. This hope is illusory. The distribution of the visible stars is extremely irregular, so that we on no account may venture to set down the mean density of star-matter in the universe as equal, let us say, to the mean density in the Milky Way. In any case, however great the space examined may be, we could not feel convinced that there were no more stars beyond that space. So it seems impossible to estimate the mean density. But there is another road, which seems to me more practicable, although it also presents great difficulties. For if we inquire into the deviations shown by the consequences of the general theory of relativity which are accessible to experience, when these are compared with the consequences of the Newtonian theory, we first of all find a deviation which shows itself in close proximity to gravitating mass, and has been confirmed in the case of the planet Mercury. But if the universe is spatially finite there is a second deviation from the Newtonian theory, which, in the language of the Newtonian theory, may be expressed thus:—The gravitational field is in its nature such as if it were produced, not only by the ponderable masses, but also by a mass-density of negative sign, distributed uniformly throughout space. Since this factitious mass-density would have to be enormously small, it could make its presence felt only in gravitating systems of very great extent.

Assuming that we know, let us say, the statistical distribution of the stars in the Milky Way, as well as their masses, then by Newton's law we can calculate the gravitational field and the mean velocities which the stars must have, so that the Milky Way should not collapse under the mutual attraction of its stars, but should maintain its actual extent. Now if the actual velocities of the stars, which can, of course, be measured, were smaller than the calculated velocities, we should have a proof that the actual attractions at great distances are smaller than by Newton's law. From such a deviation it could be proved indirectly that the universe is finite. It would even be possible to estimate its spatial magnitude.

Can we picture to ourselves a three-dimensional universe which is finite, yet unbounded?

The usual answer to this question is "No," but that is not the right answer. The purpose of the following remarks is to show that the answer should be "Yes." I want to show that without any extraordinary difficulty we can illustrate the theory of a finite universe by means of a mental image to which, with some practice, we shall soon grow accustomed.

First of all, an obervation of epistemological nature. A geometrical-physical theory as such is incapable of being directly pictured, being merely a system of concepts. But these concepts serve the purpose of bringing a multiplicity of real or imaginary sensory experiences into connection in the mind. To "visualise" a theory, or bring it home to one's mind, therefore means to give a representation to that abundance of experiences for which the theory supplies the schematic arrangement. In the present case we have to ask ourselves how we can represent that relation of solid bodies with respect to their reciprocal disposition (contact) which corresponds to the theory of a finite universe. There is really nothing new in what I have to say about this; but innumerable questions addressed to me prove that the requirements of those who thirst for knowledge of these matters have not yet been completely satisfied.

So, will the initiated please pardon me, if part of what I shall bring forward has long been known?

What do we wish to express when we say that our space is infinite? Nothing more than that we might lay any number whatever of bodies of equal sizes side by side without ever filling space. Suppose that we are provided with a great many wooden cubes all of the same size. In accordance with Euclidean geometry we can place them above, beside, and behind one another so as to fill a part of space of any dimensions; but this construction would never be finished; we could go on adding more and more cubes without ever finding that there was no more room. That is what we wish to express when we say that space is infinite. It would be better to say that space is infinite in relation to practically-rigid bodies, assuming that the laws of disposition for these bodies are given by Euclidean geometry.

Another example of an infinite continuum is the plane. On a plane surface we may lay squares of cardboard so that each side of any square has the side of another square adjacent to it. The construction is never finished; we can always go on laying squares—if their laws of disposition correspond to those of plane figures of Euclidean geometry. The plane is therefore infinite in relation to the cardboard squares. Accordingly we say that the plane is an infinite continuum of two dimensions, and space an infinite continuum of three dimensions. What is here meant by the number of dimensions, I think I may assume to be known.

Now we take an example of a two-dimensional continuum which is finite, but unbounded. We imagine the surface of a large globe and a quantity of small paper discs, all of the same size. We place one of the discs anywhere on the surface of the globe. If we move the disc about, anywhere we like, on the surface of the globe, we do not come upon a limit or boundary anywhere on the journey. Therefore we say that the spherical surface of the globe is an unbounded continuum. Moreover, the spherical surface is a finite continuum. For if we stick the paper discs on the globe, so that no disc overlaps another, the surface of the globe will finally become so full that there is no room for another disc. This simply means that the spherical surface of the globe is finite in relation to the paper discs. Further, the spherical surface is a non-Euclidean continuum of two dimensions, that is to say, the laws of disposition for the rigid figures lying in it do not agree with those of the Euclidean plane. This can be shown in the following way. Place a paper disc on the spherical surface, and around it in a circle place six more discs, each of which is to be surrounded in turn by six discs, and so on. If this construction is made on a plane surface, we have an uninterrupted disposition in which there are six discs touching every disc except those which lie on the outside.

On the spherical surface the construction also seems to promise success at the outset, and the smaller the radius of the disc in proportion to that of the sphere, the more promising it seems. But as the construction progresses it becomes more and more patent that the disposition of the discs in the manner indicated, without interruption, is not possible, as it should be possible by Euclidean geometry of the the plane surface. In this way creatures which cannot leave the spherical surface, and cannot even peep out from the spherical surface into three-dimensional space, might discover, merely by experimenting with discs, that their two-dimensional "space" is not Euclidean, but spherical space.

From the latest results of the theory of relativity it is probable that our three-dimensional space is also approximately spherical, that is, that the laws of disposition of rigid bodies in it are not given by Euclidean geometry, but approximately by spherical geometry, if only we consider parts of space which are sufficiently great. Now this is the place where the reader's imagination boggles. "Nobody can imagine this thing," he cries indignantly. "It can be said, but cannot be thought. I can represent to myself a spherical surface well enough, but nothing analogous to it in three dimensions."

We must try to surmount this barrier in the mind, and the patient reader will see that it is by no means a particularly difficult task. For this purpose we will first give our attention once more to the geometry of two-dimensional spherical surfaces. In the adjoining figure let $K$ be the spherical surface, touched at $S$ by a plane, $E$, which, for facility of presentation, is shown in the drawing as a bounded surface. Let $L$ be a disc on the spherical surface. Now let us imagine that at the point $N$ of the spherical surface, diametrically opposite to $S$, there is a luminous point, throwing a shadow $L'$ of the disc $L$ upon the plane $E$. Every point on the sphere has its shadow on the plane. If the disc on the sphere $K$ is moved, its shadow $L'$ on the plane $E$ also moves. When the disc $L$ is at $S$, it almost exactly coincides with its shadow. If it moves on the spherical surface away from $S$ upwards, the disc shadow $L'$ on the plane also moves away from $S$ on the plane outwards, growing bigger and bigger. As the disc $L$ approaches the luminous point $N$, the shadow moves off to infinity, and becomes infinitely great.

Now we put the question. What are the laws of disposition of the disc-shadows $L'$ on the plane $E$? Evidently they are exactly the same as the laws of disposition of the discs $L$ on the spherical surface. For to each original figure on $K$ there is a corresponding shadow figure on $E$. If two discs on $K$ are touching, their shadows on $E$ also touch. The shadow-geometry on the plane agrees with the the disc-geometry on the sphere. If we call the disc-shadows rigid figures, then spherical geometry holds good on the plane $E$ with respect to these rigid figures. Moreover, the plane is finite with respect to the disc-shadows, since only a finite number of the shadows can find room on the plane.

At this point somebody will say, "That is nonsense. The disc-shadows are *not* rigid figures. We have only to move a two-foot rule about on the plane $E$ to convince ourselves that the shadows constantly increase in size as they move away from $S$ on the plane towards infinity." But what if the two-foot rule were to behave on the plane $E$ in the same way as the disc-shadows $L'$? It would then be impossible to show that the shadows increase in size as they move away from $S$; such an assertion would then no longer have any meaning whatever. In fact the only objective assertion that can be made about the disc-shadows is just this, that they are related in exactly the same way as are the rigid discs on the spherical surface in the sense of Euclidean geometry.

We must carefully bear in mind that our statement as to the growth of the disc-shadows, as they move away from $S$ towards infinity, has in itself no objective meaning, as long as we are unable to employ Euclidean rigid bodies which can be moved about on the plane $E$ for the purpose of comparing the size of the disc-shadows. In respect of the laws of disposition of the shadows $L'$, the point $S$ has no special privileges on the plane any more than on the spherical surface.

The representation given above of spherical geometry on the plane is important for us, because it readily allows itself to be transferred to the three-dimensional case.

Let us imagine a point $S$ of our space, and a great number of small spheres, $L'$, which can all be brought to coincide with one another. But these spheres are not to be rigid in the sense of Euclidean geometry; their radius is to increase (in the sense of Euclidean geometry) when they are moved away from $S$ towards infinity, and this increase is to take place in exact accordance with the same law as applies to the increase of the radii of the disc-shadows $L'$ on the plane.

After having gained a vivid mental image of the geometrical behaviour of our $L'$ spheres, let us assume that in our space there are no rigid bodies at all in the sense of Euclidean geometry, but only bodies having the behaviour of our $L'$ spheres. Then we shall have a vivid representation of three-dimensional spherical space, or, rather of three-dimensional spherical geometry. Here our spheres must be called "rigid" spheres. Their increase in size as they depart from $S$ is not to be detected by measuring with measuring-rods, any more than in the case of the disc-shadows on $E$, because the standards of measurement behave in the same way as the spheres. Space is homogeneous, that is to say, the same spherical configurations are possible in the environment of all points.* Our space is finite, because, in consequence of the "growth" of the spheres, only a finite number of them can find room in space.

*This is intelligible without calculation—but only for the two-dimensional case—if we revert once more to the case of the disc on the surface of the sphere.

In this way, by using as stepping-stones the practice in thinking and visualisation which Euclidean geometry gives us, we have acquired a mental picture of spherical geometry. We may without difficulty impart more depth and vigour to these ideas by carrying out special imaginary constructions. Nor would it be difficult to represent the case of what is called elliptical geometry in an analogous manner. My only aim to-day has been to show that the human faculty of visualisation is by no means bound to capitulate to non-Euclidean geometry.

# Selections from *The Meaning of Relativity*

T hree hundred years before Einstein, Galileo Galilei developed a theory of relativity that ultimately formed one of the pillars of Sir Isaac Newton's mechanics. In *The Meaning of Relativity*, Einstein presents Galilean relativity as a precursor not only to Newton's work, but to his own as well.

Galilean relativity is based on the simple and intuitive idea that time flows constantly for all observers regardless of their state of motion. Galileo thus anticipated Newton's first law of motion: objects in motion will maintain a constant speed and direction unless acted upon by an outside force.

For all of the apparent mathematical complexity in *The Meaning of Relativity*, Einstein's goals are quite modest. He is simply demonstrating that measurements on either stationary or moving frames will both yield results that satisfy Newton's laws of motion. He proceeds to show at the end of the chapter that a different set of transformations, those suggested by Hendrik Lorentz, are required to make Maxwell's equations of electricity and magnetism work in moving frames of reference.

The difference between Galileo's transformations and those of Lorentz is that the former result in a constant flow of time for all observers, while the latter yield different rates of time for observers in different states of motion.

But what do we mean by a "constant flow of time"? It is simple to say that one event precedes another, or that they occur simultaneously, but how do we measure time, except through time itself? Einstein suggests using the reflection of an emitted light beam as a clock, postulating that light travels at the same speed regardless of the state of motion of an observer. This simple thought experiment yields some very surprising results.

For example, he finds that an observer measuring the events in a train moving past him will see its inhabitants moving slowly: a slowed heartbeat, a slowed wall-clock, and all other times slowed as well. Likewise, the person on the train thinks of himself as completely normal, but sees the clock at the train station running slowly. But because light must always travel at a constant speed, if time is related to the speed of travel, then lengths along the direction of motion must be related as well. In everyday life, these effects are not evident, only becoming manifest when the speeds involved approach that of light. Thus, at normal speeds, Einstein and Galileo's relativity behave exactly the same.

In this work, Einstein argues effectively that although virtually all of our experiences suggest that Galileo and Newton were correct, unification of different branches of physics demand a new theory: special relativity.

# SPACE AND TIME IN PRE-RELATIVITY PHYSICS

The theory of relativity is intimately connected with the theory of space and time. I shall therefore begin with a brief investigation of the origin of our ideas of space and time, although in doing so I know that I introduce a controversial subject. The object of all science, whether natural science or psychology, is to co-ordinate our experiences and to bring them into a logical system. How are our customary ideas of space and time related to the character of our experiences?

The experiences of an individual appear to us arranged in a series of events; in this series the single events which we remember appear to be ordered according to the criterion of "earlier" and "later," which cannot be analysed further. There exists, therefore, for the individual, an I-time, or subjective time. This in itself is not measurable. I can, indeed, associate numbers with the events, in such a way that a greater number is associated with the later event than with an earlier one; but the nature of this association may be quite arbitrary. This association I can define by means of a clock by comparing the order of events furnished by the clock with the order of the given series of events. We understand by a clock something which provides a series of events which can be counted, and which has other properties of which we shall speak later.

By the aid of language different individuals can, to a certain extent, compare their experiences. Then it turns out that certain sense perceptions of different individuals correspond to each other, while for other sense perceptions no such correspondence can be established. We are accustomed to regard as real those sense perceptions which are common to different individuals, and which therefore are, in a measure, impersonal. The natural sciences, and in particular, the most fundamental of them, physics, deal with such sense perceptions. The conception of physical bodies, in particular of rigid bodies, is a relatively

Reprinted courtesy of Princeton University Press

constant complex of such sense perceptions. A clock is also a body, or a system, in the same sense, with the additional property that the series of events which it counts is formed of elements all of which can be regarded as equal.

The only justification for our concepts and system of concepts is that they serve to represent the complex of our experiences; beyond this they have no legitimacy. I am convinced that the philosophers have had a harmful effect upon the progress of scientific thinking in removing certain fundamental concepts from the domain of empiricism, where they are under our control, to the intangible heights of the *a priori*. For even if it should appear that the universe of ideas cannot be deduced from experience by logical means, but is, in a sense, a creation of the human mind, without which no science is possible, nevertheless this universe of ideas is just as little independent of the nature of our experiences as clothes are of the form of the human body. This is particularly true of our concepts of time and space, which physicists have been obliged by the facts to bring down from the Olympus of the *a priori* in order to adjust them and put them in a serviceable condition.

We now come to our concepts and judgments of space. It is essential here also to pay strict attention to the relation of experience to our concepts. It seems to me that Poincaré clearly recognized the truth in the account he gave in his book, "La Science et l'Hypothèse." Among all the changes which we can perceive in a rigid body those which can be cancelled by a voluntary motion of our body are marked by their simplicity; Poincaré calls these, changes in position. By means of simple changes in position we can bring two bodies into contact. The theorems of congruence, fundamental in geometry, have to do with the laws that govern such changes in position. For the concept of space the following seems essential. We can form new bodies by bringing bodies B, C, . . . up to body A; we say that we *continue* body A. We can continue body A in such a way that it comes into contact with any other body, X. The ensemble of all continuations of body A we can designate as the "space of the body A." Then it is true that all bodies are in the "space of the (arbitrarily chosen) body A." In this sense we cannot speak of space in the abstract, but only of the "space belonging to a body *A*." The earth's crust plays such a dominant rôle in our daily life in judging the relative positions of bodies that it has led to an abstract conception of space which certainly cannot be defended. In order to free ourselves from this fatal error we shall speak only of "bodies of reference," or "space of reference." It was only through the theory of general relativity that refinement of these concepts became necessary, as we shall see later.

I shall not go into detail concerning those properties of the space of reference which lead to our conceiving points as elements of space, and space as a continuum. Nor shall I attempt to analyse further the properties of space which justify the conception of continuous series of points, or lines. If these concepts are assumed, together with their relation to the solid bodies of experience, then it is easy to say what we mean by the three-dimensionality of space; to each point three numbers, $x_1, x_2, x_3$ (co-ordinates), may be associated, in such a way that this association is uniquely reciprocal, and that $x_1$, $x_2$, and $x_3$ vary continuously when the point describes a continuous series of points (a line).

It is assumed in pre-relativity physics that the laws of the configuration of ideal rigid bodies are consistent with Euclidean geometry. What this means may be expressed as follows: Two points marked on a rigid body form an *interval*. Such an interval can be oriented at rest, relatively to our space of reference, in a multiplicity of ways. If, now, the points of this space can be referred to co-ordinates $x_1, x_2, x_3$, in such a way that the differences of the co-ordinates, $\Delta x_1, \Delta x_2, \Delta x_3$, of the two ends of the interval furnish the same sum of squares,

$$s^2 = \Delta x_1^2 + \Delta x_2^2 + \Delta x_3^2 \tag{1}$$

for every orientation of the interval, then the space of reference is called Euclidean, and the co-ordinates Cartesian.* It is sufficient, indeed, to make this assumption in the limit for an infinitely small interval. Involved in this assumption there are some which are rather

---

* This relation must hold for an arbitrary choice of the origin and of the direction (ratios $\Delta x_1 : \Delta x_2 : \Delta x_3$) of the interval.

less special, to which we must call attention on account of their fundamental significance. In the first place, it is assumed that one can move an ideal rigid body in an arbitrary manner. In the second place, it is assumed that the behaviour of ideal rigid bodies towards orientation is independent of the material of the bodies and their changes of position, in the sense that if two intervals can once be brought into coincidence, they can always and everywhere be brought into coincidence. Both of these assumptions, which are of fundamental importance for geometry and especially for physical measurements, naturally arise from experience; in the theory of general relativity their validity needs to be assumed only for bodies and spaces of reference which are infinitely small compared to astronomical dimensions.

The quantity $s$ we call the length of the interval. In order that this may be uniquely determined it is necessary to fix arbitrarily the length of a definite interval; for example, we can put it equal to 1 (unit of length). Then the lengths of all other intervals may be determined. If we make the $x_\nu$ linearly dependent upon a parameter $\lambda$,

$$x_\nu = a_\nu + \lambda b_\nu,$$

we obtain a line which has all the properties of the straight lines of the Euclidean geometry. In particular, it easily follows that by laying off $n$ times the interval $s$ upon a straight line, an interval of length $n \cdot s$ is obtained. A length, therefore, means the result of a measurement carried out along a straight line by means of a unit measuring rod. It has a significance which is as independent of the system of co-ordinates as that of a straight line, as will appear in the sequel.

We come now to a train of thought which plays an analogous rôle in the theories of special and general relativity. We ask the question: besides the Cartesian co-ordinates which we have used are there other equivalent co-ordinates? An interval has a physical meaning which is independent of the choice of co-ordinates; and so has the spherical surface which we obtain as the locus of the end points of all equal intervals that we lay off from an arbitrary point of our space of reference. If $x_\nu$ as well as $x_\nu'$ ($\nu$ from 1 to 3) are Cartesian co-ordinates of our space of reference, then the spherical surface will be expressed in our two systems of co-ordinates by the equations

$$\sum \Delta x_\nu^2 = \text{const.} \tag{2}$$

$$\sum \Delta x_\nu'^2 = \text{const.} \tag{2a}$$

How must the $x_\nu'$ be expressed in terms of the $x_\nu$ in order that equations (2) and (2a) may be equivalent to each other? Regarding the $x_\nu'$ expressed as functions of the $x_\nu$, we can write, by Taylor's theorem, for small values of the $\Delta x_\nu$,

$$\Delta x_\nu' = \sum_\alpha \frac{\partial x_\nu'}{\partial x_\alpha} \Delta x_\alpha + \tfrac{1}{2} \sum_{\alpha\beta} \frac{\partial^2 x_\nu'}{\partial x_\alpha \partial x_\beta} \Delta x_\alpha \Delta x_\beta \dots$$

If we substitute (2a) in this equation and compare with (1), we see that the $x_\nu'$ must be linear functions of the $x_\nu$. If we therefore put

$$x_\nu' = \alpha_\nu + \sum_\alpha b_{\nu\alpha} x_\alpha \tag{3}$$

or

$$\Delta x_\nu' = \sum_\alpha b_{\nu\alpha} \Delta x_\alpha \tag{3a}$$

then the equivalence of equations (2) and (2a) is expressed in the form

$$\sum \Delta x_\nu'^2 = \lambda \sum \Delta x_\nu^2 \quad (\lambda \text{ independent of } \Delta x_\nu) \tag{2b}$$

It therefore follows that $\lambda$ must be a constant. If we put $\lambda = 1$, (2b) and (3a) furnish the conditions

$$\sum_\nu b_{\nu\alpha} b_{\nu\beta} = \delta_{\alpha\beta} \tag{4}$$

in which $\delta_{\alpha\beta} = 1$, or $\delta_{\alpha\beta} = 0$, according as $\alpha = \beta$ or $\alpha \neq \beta$. The conditions (4) are called the conditions of orthogonality, and the transformations (3), (4), linear orthogonal transformations. If we stipulate that $s^2 = \sum \Delta x_\nu^2$ shall be equal to the square of the length in every system of co-ordinates, and if we always measure with the same unit scale, then $\lambda$ must be equal to 1. Therefore the linear orthogonal transformations are the only ones by means of which we can pass from one Cartesian system of co-ordinates in our space of reference to another. We see that in applying such transformations the equations of a straight line become equations of a straight line. Reversing equations

(3a) by multiplying both sides by $b_{\nu\beta}$ and summing for all the $\nu$'s, we obtain

$$\sum_{\nu\alpha} b_{\nu\beta}\Delta x_\nu' = \sum b_{\nu\alpha}b_{\nu\beta}\Delta x_\alpha = \sum_\alpha \delta_{\alpha\beta}\Delta x_\alpha = \Delta x_\beta \qquad (5)$$

The same coefficients, $b$, also determine the inverse substitution of $\Delta x_\nu$. Geometrically, $b_{\nu\alpha}$ is the cosine of the angle between the $x_\nu'$ axis and the $x_\alpha$ axis.

To sum up, we can say that in the Euclidean geometry there are (in a given space of reference) preferred systems of co-ordinates, the Cartesian systems, which transform into each other by linear orthogonal transformations. The distance $s$ between two points of our space of reference, measured by a measuring rod, is expressed in such co-ordinates in a particularly simple manner. The whole of geometry may be founded upon this conception of distance. In the present treatment, geometry is related to actual things (rigid bodies), and its theorems are statements concerning the behaviour of these things, which may prove to be true or false.

One is ordinarily accustomed to study geometry divorced from any relation between its concepts and experience. There are advantages in isolating that which is purely logical and independent of what is, in principle, incomplete empiricism. This is satisfactory to the pure mathematician. He is satisfied if he can deduce his theorems from axioms correctly, that is, without errors of logic. The questions as to whether Euclidean geometry is true or not does not concern him. But for our purpose it is necessary to associate the fundamental concepts of geometry with natural objects; without such an association geometry is worthless for the physicist. The physicist is concerned with the question as to whether the theorems of geometry are true or not. That Euclidean geometry, from this point of view, affirms something more than the mere deductions derived logically from definitions may be seen from the following simple consideration.

Between $n$ points of space there are $\dfrac{n(n-1)}{2}$ distances, $s_{\mu\nu}$; between these and the $3n$ co-ordinates we have the relations

$$s_{\mu\nu}{}^2 = \left(x_{1(\mu)} - x_{1(\nu)}\right)^2 + \left(x_{2(\mu)} - x_{2(\nu)}\right)^2 + \cdots$$

From these $\dfrac{n(n-1)}{2}$ equations the $3n$ co-ordinates may be elim-

inated, and from this elimination at least $\dfrac{n(n-1)}{2} - 3n$ equations in

the $s_{\mu\nu}$ will result.* Since the $s_{\mu\nu}$ are measurable quantities, and by definition are independent of each other, these relations between the $s_{\mu\nu}$
are not necessary *a priori.*

From the foregoing it is evident that the equations of transformation (3), (4) have a fundamental significance in Euclidean geometry, in that they govern the transformation from one Cartesian system
of co-ordinates to another. The Cartesian systems of co-ordinates are
characterized by the property that in them the measurable distance
between two points, $s$, is expressed by the equation

$$s^2 = \sum \Delta x_\nu^{\,2}.$$

If $K_{(x_\nu)}$ and $K'_{(x_\nu)}$ are two Cartesian systems of co-ordinates, then

$$\sum \Delta x_\nu^{\,2} = \sum \Delta x_\nu'^{\,2}.$$

The right-hand side is identically equal to the left-hand side on
account of the equations of the linear orthogonal transformation, and
the right-hand side differs from the left-hand side only in that the $x_\nu$
are replaced by the $x'_\nu$. This is expressed by the statement that $\sum \Delta x_\nu^2$
is an invariant with respect to linear orthogonal transformations. It is
evident that in the Euclidean geometry only such, and all such, quantities have an objective significance, independent of the particular
choice of the Cartesian co-ordinates, as can be expressed by an invariant with respect to linear orthogonal transformations. This is the reason that the theory of invariants, which has to do with the laws that
govern the form of invariants, is so important for analytical geometry.

As a second example of a geometrical invariant, consider a volume. This is expressed by

$$V = \int \int \int dx_1\, dx_2\, dx_3.$$

---

*In reality there are $\dfrac{n(n-1)}{2} - 3n + 6$ equations.

By means of Jacobi's theorem we may write

$$\int\int\int dx'_1\, dx'_2\, dx'_3 = \int\int\int \frac{\partial(x'_1, x'_2, x'_3)}{\partial(x_1, x_2, x_3)}\, dx_1\, dx_2\, dx_3$$

where the integrand in the last integral is the functional determinant of the $x'_\nu$ with respect to the $x_\nu$, and this by (3) is equal to the determinant $|b_{\mu\nu}|$ of the coefficients of substitution, $b_{\nu\alpha}$. If we form the determinant of the $\delta_{\mu\alpha}$ from equation (4), we obtain, by means of the theorem of multiplication of determinants,

$$1 = |\delta_{\alpha\beta}| = \left|\sum_\nu b_{\nu\alpha}\, b_{\nu\beta}\right| = |b_{\mu\nu}|^2; |b_{\mu\nu}| = \pm 1 \qquad (6)$$

If we limit ourselves to those transformations which have the determinant $+1^*$ (and only these arise from continuous variations of the systems of co-ordinates) then $V$ is an invariant.

Invariants, however, are not the only forms by means of which we can give expression to the independence of the particular choice of the Cartesian co-ordinates. Vectors and tensors are other forms of expression. Let us express the fact that the point with the current co-ordinates $x_\nu$ lies upon a straight line. We have

$$x_\nu - A_\nu = \lambda B_\nu \text{ ($\nu$ from 1 to 3).}$$

Without limiting the generality we can put

$$\sum B_\nu^2 = 1.$$

If we multiply the equations by $b_{\beta\nu}$ (compare (3a) and (5)) and sum for all the $\nu$'s, we get

$$x'_\beta - A'_\beta = \lambda B'_\beta$$

where we have written

$$B'_\beta = \sum_\nu b_{\beta\nu}\, B_\nu; A'_\beta = \sum_\nu b_{\beta\nu}\, A_\nu.$$

These are the equations of straight lines with respect to a second Cartesian system of co-ordinates $K'$. They have the same form as the equations with respect to the original system of co-ordinates. It is therefore evident that straight lines have a significance which is independent

---

* There are thus two kinds of Cartesian systems which are designated as "right-handed" and "left-handed" systems. The difference between these is familiar to every physicist and engineer. It is interesting to note that these two kinds of systems cannot be defined geometrically, but only the contrast between them.

of the system of co-ordinates. Formally, this depends upon the fact that the quantities $(x_\nu - A_\nu) - \lambda B_\nu$ are transformed as the components of an interval, $\Delta x_\nu$. The ensemble of three quantities, defined for every system of Cartesian co-ordinates, and which transform as the components of an interval, is called a vector. If the three components of a vector vanish for one system of Cartesian co-ordinates, they vanish for all systems, because the equations of transformation are homogeneous. We can thus get the meaning of the concept of a vector without referring to a geometrical representation. This behaviour of the equations of a straight line can be expressed by saying that the equation of a straight line is co-variant with respect to linear orthogonal transformations.

We shall now show briefly that there are geometrical entities which lead to the concept of tensors. Let $P_0$ be the centre of a surface of the second degree, $P$ any point on the surface, and $\xi_\nu$ the projections of the interval $P_0 P$ upon the co-ordinate axes. Then the equation of the surface is

$$\sum a_{\mu\nu} \xi_\mu \xi_\nu = 1.$$

In this, and in analogous cases, we shall omit the sign of summation, and understand that the summation is to be carried out for those indices that appear twice. We thus write the equation of the surface

$$a_{\mu\nu} \xi_\mu \xi_\nu = 1.$$

The quantities $a_{\mu\nu}$ determine the surface completely, for a given position of the centre, with respect to the chosen system of Cartesian co-ordinates. From the known law of transformation for the $\xi_\nu$ (3a) for linear orthogonal transformations, we easily find the law of transformation for the $a_{\mu\nu}$*:

$$a'_{\sigma r} = b_{\sigma\mu} b_{r\nu} a_{\mu\nu}.$$

This transformation is homogeneous and of the first degree in the $a_{\mu\nu}$. On account of this transformation, the $a_{\mu\nu}$ are called components of a tensor of the second rank (the latter on account of the double index). If all the components, $a_{\mu\nu}$, of a tensor with respect to any system of

---

*The equation $a'_{\sigma r} \xi'_\sigma \xi'_r = 1$ may, by (5), be replaced by $a'_{\sigma r} b_{\mu\sigma} b_{\nu r} \xi_\sigma \xi_r = 1$, from which the result stated immediately follows.

Cartesian co-ordinates vanish, they vanish with respect to every other Cartesian system. The form and the position of the surface of the second degree is described by this tensor $(a)$.

Tensors of higher rank (number of indices) may be defined analytically. It is possible and advantageous to regard vectors as tensors of rank 1, and invariants (scalars) as tensors of rank 0. In this respect, the problem of the theory of invariants may be so formulated: according to what laws may new tensors be formed from given tensors? We shall consider these laws now, in order to be able to apply them later. We shall deal first only with the properties of tensors with respect to the transformation from one Cartesian system to another in the same space of reference, by means of linear orthogonal transformations. As the laws are wholly independent of the number of dimensions, we shall leave this number, $n$, indefinite at first.

*Definition.* If an object is defined with respect to every system of Cartesian co-ordinates in a space of reference of $n$ dimensions by the $n^\alpha$ numbers $A_{\mu\nu\rho}\ldots$ ($\alpha$ = number of indices), then these numbers are the components of a tensor of rank $\alpha$ if the transformation law is

$$A'_{\mu'\nu'\rho'}\ldots = b_{\mu'\mu}b_{\nu'\nu}b_{\rho'\rho}\ldots A_{\mu\nu\rho}\ldots \tag{7}$$

*Remark.* From this definition it follows that

$$A_{\mu\nu\rho}\ldots B_\mu C_\nu D_\rho\ldots \tag{8}$$

is an invariant, provided that $(B)$, $(C)$, $(D)$ ... are vectors. Conversely, the tensor character of $(A)$ may be inferred, if it is known that the expression (8) leads to an invariant for an arbitrary choice of the vectors $(B)$, $(C)$, etc.

*Addition and Subtraction.* By addition and subtraction of the corresponding components of tensors of the same rank, a tensor of equal rank results:

$$A_{\mu\nu\rho}\ldots \pm B_{\mu\nu\rho}\ldots = C_{\mu\nu\rho}\ldots \tag{9}$$

The proof follows from the definition of a tensor given above.

*Multiplication.* From a tensor of rank $\alpha$ and a tensor of rank $\beta$ we may obtain a tensor of rank $\alpha + \beta$ by multiplying all the components of the first tensor by all the components of the second tensor:

$$T_{\mu\nu\rho}\ldots{}_{\alpha\beta\gamma}\ldots = A_{\mu\nu\rho}\ldots B_{\alpha\beta\gamma}\ldots \tag{10}$$

*Contraction.* A tensor of rank $\alpha - 2$ may be obtained from one of rank $\alpha$ by putting two definite indices equal to each other and then summing for this single index:

$$T_\rho \ldots = A_{\mu\mu\rho} \ldots \left( = \sum_\mu A_{\mu\mu\rho} \ldots \right) \tag{11}$$

The proof is

$$A'_{\mu\mu\rho} \ldots = b_{\mu\alpha} b_{\mu\beta} b_{\rho\gamma} \ldots A_{\alpha\beta\gamma} \ldots = \delta_{\alpha\beta} b_{\rho\gamma} \ldots A_{\alpha\beta\gamma} \ldots$$
$$= b_{\rho\gamma} \ldots A_{\alpha\alpha\gamma} \ldots$$

In addition to these elementary rules of operation there is also the formation of tensors by differentiation ("Erweiterung"):

$$T_{\mu\nu\rho} \cdots _\alpha = \frac{\partial A_{\mu\nu\rho} \cdots}{\partial x_\alpha} \tag{12}$$

New tensors, in respect to linear orthogonal transformations, may be formed from tensors according to these rules of operation.

*Symmetry Properties of Tensors.* Tensors are called symmetrical or skew-symmetrical in respect to two of their indices, $\mu$ and $\nu$, if both the components which result from interchanging the indices $\mu$ and $\nu$ are equal to each other or equal with opposite signs.

Condition for symmetry: $\quad A_{\mu\nu\rho} = A_{\nu\mu\rho}.$

Condition for skew-symmetry: $\quad A_{\mu\nu\rho} = -A_{\nu\mu\rho}.$

*Theorem.* The character of symmetry or skew-symmetry exists independently of the choice of co-ordinates, and in this lies its importance. The proof follows from the equation defining tensors.

*Special Tensors.*

I. The quantities $\delta_{\rho\sigma}$ (4) are tensor components (fundamental tensor).

*Proof.* If in the right-hand side of the equation of transformation $A'_{\mu\nu} = b_{\mu\alpha} b_{\nu\beta} A_{\alpha\beta}$, we substitute for $A_{\alpha\beta}$ the quantities $\delta_{\alpha\beta}$ (which are equal to 1 or 0 according as $\alpha = \beta$ or $\alpha \neq \beta$), we get

$$A'_{\mu\nu} = b_{\mu\alpha} b_{\nu\alpha} = \delta_{\mu\nu}.$$

The justification for the last sign of equality becomes evident if one applies (4) to the inverse substitution (5).

II. There is a tensor $(\delta_{\mu\nu\rho} \ldots)$ skew-symmetrical with respect to all pairs of indices, whose rank is equal to the number of dimensions, $n$, and whose components are equal to $+1$ or $-1$ according as $\mu\nu\rho\ldots$ is an even or odd permutation of $123\ldots$

The proof follows with the aid of the theorem proved above $|b_{\rho\sigma}| = 1$.

These few simple theorems form the apparatus from the theory of invariants for building the equations of pre-relativity physics and the theory of special relativity.

We have seen that in pre-relativity physics, in order to specify relations in space, a body of reference, or a space of reference, is required, and, in addition, a Cartesian system of co-ordinates. We can fuse both these concepts into a single one by thinking of a Cartesian system of co-ordinates as a cubical frame-work formed of rods each of unit length. The co-ordinates of the lattice points of this frame are integral numbers. It follows from the fundamental relation

$$s^2 = \Delta x_1^2 + \Delta x_2^2 + \Delta x_3^2 \tag{13}$$

that the members of such a space-lattice are all of unit length. To specify relations in time, we require in addition a standard clock placed, say, at the origin of our Cartesian system of co-ordinates or frame of reference. If an event takes place anywhere we can assign to it three co-ordinates, $x_\nu$, and a time $t$, as soon as we have specified the time of the clock at the origin which is simultaneous with the event. We therefore give (hypothetically) an objective significance to the statement of the simultaneity of distant events, while previously we have been concerned only with the simultaneity of two experiences of an individual. The time so specified is at all events independent of the position of the system of co-ordinates in our space of reference, and is therefore an invariant with respect to the transformation (3).

It is postulated that the system of equations expressing the laws of pre-relativity physics is co-variant with respect to the transformation (3), as are the relations of Euclidean geometry. The isotropy and homogeneity of space is expressed in this way.* We shall now consider some of the more important equations of physics from this point of view.

---

*The laws of physics could be expressed, even in case there were a preferred direction in space, in such a way as to be co-variant with respect to the transformation (3); but such an expression would in this case be unsuitable. If there were a preferred direction in space it would simplify the description of natural phenomena to orient the system of co-ordinates in a definite way with respect to this direction. But if, on the other hand, there is no unique direction in space it is not logical to formulate the laws of nature in such a way as to conceal the equivalence of systems of co-ordinates that are oriented differently. We shall meet with this point of view again in the theories of special and general relativity.

The equations of motion of a material particle are

$$m\frac{d^2x_\nu}{dt^2} = X_\nu \qquad (14)$$

$(dx_\nu)$ is a vector; $dt$, and therefore also $\frac{1}{dt}$, an invariant; thus $\left(\frac{dx_\nu}{dt}\right)$ is a vector; in the same way it may be shown that $\left(\frac{d^2x_\nu}{dt^2}\right)$ is a vector. In general, the operation of differentiation with respect to time does not alter the tensor character. Since $m$ is an invariant (tensor of rank 0), $\left(m\frac{d^2x_\nu}{dt^2}\right)$ is a vector, or tensor of rank 1 (by the theorem of the multiplication of tensors). If the force $(X_\nu)$ has a vector character, the same holds for the difference $\left(m\frac{d^2x_\nu}{dt^2} - X_\nu\right)$. These equations of motion are therefore valid in every other system of Cartesian co-ordinates in the space of reference. In the case where the forces are conservative we can easily recognize the vector character of $(X_\nu)$. For a potential energy, $\Phi$, exists, which depends only upon the mutual distances of the particles, and is therefore an invariant. The vector character of the force, $X_\nu = -\frac{\partial \Phi}{\partial x_\nu}$, is then a consequence of our general theorem about the derivative of a tensor of rank 0.

Multiplying by the velocity, a tensor of rank 1, we obtain the tensor equation

$$\left(m\frac{d^2x_\nu}{dt^2} - X_\nu\right)\frac{dx_\mu}{dt} = 0.$$

By contraction and multiplication by the scalar $dt$ we obtain the equation of kinetic energy

$$d\left(\frac{mq^2}{2}\right) = X_\nu\, dx_\nu.$$

If $\xi_\nu$ denotes the difference of the co-ordinates of the material particle and a point fixed in space, then the $\xi_\nu$ have vector character. We evidently have $\frac{d^2x_\nu}{dt^2} = \frac{d^2\xi_\nu}{dt^2}$, so that the equations of motion of the particle may be written

$$m\frac{d^2\xi_\nu}{dt^2} - X_\nu = 0.$$

Multiplying this equation by $\xi_\mu$ we obtain a tensor equation

$$\left(m\frac{d^2\xi_\nu}{dt^2} - X_\nu\right)\xi_\mu = 0.$$

Contracting the tensor on the left and taking the time average we obtain the virial theorem, which we shall not consider further. By interchanging the indices and subsequent subtraction, we obtain, after a simple transformation, the theorem of moments,

$$\frac{d}{dt}\left[m\left(\xi_\mu\frac{d\xi_\nu}{dt} - \xi_\nu\frac{d\xi_\mu}{dt}\right)\right] = \xi_\mu X_\nu - \xi_\nu X_\mu \tag{15}$$

It is evident in this way that the moment of a vector is not a vector but a tensor. On account of their skew-symmetrical character there are not nine, but only three independent equations of this system. The possibility of replacing skew-symmetrical tensors of the second rank in space of three dimensions by vectors depends upon the formation of the vector

$$A_\mu = \tfrac{1}{2}A_{\sigma\tau}\delta_{\sigma\tau\mu}.$$

If we multiply the skew-symmetrical tensor of rank 2 by the special skew-symmetrical tensor $\delta$ introduced above, and contract twice, a vector results whose components are numerically equal to those of the tensor. These are the so-called axial vectors which transform differently, from a right-handed system to a left-handed system, from the $\Delta x_\nu$. There is a gain in picturesqueness in regarding a skew-symmetrical tensor of rank 2 as a vector in space of three dimensions, but it does not represent the exact nature of the corresponding quantity so well as considering it a tensor.

We consider next the equations of motion of a continuous medium. Let $\rho$ be the density, $u_\nu$ the velocity components considered as functions of the co-ordinates and the time, $X_\nu$ the volume forces per unit of mass, and $p_{v\sigma}$ the stresses upon a surface perpendicular to the $\sigma$-axis in the direction of increasing $x_v$. Then the equations of motion area, by Newton's law,

$$\rho \frac{du_v}{dt} = -\frac{\partial p_{v\sigma}}{\partial x_\sigma} + \rho X_v$$

in which $\dfrac{du_v}{dt}$ is the acceleration of the particle which at time $t$ has the co-ordinates $x_v$. If we express this acceleration by partial differential coefficients, we obtain, after dividing by $\rho$,

$$\frac{\partial u_v}{\partial t} + \frac{\partial u_v}{\partial x_\sigma} u_\sigma = -\frac{1}{\rho}\frac{\partial p_{v\sigma}}{\partial x_\sigma} + X_v \tag{16}$$

We must show that this equation holds independently of the special choice of the Cartesian system of co-ordinates. $(u_v)$ is a vector, and therefore $\dfrac{\partial u_v}{\partial t}$ is also a vector. $\dfrac{\partial u_v}{\partial x_\sigma}$ is a tensor of rank 2, $\dfrac{\partial u_v}{\partial x_\sigma} u_\tau$ is a tensor of rank 3. The second term on the left results from contraction in the indices $\sigma, \tau$. The vector character of the second term on the right is obvious. In order that the first term on the right may also be a vector it is necessary for $p_{v\sigma}$ to be a tensor. Then by differentiation and contraction $\dfrac{\partial p_{v\sigma}}{\partial x_\sigma}$ results, and is therefore a vector, as it also is after multiplication by the reciprocal scalar $\dfrac{1}{\rho}$. That $p_{v\sigma}$ is a tensor, and therefore transforms according to the equation

$$p'_{\mu v} = b_{\mu\alpha} b_{v\beta} p_{\alpha\beta},$$

is proved in mechanics by integrating this equation over an infinitely small tetrahedron. It is also proved there, by application of the theorem of moments to an infinitely small parallelepipedon, that $p_{v\sigma} = p_{\sigma v}$, and hence that the tensor of the stress is a symmetrical tensor. From what has been said it follows that, with the aid of the rules given above, the equation is co-variant with respect to orthogonal transformations in space (rotational transformations); and the rules according to which the quantities in the equation must be transformed in order that the equation may be co-variant also become evident.

The co-variance of the equation of continuity,

$$\frac{\partial \rho}{\partial t} + \frac{\partial(\rho u_\nu)}{\partial x_\nu} = 0 \tag{17}$$

requires, from the foregoing, no particular discussion.

We shall also test for co-variance the equations which express the dependence of the stress components upon the properties of the matter, and set up these equations for the case of a compressible viscous fluid with the aid of the conditions of co-variance. If we neglect the viscosity, the pressure, $p$, will be a scalar, and will depend only upon the density and the temperature of the fluid. The contribution to the stress tensor is then evidently

$$p\delta_{\mu\nu}$$

in which $\delta_{\mu\nu}$ is the special symmetrical tensor. This term will also be present in the case of a viscous fluid. But in this case there will also be pressure terms, which depend upon the space derivatives of the $u_\nu$. We shall assume that this dependence is a linear one. Since these terms must be symmetrical tensors, the only ones which enter will be

$$\alpha\left(\frac{\partial u_\mu}{\partial x_\nu} + \frac{\partial u_\nu}{\partial x_\mu}\right) + \beta\delta_{\mu\nu}\frac{\partial u_\alpha}{\partial x_\alpha}$$

$\left(\text{for } \dfrac{\partial u_\alpha}{\partial x_\alpha} \text{ is a scalar}\right)$. For physical reasons (no slipping) it is assumed that for symmetrical dilatations in all directions, i.e. when

$$\frac{\partial u_1}{\partial x_1} = \frac{\partial u_2}{\partial x_2} = \frac{\partial u_3}{\partial x_3}; \frac{\partial u_1}{\partial x_2}, \text{ etc., } = 0,$$

there are no frictional forces present, from which it follows that $\beta = -\frac{2}{3}\alpha$. If only $\dfrac{\partial u_1}{\partial x_3}$ is different from zero, let $p_{31} = -\eta\dfrac{\partial u_1}{\partial x_3}$, by which $\alpha$ is determined. We then obtain for the complete stress tensor,

$$(18) \quad p_{\mu\nu} = p\delta_{\mu\nu} - \eta\left[\left(\frac{\partial u_\mu}{\partial x_\nu} + \frac{\partial u_\nu}{\partial x_\mu}\right) - \frac{2}{3}\left(\frac{\partial u_1}{\partial x_1} + \frac{\partial u_2}{\partial x_2} + \frac{\partial u_3}{\partial x_3}\right)\delta_{\mu\nu}\right]$$

The heuristic value of the theory of invariants, which arises from the isotropy of space (equivalence of all directions), becomes evident from this example.

We consider, finally, Maxwell's equations in the form which are the foundation of the electron theory of Lorentz.

$$
\begin{cases}
\dfrac{\partial h_3}{\partial x_2} - \dfrac{\partial h_2}{\partial x_3} = \dfrac{1}{c}\dfrac{\partial e_1}{\partial t} + \dfrac{1}{c}\,i_1 \\[2ex]
\dfrac{\partial h_1}{\partial x_3} - \dfrac{\partial h_3}{\partial x_1} = \dfrac{1}{c}\dfrac{\partial e_2}{\partial t} + \dfrac{1}{c}\,i_2 \\[2ex]
\cdot \qquad\quad \cdot \qquad\qquad \cdot \qquad\quad \cdot \\[1ex]
\dfrac{\partial e_1}{\partial x_1} + \dfrac{\partial e_2}{\partial x_2} + \dfrac{\partial e_3}{\partial x_3} = \rho
\end{cases}
\tag{19}
$$

$$
\begin{cases}
\dfrac{\partial e_3}{\partial x_2} - \dfrac{\partial e_2}{\partial x_3} = -\dfrac{1}{c}\dfrac{\partial h_1}{\partial t} \\[2ex]
\dfrac{\partial e_1}{\partial x_3} - \dfrac{\partial e_3}{\partial x_1} = -\dfrac{1}{c}\dfrac{\partial h_2}{\partial t} \\[2ex]
\cdot \qquad\quad \cdot \qquad\qquad \cdot \\[1ex]
\dfrac{\partial h_1}{\partial x_1} + \dfrac{\partial h_2}{\partial x_2} + \dfrac{\partial h_3}{\partial x_3} = 0
\end{cases}
\tag{20}
$$

**i** is a vector, because the current density is defined as the density of electricity multiplied by the vector velocity of the electricity. According to the first three equations it is evident that **e** is also to be regarded as a vector. Then **h** cannot be regarded as a vector.* The equations may, however, easily be interpreted if **h** is regarded as a skew-symmetrical tensor of the second rank. Accordingly, we write $h_{23}$, $h_{31}$, $h_{12}$, in place of $h_1$, $h_2$, $h_3$ respectively. Paying attention to the skew-symmetry of $h_{\mu\nu}$, the first three equations of (19) and (20) may be written in the form

$$
\frac{\partial h_{\mu\nu}}{\partial x_\nu} = \frac{1}{c}\frac{\partial e_\mu}{\partial t} + \frac{1}{c}\,i_\mu
\tag{19a}
$$

$$
\frac{\partial e_\mu}{\partial x_\nu} - \frac{\partial e_\nu}{\partial x_\mu} = +\frac{1}{c}\frac{\partial h_{\mu\nu}}{\partial t}
\tag{20a}
$$

In contrast to **e, h** appears as a quantity which has the same type of symmetry as an angular velocity. The divergence equations then take

---

* These considerations will make the reader familiar with tensor operations without the special difficulties of the four-dimensional treatment; corresponding considerations in the theory of special relativity (Minkowski's interpretation of the field) will then offer fewer difficulties.

the form

$$\frac{\partial e_\nu}{\partial x_\nu} = \rho \tag{19b}$$

$$\frac{\partial h_{\mu\nu}}{\partial x_\rho} + \frac{\partial h_{\nu\rho}}{\partial x_\mu} + \frac{\partial h_{\rho\mu}}{\partial x_\nu} = 0 \tag{20b}$$

The last equation is a skew-symmetrical tensor equation of the third rank (the skew-symmetry of the left-hand side with respect to every pair of indices may easily be proved, if attention is paid to the skew-symmetry of $h_{\mu\nu}$). This notation is more natural than the usual one, because, in contrast to the latter, it is applicable to Cartesian left-handed systems as well as to right-handed systems without change of sign.

# Selections from
# *The Evolution of Physics*

During the first half of the twentieth century, quantum theory was transforming the landscape of physics, much as the theory of electromagnetism did a century earlier. In *The Evolution of Physics*, Albert Einstein and Leopold Infeld describe this revolution from the eye of the storm. Today we have grown so accustomed to the idea of nanotechnology and microelectronics, technologies which could not exist without quantum mechanics, that it is easy to forget what a monumental shift in our understanding is required to think in quantum terms.

In the continuous picture, a piece of iron, for example, may have any mass whatsoever. In the quantum picture, this is shown to be an illusion. Each lump of iron has a certain number of atoms in it, and each atom has a fixed mass. Another lump can only differ by an integer number of atoms and thus, by "quantized" masses. Atoms themselves are made of still smaller quantized elements, protons and neutrons. And that's not even the end of it! About two decades after the publication of *The Evolution of Physics*, Murray Gell-Mann and Kazuhiko Nishijima proposed that protons and neutrons were made of yet smaller quantized particles known as quarks.

The idea that particles may be only divided a finite number of times before reaching the atomic scale was not a new one; it had its origins as far back as Democritus and the early Greek atomists. The strength of modern quantum theory came, rather, from the properties ascribed to microscopic particles. While on the human scale, we normally say that a particle has a well-defined position and speed, we can make no such statements on the quantum scale. Instead, particles are defined by their probability waves. One of the strangest examples of quantum weirdness comes from the idea that prior to observation by an experiment, an electron does not have a well-defined position, but that by observing it, we "force" it into a particular state. Let's be clear that quantum theory doesn't say that we don't *know* the position prior to observations, but that such a thing as definite position really doesn't exist!

The amazing thing is that while the microscopic world is governed by statistics, the macroscopic world seems governed by Newton's laws, which are themselves deterministic. How can that be, since, in the end, macroscopic objects are made of protons, neutrons, and electrons? We see the same effect when we think about the air in a room. While the individual molecules fly around in a haphazard manner, on the human scale, they normally seem much steadier. In a sense, the distinction between the wave properties and particle properties of matter are really just a function of physical scale. The quantum theory shows that on the smallest scales, particles look more and more wave-like, and are governed more and more by statistics.

This wave-particle duality doesn't just exist for objects like electrons and protons, however. Isaac Newton originally proposed that light must have particle properties, a theory which was rejected in the nineteenth century when light was observed to exhibit interference patterns, a property of waves. Ultimately, light was understood to have both wave properties, as with radio waves, and quantized particle properties, which came to be known as photons. Modesty must have prevented Einstein from noting that it was his own interpretation of the photoelectric effect that ultimately gave rise to the modern particle picture of light. In this experiment, an ultraviolet beam is shined on metal, and electrons are ejected, a very particle-like behavior. His 1905 paper describing this effect earned him the Nobel Prize in 1921.

Einstein's *Evolution of Physics* gives us an insight into the state of science in the early twentieth century, including glimpses into his own considerable contributions. Nearly seventy years later, though they have refined their models considerably, physicists are still dealing with the fallout of the weirdness born from the quantum picture of the universe.

# Field, Relativity

*The field as representation . . . The two pillars of the field theory . . . The reality of the field . . . Field and ether . . . The mechanical scaffold . . . Ether and motion . . . Time, distance, relativity . . . Relativity and mechanics . . . The time-space continuum . . . General relativity . . . Outside and inside the elevator . . . Geometry and experiment . . . General relativity and its verification . . . Field and matter*

## THE FIELD AS REPRESENTATION

DURING the second half of the nineteenth century new and revolutionary ideas were introduced into physics; they opened the way to a new philosophical view, differing from the mechanical one. The results of the work of Faraday, Maxwell, and Hertz led to the development of modern physics, to the creation of new concepts, forming a new picture of reality.

Our task now is to describe the break brought about in science by these new concepts and to show how they gradually gained clarity and strength. We shall try to reconstruct the line of progress logically, without bothering too much about chronological order.

The new concepts originated in connection with the phenomena of electricity, but it is simpler to introduce them, for the first time, through mechanics. We know that two particles attract each other and that this force of attraction decreases with the square of the distance. We can represent this fact in a new way, and shall do so even though it is difficult to understand the advantage of this. The small circle in our drawing on page 286 represents an attracting body, say, the sun. Actually, our diagram should be imagined as a model in space and not as a drawing on a plane. Our small circle, then, stands for a sphere in space, say, the sun. A body, the so-called *test body*, brought somewhere within the vicinity of the sun will be attracted along the line

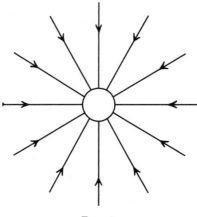

FIG. 1.

connecting the centers of the two bodies. Thus the lines in our drawing indicate the direction of the attracting force of the sun for different positions of the test body. The arrow on each line shows that the
force is directed toward the sun; this means the force is an attraction.
These are the *lines of force of the gravitational field.* For the moment,
this is merely a name and there is no reason for stressing it further.
There is one characteristic feature of our drawing which will be emphasized later. The lines of force are constructed in space, where no matter
is present. For the moment, all the lines of force, or briefly speaking,
the *field,* indicate only how a test body would behave if brought into
the vicinity of the sphere for which the field is constructed.

The lines in our space model are always perpendicular to the surface of the sphere. Since they diverge from one point, they are dense
near the sphere and become less and less so farther away. If we increase
the distance from the sphere twice or three times, then the density of
the lines, in our space-model, though not in the drawing, will be four
or nine times less. Thus the lines serve a double purpose. On the one
hand they show the direction of the force acting on a body brought
into the neighborhood of the sphere-sun. On the other hand the density of the lines of force in space shows how the force varies with the
distance. The drawing of the field, correctly interpreted, represents the
direction of the gravitational force and its dependence on distance. One

can read the law of gravitation from such a drawing just as well as from a description of the action in words, or in the precise and economical language of mathematics. This *field representation*, as we shall call it, may appear clear and interesting but there is no reason to believe that it marks any real advance. It would be quite difficult to prove its usefulness in the case of gravitation. Some may, perhaps, find it helpful to regard these lines as something more than drawings, and to imagine the real actions of force passing through them. This may be done, but then the speed of the actions along the lines of force must be assumed as infinitely great! The force between two bodies, according to Newton's law, depends only on distance; time does not enter the picture. The force has to pass from one body to another in no time! But, as motion with infinite speed cannot mean much to any reasonable person, an attempt to make our drawing something more than a model leads nowhere.

We do not intend, however, to discuss the gravitational problem just now. It served only as an introduction, simplifying the explanation of similar methods of reasoning in the theory of electricity.

We shall begin with a discussion of the experiment which created serious difficulties in our mechanical interpretation. We had a current flowing through a wire circuit in the form of a circle. In the middle of the circuit was a magnetic needle. The moment the current began to flow a new force appeared, acting on the magnetic pole, and perpendicular to any line connecting the wire and the pole. This force, if caused by a circulating charge, depended, as shown by Rowland's experiment, on the velocity of the charge. These experimental facts contradicted the philosophical view that all forces must act on the line connecting the particles and can depend only upon distance.

The exact expression for the force of a current acting on a magnetic pole is quite complicated, much more so, indeed, than the expression for gravitational forces. We can, however, attempt to visualize the actions just as we did in the case of a gravitational force. Our question is: with what force does the current act upon a magnetic pole placed somewhere in its vicinity? It would be rather difficult to describe this force in words. Even a mathematical formula would be complicated and awkward. It is best to represent all we know about the acting forces by a drawing, or rather by a spatial model, with lines of force. Some difficulty is caused by the fact that a magnetic pole exists only in connection with another magnetic pole, forming a dipole. We can, however, always imagine the magnetic needle of such length that only the force acting upon the pole nearer the current has to be taken into account. The other pole is far enough away for the force acting upon it to be negligible. To avoid ambiguity we shall say that the magnetic pole brought nearer to the wire is the *positive* one.

The character of the force acting upon the positive magnetic pole can be read from our drawing.

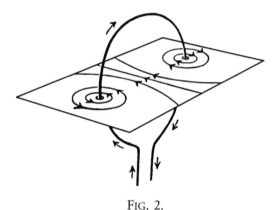

FIG. 2.

First we notice an arrow near the wire indicating the direction of the current, from higher to lower potential. All other lines are just fines of force belonging to this current and lying on a certain plane. If drawn properly, they tell us the direction of the force vector representing the action of the current on a given positive magnetic pole as well as something about the length of this vector. Force, as we know, is a vector and to determine it we must know its direction as well as its length. We are chiefly concerned with the problem of the direction of the force acting upon a pole. Our question is: how can we find, from the drawing, the direction of the force, at any point in space?

The rule for reading the direction of a force from such a model is not as simple as in our previous example, where the lines of force

FIG. 3.

were straight. In our next diagram only one line of force is drawn in order to clarify the procedure. The force vector lies on the tangent to the line of force, as indicated. The arrow of the force vector and the arrows on the line of force point in the same direction. Thus this is the direction in which the force acts on a magnetic pole at this point. A good drawing, or rather a good model, also tells us something about the length of the force vector at any point. This vector has to be longer where the lines are denser, i.e., near the wire, shorter where the lines are less dense, i.e., far from the wire.

In this way, the lines of force, or in other words, the field, enable us to determine the forces acting on a magnetic pole at any point in space. This, for the time being, is the only justification for our elaborate construction of the field. Knowing what the field expresses, we shall examine with a far deeper interest the lines of force corresponding to the current. These lines are circles surrounding the wire and lying on the plane perpendicular to that in which the wire is situated. Reading the character of the force from the drawing we come once more to the conclusion that the force acts in a direction perpendicular to any line connecting the wire and the pole, for the tangent to a circle is always perpendicular to its radius. Our entire knowledge of the acting forces can be summarized in the construction of the field. We sandwich the concept of the field between that of the current and that of the magnetic pole in order to represent the acting forces in a simple way.

Every current is associated with a magnetic field, i.e., a force always acts on a magnetic pole brought near the wire through which a current flows. We may remark in passing that this property enables us to construct sensitive apparatus for detecting the existence of a current. Once having learned how to read the character of the magnetic forces from the field model of a current, we shall always draw the field surrounding the wire through which the current flows, in order to represent the action of the magnetic forces at any point in space. Our first example is the so-called solenoid. This is, in fact, a coil of wire as shown in the drawing. Our aim is to learn, by experiment, all we can about the magnetic field associated with the current flowing through a solenoid and to incorporate this knowledge in the construction of a field. A drawing represents our result. The curved lines of force are closed, and surround the solenoid in a way characteristic of the magnetic field of a current.

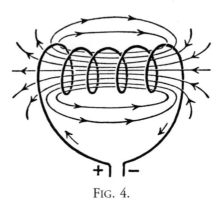

FIG. 4.

The field of a bar magnet can be represented in the same way as that of a current. Another drawing shows this. The lines of force are directed from the positive to the negative pole. The force vector always lies on the tangent to the line of force and is longest near the poles because the density of the lines is greatest at these points. The force vector represents the action of the magnet on a positive magnetic pole. In this case the magnet and not the current is the "source" of the field.

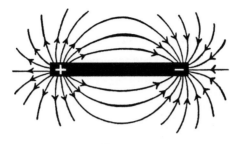

FIG. 5.

Our last two drawings should be carefully compared. In the first, we have the magnetic field of a current flowing through a solenoid; in the second, the field of a bar magnet. Let us ignore both the solenoid and the bar and observe only the two outside fields. We immediately notice that they are of exactly the same character; in each case the lines of force lead from one end of the solenoid or bar to the other.

The field representation yields its first fruit! It would be rather difficult to see any strong similarity between the current flowing through a solenoid and a bar magnet if this were not revealed by our construction of the field.

The concept of field can now be put to a much more severe test. We shall soon see whether it is anything more than a new representation of the acting forces. We could reason: assume, for a moment, that the field characterizes all actions determined by its sources in a unique way. This is only a guess. It would mean that if a solenoid and a bar magnet have the same field, then all their influences must also be the same. It would mean that two solenoids, carrying electric currents, behave like two bar magnets, that they attract or repel each other depending, exactly as in the case of bars, on their relative positions. It would also mean that a solenoid and a bar attract or repel each other in the same way as two bars. Briefly speaking, it would mean that all actions of a solenoid through which a current flows, and of a corresponding bar magnet are the same, since the field alone is responsible for them, and the field in both cases is of the same character. Experiment fully confirms our guess!

291

How difficult it would be to find those facts without the concept of field! The expression for a force acting between a wire through which a current flows and a magnetic pole is very complicated. In the case of two solenoids we should have to investigate the forces with which two currents act upon each other. But if we do this, with the help of the field, we immediately notice the character of all those actions at the moment when the similarity between the field of a solenoid and that of a bar magnet is seen.

We have the right to regard the field as something much more than we did at first. The properties of the field alone appear to be essential for the description of phenomena; the differences in source do not matter. The concept of field reveals its importance by leading to new experimental facts.

The field proved a very helpful concept. It began as something placed between the source and the magnetic needle in order to describe the acting force. It was thought of as an "agent" of the current, through which all action of the current was performed. But now the agent also acts as an interpreter, one who translates the laws into a simple, clear language, easily understood.

The first success of the field description suggests that it may be convenient to consider all actions of currents, magnets and charges indirectly, i.e., with the help of the field as an interpreter. A field may be regarded as something always associated with a current. It is there even in the absence of a magnetic pole to test its existence. Let us try to follow this new clew consistently.

The field of a charged conductor can be introduced in much the same way as the gravitational field, or the field of a current or magnet. Again only the simplest example! To design the field of a positively charged sphere, we must ask what kind of forces are acting on a small positively charged test body brought near the source of the field, the charged sphere. The fact that we use a positively and not a negatively charged test body is merely a convention, indicating in which direction the arrows on the line of force should be drawn. The model is analogous to that of a gravitational field (figure 1) because of

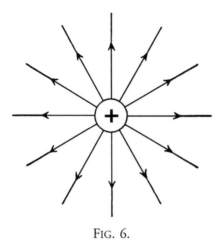

Fig. 6.

the similarity between Coulomb's law and Newton's. The only difference between the two models is that the arrows point in opposite
directions. Indeed, we have repulsion of two positive charges and
attraction of two masses. However, the field of a sphere with a negative charge will be identical with a gravitational field since the small
positive testing charge will be attracted by the source of the field.

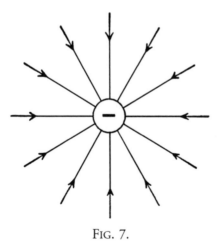

Fig. 7.

If both electric and magnetic poles are at rest, there is no action
between them, neither attraction nor repulsion. Expressing the same
fact in the field language we can say: an electrostatic field does not

influence a magnetostatic one and vice versa. The words "static field" mean a field that does not change with time. The magnets and charges would rest near one another for an eternity if no external forces disturbed them. Electrostatic, magnetostatic and gravitational fields are all of different character. They do not mix; each preserves its individuality regardless of the others.

Let us return to the electric sphere which was, until now, at rest, and assume that it begins to move due to the action of some external force. The charged sphere moves. In the field language this sentence reads: the field of the electric charge changes with time. But the motion of this charged sphere is, as we already know from Rowland's experiment, equivalent to a current. Further, every current is accompanied by a magnetic field. Thus the chain of our argument is:

$$\text{motion of charge} \rightarrow \text{change of an electric field}$$
$$\downarrow$$
$$\text{current} \rightarrow \text{associated magnetic field.}$$

We, therefore, conclude: *The change of an electric field produced by the motion of a charge is always accompanied by a magnetic field.*

Our conclusion is based on Oersted's experiment but it covers much more. It contains the recognition that the association of an electric field, changing in time, with a magnetic field is essential for our further argument.

As long as a charge is at rest there is only an electrostatic field. But a magnetic field appears as soon as the charge begins to move. We can say more. The magnetic field created by the motion of the charge will be stronger if the charge is greater and if it moves faster. This also is a consequence of Rowland's experiment. Once again using the field language, we can say: the faster the electric field changes, the stronger the accompanying magnetic field.

We have tried here to translate familiar facts from the language of fluids, constructed according to the old mechanical view, into the new language of fields. We shall see later how clear, instructive, and far-reaching our new language is.

# THE TWO PILLARS OF THE FIELD THEORY

"The change of an electric field is accompanied by a magnetic field." If we interchange the words "magnetic" and "electric," our sentence reads: "The change of a magnetic field is accompanied by an electric field." Only an experiment can decide whether or not this statement is true. But the idea of formulating this problem is suggested by the use of the field language.

Just over a hundred years ago, Faraday performed an experiment which led to the great discovery of induced currents.

The demonstration is very simple. We need only a solenoid or some other circuit, a bar magnet, and one of the many types of apparatus for detecting the existence of an electric current. To begin with, a bar magnet is kept at rest near a solenoid which forms a closed circuit. No current flows through the wire, for no source is present. There is only the magnetostatic field of the bar magnet which does not change with time. Now, we quickly change the position of the magnet either by removing it or by bringing it nearer the solenoid, whichever we prefer. At this moment, a current will appear for a very short time and then vanish.

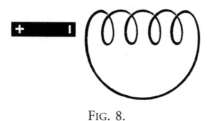

FIG. 8.

Whenever the position of the magnet is changed, the current reappears, and can be detected by a sufficiently sensitive apparatus. But a current—from the point of view of the field theory—means the existence of an electric field forcing the flow of the electric fluids through the wire. The current, and therefore the electric field, too, vanishes when the magnet is again at rest.

Imagine for a moment that the field language is unknown and the results of this experiment have to be described, qualitatively and quantitatively, in the language of old mechanical concepts. Our experiment then shows: by the motion of a magnetic dipole a new force was created, moving the electric fluid in the wire. The next question would be: upon what does this force depend? This would be very difficult to answer. We should have to investigate the dependence of the force upon the velocity of the magnet, upon its shape, and upon the shape of the circuit. Furthermore, this experiment, if interpreted in the old language, gives us no hint at all as to whether an induced current can be excited by the motion of another circuit carrying a current, instead of by motion of a bar magnet.

It is quite a different matter if we use the field language and again trust our principle that the action is determined by the field. We see at once that a solenoid through which a current flows would serve as well as a bar magnet. The drawing shows two solenoids: one, small, through which a current flows, and the other, in which the induced current is detected, larger. We could move the small solenoid, as we previously moved the bar magnet, creating an induced current in the larger

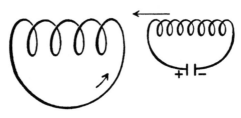

FIG. 9.

solenoid. Furthermore, instead of moving the small solenoid, we could create and destroy a magnetic field by creating and destroying the current, that is, by opening and closing the circuit. Once again, new facts suggested by the field theory are confirmed by experiment!

Let us take a simpler example. We have a closed wire without any source of current. Somewhere in the vicinity is a magnetic field. It means nothing to us whether the source of this magnetic field is another circuit through which an electric current flows, or a bar magnet. Figure 10 shows the closed circuit and the magnetic lines of force.

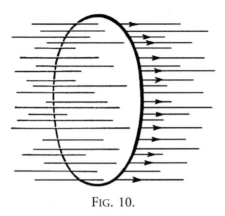

FIG. 10.

The qualitative and quantitative description of the induction phenomena is very simple in terms of the field language. As marked on the drawing, some lines of force go through the surface bounded by the wire. We have to consider the lines of force cutting that part of the plane which has the wire for a rim. No electric current is present so long as the field does not change, no matter how great its strength. But a current begins to flow through the rim-wire as soon as the number of lines passing through the surface surrounded by wire changes. The current is determined by the change, however it may be caused, of the number of lines passing the surface. This change in the number of lines of force is the only essential concept for both the qualitative and the quantitative descriptions of the induced current. "The number of lines changes" means that the density of the lines changes and this, we remember, means that the field strength changes.

These then are the essential points in our chain of reasoning: change of magnetic field → induced current → motion of charge → existence of an electric field.

Therefore: *a changing magnetic field is accompanied by an electric field.*

Thus we have found the two most important pillars of support for the theory of the electric and magnetic field. The first is the connection between the changing electric field and the magnetic field. It arose from Oersted's experiment on the deflection of a magnetic needle and led to the conclusion: *a changing electric field is accompanied by a magnetic field.*

The second connects the changing magnetic field with the induced current and arose from Faraday's experiment. Both formed a basis for quantitative description.

Again the electric field accompanying the changing magnetic field appears as something real. We had to imagine, previously, the magnetic field of a current existing without the testing pole. Similarly, we must claim here that the electric field exists without the wire testing the presence of an induced current.

In fact, our two-pillar structure could be reduced to only one, namely, to that based on Oersted's experiment. The result of Faraday's experiment could be deduced from this with the law of conservation of energy. We used the two-pillar structure only for the sake of clearness and economy.

One more consequence of the field description should be mentioned. There is a circuit through which a current flows, with for instance, a voltaic battery as the source of the current. The connection between the wire and the source of the current is suddenly broken. There is, of course, no current now! But during this short interruption an intricate process takes place, a process which could again have been foreseen by the field theory. Before the interruption of the current there was a magnetic field surrounding the wire. This ceased to exist the moment the current was interrupted. Therefore, through the interruption of a current, a magnetic field disappeared. The number of lines of force passing through the surface surrounded by the wire changed very rapidly. But such a rapid change, however it is produced, must create an induced current. What really matters is the change of the magnetic field making the induced current stronger if the change is greater. This consequence is another test for the theory. The disconnection of a current must be accompanied by the appearance of a strong, momentary induced current. Experiment again confirms the prediction. Anyone who has ever disconnected a current must have noticed that a spark appears. This spark reveals the strong potential differences caused by the rapid change of the magnetic field.

The same process can be looked at from a different point of view, that of energy. A magnetic field disappeared and a spark was created. A spark represents energy, therefore, so also must the magnetic field. To use the field concept and its language consistently, we must regard the magnetic field as a store of energy. Only in this way shall we be able to describe the electric and magnetic phenomena in accordance with the law of conservation of energy.

Starting as a helpful model the field became more and more real. It helped us to understand old facts and led us to new ones. The attribution of energy to the field is one step further in the development in which the field concept was stressed more and more, and the concepts of substances, so essential to the mechanical point of view, were more and more suppressed.

# QUANTA

*Continuity—discontinuity . . . Elementary quanta of matter and electricity . . . The quanta of light . . . Light spectra . . . The waves of matter . . . Probability waves . . . Physics and reality*

## CONTINUITY—DISCONTINUITY

A MAP of New York City and the surrounding country is spread before us. We ask: which points on this map can be reached by train? After looking up these points in a railway timetable, we mark them on the map. We now change our question and ask: which points can be reached by car? If we draw lines on the map representing all the roads starting from New York, every point on these roads can, in fact, be reached by car. In both cases we have sets of points. In the first they are separated from each other and represent the different railway stations, and in the second they are the points along the lines representing the roads. Our next question is about the distance of each of these points from New York, or, to be more rigorous, from a certain spot in that city. In the first case, certain numbers correspond to the points on our map. These numbers change by irregular, but always finite, leaps and bounds. We say: the distances from New York of the places which can be reached by train change only in a *discontinuous* way. Those of the places which can be reached by car, however, may change by steps as small as we wish, they can vary in a *continuous* way. The changes in distance can be made arbitrarily small in the case of a car, but not in the case of a train.

The output of a coal mine can change in a continuous way. The amount of coal produced can be decreased or increased by arbitrarily small steps. But the number of miners employed can change only discontinuously. It would be pure nonsense to say: "Since yesterday, the number of employees has increased by 3.783."

Asked about the amount of money in his pocket, a man can give a number containing only two decimals. A sum of money can change only by jumps, in a discontinuous way. In America the smallest permissible change or, as we shall call it, the "elementary quantum" for American money, is one cent. The elementary quantum for English money is one farthing, worth only half the American elementary quantum. Here we have an example of two elementary quanta whose mutual values can be compared. The ratio of their values has a definite sense since one of them is worth twice as much as the other.

We can say: some quantities can change continuously and others can change only discontinuously, by steps which cannot be further decreased. These indivisible steps are called the *elementary quanta* of the particular quantity to which they refer.

We can weigh large quantities of sand and regard its mass as continuous even though its granular structure is evident. But if the sand were to become very precious and the scales used very sensitive, we should have to consider the fact that the mass always changes by a multiple number of one grain. The mass of this one grain would be our elementary quantum. From this example we see how the discontinuous character of a quantity, so far regarded as continuous, can be detected by increasing the precision of our measurements.

If we had to characterize the principal idea of the quantum theory in one sentence, we could say: *it must be assumed that some physical quantities so far regarded as continuous are composed of elementary quanta.*

The region of facts covered by the quantum theory is tremendously great. These facts have been disclosed by the highly developed technique of modern experiment. As we can neither show nor describe even the basic experiments, we shall frequently have to quote their results dogmatically. Our aim is to explain the principal underlying ideas only.

## ELEMENTARY QUANTA OF MATTER AND ELECTRICITY

In the picture of matter drawn by the kinetic theory, all elements are built of molecules. Take the simplest case of the lightest element, that is hydrogen. . . . Its value is:

0.000 000 000 000 000 000 000 0033 grams.

This means that mass is discontinuous. The mass of a portion of hydrogen can change only by a whole number of small steps each corresponding to the mass of one hydrogen molecule. But chemical processes show that the hydrogen molecule can be broken up into two parts, or, in other words, that the hydrogen molecule is composed of two atoms. In chemical processes it is the atom and not the molecule which plays the role of an elementary quantum. Dividing the above number by two, we find the mass of a hydrogen atom. This is about

0.000 000 000 000 000 000 000 0017 grams.

Mass is a discontinuous quantity. But, of course, we need not bother about this when determining weight. Even the most sensitive scales are far from attaining the degree of precision by which the discontinuity in mass variation could be detected.

Let us return to a well-known fact. A wire is connected with the source of a current. The current is flowing through the wire from higher to lower potential. We remember that many experimental facts were explained by the simple theory of electric fluids flowing through the wire. We also remember that the decision as to whether the positive fluid flows from higher to lower potential, or the negative fluid flows from lower to higher potential, was merely a matter of convention. For the moment we disregard all the further progress resulting from the field concepts. Even when thinking in the simple terms of electric fluids, there still remain some questions to be settled. As the name "fluid" suggests, electricity was regarded, in the early days, as a continuous quantity. The amount of charge could be changed, according to these old views, by arbitrarily small steps. There was no need to assume elementary electric quanta. The achievements of the kinetic theory of matter prepared us for a new question: do elementary quanta of electric fluids exist? The other question to be settled is: does the current consist of a flow of positive, negative or perhaps of both fluids?

The idea of all the experiments answering these questions is to tear the electric fluid from the wire, to let it travel through empty space, to deprive it of any association with matter and then to investigate its properties, which must appear most clearly under these conditions. Many experiments of this kind were performed in the late nineteenth century. Before explaining the idea of these experimental arrangements, at least in one case, we shall quote the results. The electric fluid flowing through the wire is a negative one, directed, therefore, from lower to higher potential. Had we known this from the start, when the theory of electric fluids was first formed, we should certainly have interchanged the words, and called the electricity of the rubber rod positive, that of the glass rod negative. It would then have been more convenient to regard the flowing fluid as the positive one. Since our first guess was wrong we now have to put up with the inconvenience. The next important question is whether the structure of this negative fluid is "granular," whether or not it is composed of electric quanta. Again a number of independent experiments show that there is no doubt as to the existence of an elementary quantum of this negative electricity. The negative electric fluid is constructed of grains, just as the beach is composed of grains of sand, or a house built of bricks. This result was formulated most clearly by J. J. Thomson, about forty years ago. The elementary quanta of negative electricity are called *electrons*. Thus every negative electric charge is composed of a multitude of elementary charges represented by electrons. The negative charge can, like mass, vary only discontinuously. The elementary electric charge is, however, so small that in many investigations it is equally possible and sometimes even more convenient to regard it as a continuous quantity. Thus the atomic and electron theories introduce into science discontinuous physical quantities which can vary only by jumps.

Imagine two parallel metal plates in some place from which all air has been extracted. One of the plates has a positive, the other a negative charge. A positive test charge brought between the two plates will be repelled by the positively charged and attracted by the negatively charged plate. Thus the lines of force of the electric field will be directed from the positively to the negatively charged plate. A force acting on a negatively charged test body would have the opposite direction. If the plates are sufficiently large, the lines of force between them will be equally dense everywhere; it is immaterial where the test body is placed, the force and, therefore, the density of the lines of force

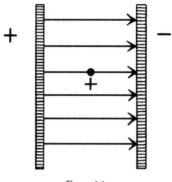

FIG. 11.

will be the same. Electrons brought somewhere between the plates would behave like raindrops in the gravitational field of the earth, moving parallel to each other from the negatively to the positively charged plate. There are many known experimental arrangements for bringing a shower of electrons into such a field which directs them all in the same way. One of the simplest is to bring a heated wire between the charged plates. Such a heated wire emits electrons which are afterwards directed by the lines of force of the external field. For instance, radio tubes, familiar to everyone, are based on this principle.

Many very ingenious experiments have been performed on a beam of electrons. The changes of their path in different electric and magnetic external fields have been investigated. It has even been possible to isolate a single electron and to determine its elementary charge and its mass, that is, its inertial resistance to the action of an external force. Here we shall only quote the value of the mass of an electron. It turned out to be about *two thousand times smaller* than the mass of a hydrogen atom. Thus the mass of a hydrogen atom, small as it is, appears great in comparison with the mass of an electron. From the point of view of a consistent field theory, the whole mass, that is, the whole energy, of an electron is the energy of its field; the bulk of its strength is within a very small sphere, and away from the "center" of the electron it is weak.

We said before that the atom of any element is its smallest elementary quantum. This statement was believed for a very long time. Now, however, it is no longer believed! Science has formed a new view showing the limitations of the old one. There is scarcely any statement in physics more firmly founded on facts than the one about the complex structure of the atom. First came the realization that the electron, the elementary quantum of the negative electric fluid, is also one of the components of the atom, one of the elementary bricks from which all matter is built. The previously quoted example of a heated wire emitting electrons is only one of the numerous instances of the extraction of these particles from matter. This result closely connecting the problem of the structure of matter with that of electricity follows, beyond any doubt, from very many independent experimental facts.

It is comparatively easy to extract from an atom some of the electrons from which it is composed. This can be done by heat, as in our example of a heated wire, or in a different way, such as by bombarding atoms with other electrons.

Suppose a thin, red-hot, metal wire is inserted into rarefied hydrogen. The wire will emit electrons in all directions. Under the action of a foreign electric field a given velocity will be imparted to them. An electron increases its velocity just as a stone falling in the gravitational field. By this method we can obtain a beam of electrons rushing along with a definite speed in a definite direction. Nowadays, we can reach velocities comparable to that of light by submitting electrons to the action of very strong fields. What happens, then, when a beam of electrons of a definite velocity impinges on the molecules of rarefied hydrogen? The impact of a sufficiently speedy electron will not only disrupt the hydrogen molecule into its two atoms but will also extract an electron from one of the atoms.

Let us accept the fact that electrons are constituents of matter. Then, an atom from which an electron has been torn out cannot be electrically neutral. If it was previously neutral, then it cannot be so now, since it is poorer by one elementary charge. That which remains must have a positive charge. Furthermore, since the mass of an electron is so much smaller than that of the lightest atom, we can safely conclude that by far the greater part of the mass of the atom is not represented by electrons but by the remainder of the elementary particles which are much heavier than the electrons. We call this heavy part of the atom its *nucleus*.

Modern experimental physics has developed methods of breaking up the nucleus of the atom, of changing atoms of one element into those of another, and of extracting from the nucleus the heavy elementary particles of which it is built. This chapter of physics, known as "nuclear physics," to which Rutherford contributed so much, is, from the experimental point of view, the most interesting. But a theory, simple in its fundamental ideas and connecting the rich variety of facts in the domain of nuclear physics, is still lacking. Since, in these pages, we are interested only in general physical ideas, we shall omit this chapter in spite of its great importance in modern physics.

## THE QUANTA OF LIGHT

Let us consider a wall built along the seashore. The waves from the sea continually impinge on the wall, wash away some of its surface, and retreat, leaving the way clear for the incoming waves. The mass of the wall decreases and we can ask how much is washed away in, say, one year. But now let us picture a different process. We want to diminish the mass of the wall by the same amount as previously but in a different way. We shoot at the wall and split it at the places where the bullets hit. The mass of the wall will be decreased and we can well imagine that the same reduction in mass is achieved in both cases. But from the appearance of the wall we could easily detect whether the continuous sea wave or the discontinuous shower of bullets has been acting. It will be helpful in understanding the phenomena which we are about to describe, to bear in mind the difference between sea waves and a shower of bullets.

We said, previously, that a heated wire emits electrons. Here we shall introduce another way of extracting electrons from metal. Homogeneous light, such as violet light, which is, as we know, light of a definite wave-length, is impinging on a metal surface. The light extracts electrons from the metal. The electrons are torn from the metal and a shower of them speeds along with a certain velocity. From the point of view of the energy principle we can say: the energy of light is partially transformed into the kinetic energy of expelled electrons. Modern experimental technique enables us to register these electron-bullets, to determine their velocity and thus their energy. This extraction of electrons by light falling upon metal is called the *photoelectric effect.*

Our starting point was the action of a homogeneous light wave, with some definite intensity. As in every experiment, we must now change our arrangements to see whether this will have any influence on the observed effect.

Let us begin by changing the intensity of the homogeneous violet light falling on the metal plate and note to what extent the energy of the emitted electrons depends upon the intensity of the light. Let us try to find the answer by reasoning instead of by experiment. We could argue: in the photoelectric effect a certain definite portion of the energy of radiation is transformed into energy of motion of the electrons. If we again illuminate the metal with light of the same wavelength but from a more powerful source, then the energy of the emitted electrons should be greater, since the radiation is richer in energy. We should, therefore, expect the velocity of the emitted electrons to increase if the intensity of the light increases. But experiment again contradicts our prediction. Once more we see that the laws of nature are not as we should like them to be. We have come upon one of the experiments which, contradicting our predictions, breaks the theory on which they were based. The actual experimental result is, from the point of view of the wave theory, astonishing. The observed electrons all have the same speed, the same energy, which does not change when the intensity of the light is increased.

This experimental result could not be predicted by the wave theory. Here again a new theory arises from the conflict between the old theory and experiment.

Let us be deliberately unjust to the wave theory of light, forgetting its great achievements, its splendid explanation of the bending of light around very small obstacles. With our attention focused on the photoelectric effect, let us demand from the theory an adequate explanation of this effect. Obviously, we cannot deduce from the wave theory the independence of the energy of electrons from the intensity of light by which they have been extracted from the metal plate. We shall, therefore, try another theory. We remember that Newton's corpuscular theory, explaining many of the observed phenomena of light, failed to account for the bending of light, which we are now deliberately disregarding. In Newton's time the concept of energy did not exist. Light corpuscles were, according to him, weightless; each color preserved its own substance character. Later, when the concept of energy was created and it was recognized that light carries energy, no one thought of applying these concepts to the corpuscular theory of light. Newton's theory was dead and, until our own century, its revival was not taken seriously.

To keep the principal idea of Newton's theory, we must assume that homogeneous light is composed of energy-grains and replace the old light corpuscles by light quanta, which we shall call *photons*, small portions of energy, traveling through empty space with the velocity of light. The revival of Newton's theory in this new form leads to the *quantum theory of light*. Not only matter and electric charge, but also energy of radiation has a granular structure, i.e., is built up of light quanta. In addition to quanta of matter and quanta of electricity there are also quanta of energy.

The idea of energy quanta was first introduced by Planck at the beginning of this century in order to explain some effects much more complicated than the photoelectric effect. But the photo-effect shows most clearly and simply the necessity for changing our old concepts.

It is at once evident that this quantum theory of light explains the photoelectric effect. A shower of photons is falling on a metal plate. The action between radiation and matter consists here of very many single processes in which a photon impinges on the atom and tears out an electron. These single processes are all alike and the extracted electron will have the same energy in every case. We also understand that increasing the intensity of the light means, in our new language, increasing the number of falling photons. In this case, a different number of electrons would be thrown out of the metal plate, but the energy of any single one would not change. Thus we see that this theory is in perfect agreement with observation.

What will happen if a beam of homogeneous light of a different color, say, red instead of violet, falls on the metal surface? Let us leave experiment to answer this question. The energy of the extracted electrons must be measured and compared with the energy of electrons thrown out by violet light. The energy of the electron extracted by red light turns out to be smaller than the energy of the electron extracted by violet light. This means that the energy of the light quanta is different for different colors. The photons belonging to the color red have half the energy of those belonging to the color violet. Or, more rigorously: the energy of a light quantum belonging to a homogeneous color decreases proportionally as the wave-length increases. There is an essential difference between quanta of energy and quanta of electricity. Light quanta differ for every wave-length, whereas quanta of electricity are always the same. If we were to use one of our previous analogies, we should compare light quanta to the smallest monetary quanta, differing in each country.

Let us continue to discard the wave theory of light and assume that the structure of light is granular and is formed by light quanta, that is, photons speeding through space with the velocity of light. Thus, in our new picture, light is a shower of photons, and the photon is the elementary quantum of light energy. If, however, the wave theory is discarded, the concept of a wave-length disappears. What new concept takes its place? The energy of the light quanta! Statements expressed in the terminology of the wave theory can be translated into statements of the quantum theory of radiation. For example:

| TERMINOLOGY OF THE WAVE THEORY | TERMINOLOGY OF THE QUANTUM THEORY |
|---|---|
| Homogeneous light has a definite wave-length. The wave-length of the red end of the spectrum is twice that of the violet end. | Homogeneous light contains photons of a definite energy. The energy of the photon for the red end of the spectrum is half that of the violet end. |

The state of affairs can be summarized in the following way: there are phenomena which can be explained by the quantum theory but not by the wave theory. Photo-effect furnishes an example, though other phenomena of this kind are known. There are phenomena which can be explained by the wave theory but not by the quantum theory. The bending of light around obstacles is a typical example. Finally, there are phenomena, such as the rectilinear propagation of light, which can be equally well explained by the quantum and the wave theory of light.

But what is light really? Is it a wave or a shower of photons? Once before we put a similar question when we asked: is light a wave or a shower of light corpuscles? At that time there was every reason for discarding the corpuscular theory of light and accepting the wave theory, which covered all phenomena. Now, however, the problem is much more complicated. There seems no likelihood of forming a consistent description of the phenomena of light by a choice of only one of the two possible languages. It seems as though we must use sometimes the one theory and sometimes the other, while at times we may use either. We are faced with a new kind of difficulty. We have two contradictory pictures of reality; separately neither of them fully explains the phenomena of light, but together they do!

How is it possible to combine these two pictures? How can we understand these two utterly different aspects of light? It is not easy to account for this new difficulty. Again we are faced with a fundamental problem.

For the moment let us accept the photon theory of light and try, by its help, to understand the facts so far explained by the wave theory. In this way we shall stress the difficulties which make the two theories appear, at first sight, irreconcilable.

We remember: a beam of homogeneous light passing through a pinhole gives light and dark rings. How is it possible to understand this phenomena by the help of the quantum theory of light, disregarding the wave theory? A photon passes through the hole. We could expect the screen to appear light if the photon passes through and dark if it does not. Instead, we find fight and dark rings. We could try to account for it as follows: perhaps there is some interaction between the rim of the hole and the photon which is responsible for the appearance of the diffraction rings. This sentence can, of course, hardly be regarded as an explanation. At best, it outlines a program for an explanation holding out at least some hope of a future understanding of diffraction by interaction between matter and photons.

But even this feeble hope is dashed by our previous discussion of another experimental arrangement. Let us take two pinholes. Homogeneous light passing through the two holes gives light and dark stripes on the screen. How is this effect to be understood from the point of view of the quantum theory of light? We could argue: a photon passes through either one of the two pinholes. If a photon of homogeneous light represents an elementary light particle, we can hardly imagine its division and its passage through the two holes. But then the effect should be exactly as in the first case, light and dark rings and not light and dark stripes. How is it possible then that the presence of another pinhole completely changes the effect? Apparently the hole through which the photon does not pass, even though it may be at a fair distance, changes the rings into stripes! If the photon behaves like a corpuscle in classical physics it must pass through one of the two holes.

But in this case, the phenomena of diffraction seem quite incomprehensible.

Science forces us to create new ideas, new theories. Their aim is to break down the wall of contradictions which frequently blocks the way of scientific progress. All the essential ideas in science were born in a dramatic conflict between reality and our attempts at understanding. Here again is a problem for the solution of which new principles are needed. Before we try to account for the attempts of modern physics to explain the contradiction between the quantum and the wave aspects of light, we shall show that exactly the same difficulty appears when dealing with quanta of matter instead of quanta of light.

# LIGHT SPECTRA

We already know that all matter is built of only a few kinds of particles. Electrons were the first elementary particles of matter to be discovered. But electrons are also the elementary quanta of negative electricity. We learned furthermore that some phenomena force us to assume that light is composed of elementary light quanta, differing for different wave-lengths. Before proceeding we must discuss some physical phenomena in which matter as well as radiation plays an essential role.

The sun emits radiation which can be split into its components by a prism. The continuous spectrum of the sun can thus be obtained. Every wave-length between the two ends of the visible spectrum is represented. Let us take another example. It was previously mentioned that sodium when incandescent emits homogeneous light, light of one color or one wave-length. If incandescent sodium is placed before the prism we see only one yellow line. In general, if a radiating body is placed before the prism, then the light it emits is split up into its components, revealing the spectrum characteristic of the emitting body.

The discharge of electricity in a tube containing gas produces a source of light such as seen in the neon tubes used for luminous advertisements. Suppose such a tube is placed before a spectroscope. The spectroscope is an instrument which acts like a prism, but with much greater accuracy and sensitiveness; it splits light into its components, that is, it analyzes it. Light from the sun, seen through a spectroscope, gives a continuous spectrum; all wave-lengths are represented in it. If, however, the source of light is a gas through which a current of electricity passes, the spectrum is of a different character. Instead of the continuous, multi-colored design of the sun's spectrum, bright, separated stripes appear on a continuous dark background. Every stripe, if it is very narrow, corresponds to a definite color or, in the language of the wave theory, to a definite wave-length. For example, if twenty lines are visible in the spectrum, each of them will be designated by one of twenty numbers expressing the corresponding wave-length. The vapors of the various elements possess different systems of lines, and thus different combinations of numbers designating the wave-lengths composing the emitted light spectrum. No two elements have identical systems of stripes in their characteristic spectra, just as no two persons have exactly identical fingerprints. As a catalogue of these lines was worked out by physicists, the existence of laws gradually became evident, and it was possible to replace some of the columns of seemingly disconnected numbers expressing the length of the various waves by one simple mathematical formula.

All that has just been said can now be translated into the photon language. The stripes correspond to certain definite wave-lengths or, in other words, to photons with a definite energy. Luminous gases do not, therefore, emit photons with all possible energies, but only those characteristic of the substance. Reality again limits the wealth of possibilities.

Atoms of a particular element, say, hydrogen, can emit only photons with definite energies. Only the emission of definite energy quanta is permissible, all others being prohibited. Imagine, for the sake of simplicity, that some element emits only one line, that is, photons of a quite definite energy. The atom is richer in energy before the emission and poorer afterwards. From the energy principle it must follow that the *energy level* of an atom is higher before emission and lower afterwards, and that the difference between the two levels must be equal to the energy of the emitted photon. Thus the fact that an atom of a certain element emits radiation of one wave-length only, that is photons of a definite energy only, could be expressed differently: only two energy levels are permissible in an atom of this element and the emission of a photon corresponds to the transition of the atom from the higher to the lower energy level.

But more than one line appears in the spectra of the elements, as a rule. The photons emitted correspond to many energies and not to one only. Or, in other words, we must assume that many energy levels are allowed in an atom and that the emission of a photon corresponds to the transition of the atom from a higher energy level to a lower one. But it is essential that not every energy level should be permitted, since not every wave-length, not every photon-energy, appears in the spectra of an element. Instead of saying that some definite lines, some definite wave-lengths, belong to the spectrum of every atom, we can say that every atom has some definite energy levels, and that the emission of light quanta is associated with the transition of the atom from one energy level to another. The energy levels are, as a rule, not continuous but discontinuous. Again we see that the possibilities are restricted by reality.

It was Bohr who showed for the first time why just these and no other lines appear in the spectra. His theory, formulated twenty-five years ago, draws a picture of the atom from which, at any rate in simple cases, the spectra of the elements can be calculated and the apparently dull and unrelated numbers are suddenly made coherent in the light of the theory.

Bohr's theory forms an intermediate step toward a deeper and more general theory, called the wave or quantum mechanics. It is our aim in these last pages to characterize the principal ideas of this theory. Before doing so, we must mention one more theoretical and experimental result of a more special character.

Our visible spectrum begins with a certain wave-length for the violet color and ends with a certain wave-length for the red color. Or, in other words, the energies of the photons in the visible spectrum are always enclosed within the limits formed by the photon energies of the violet and red lights. This limitation is, of course, only a property of the human eye. If the difference in energy of some of the energy levels is sufficiently great, then an *ultraviolet* photon will be sent out, giving a line beyond the visible spectrum. Its presence cannot be detected by the naked eye; a photographic plate must be used.

X rays are also composed of photons of a much greater energy than those of visible light, or in other words, their wave-lengths are much smaller, thousands of times smaller in fact, than those of visible light.

But is it possible to determine such small wavelengths experimentally? It was difficult enough to do so for ordinary light. We had to have small obstacles or small apertures. Two pinholes very near to each other, showing diffraction for ordinary light, would have to be many thousands of times smaller and closer together to show diffraction for X rays.

How then can we measure the wave-lengths of these rays? Nature herself comes to our aid.

A crystal is a conglomeration of atoms arranged at very short distances from each other on a perfectly regular plan. Our drawing shows a simple model of the structure of a crystal. Instead of minute apertures, there are extremely small obstacles formed by the atoms of the element, arranged very close to each other in absolutely regular order.

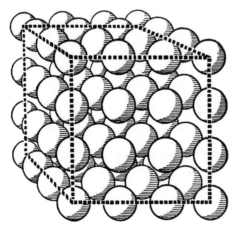

FIG. 12.

The distances between the atoms, as found from the theory of the crystal structure, are so small that they might be expected to show the effect of diffraction for X rays. Experiment proved that it is, in fact, possible to diffract the X-ray wave by means of these closely packed obstacles disposed in the regular three-dimensional arrangement occurring in a crystal.

Suppose that a beam of X rays falls upon a crystal and, after passing through it, is recorded on a photographic plate. The plate then shows the diffraction pattern. Various methods have been used to study the X-ray spectra, to deduce data concerning the wave-length from the diffraction pattern. What has been said here in a few words would fill volumes if all theoretical and experimental details were set forth. In Plate III (page 317) we give only one diffraction pattern obtained by one of the various methods. We again see the dark and light rings so characteristic of the wave theory. In the center the non-diffracted ray is visible. If the crystal were not brought between the X rays and the photographic plate, only the light spot in the center would be seen. From photographs of this kind the wave-lengths of the X-ray spectra can be calculated and, on the other hand, if the wave-length is known, conclusions can be drawn about the structure of the crystal.

## THE WAVES OF MATTER

How can we understand the fact that only certain characteristic wave-lengths appear in the spectra of the elements?

It has often happened in physics that an essential advance was achieved by carrying out a consistent analogy between apparently unrelated phenomena. In these pages we have often seen how ideas created and developed in one branch of science were afterwards successfully applied to another. The development of the mechanical and field views gives many examples of this kind. The association of solved problems with those unsolved may throw new light on our difficulties by suggesting new ideas. It is easy to find a superficial analogy which really expresses nothing. But to discover some essential common features,

## PLATE III

*(Photographed by A. G. Shenstone)*
### Spectral lines.

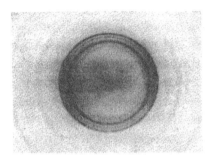

*(Photographed by Lastowiecki and Gregor)*
### Diffraction of X rays.

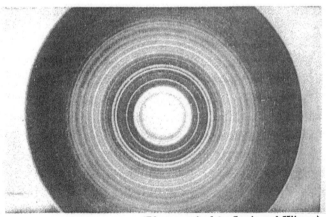

*(Photographed by Loria and Klinger)*
### Diffraction of electronic waves.

FIG. 13.

hidden beneath a surface of external differences, to form, on this basis, a new successful theory, is important creative work. The development of the so-called wave mechanics, begun by de Broglie and Schrödinger, less than fifteen years ago, is a typical example of the achievement of a successful theory by means of a deep and fortunate analogy.

317

Our starting point is a classical example having nothing to do with modern physics. We take in our hand the end of a very long flexible rubber tube, or a very long spring, and try to move it rhythmically up and down, so that the end oscillates. Then, as we have seen in many other examples, a wave is created by the oscillation which spreads through the tube with a certain velocity. If we imagine an infi-

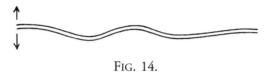

FIG. 14.

nitely long tube, then the portions of waves, once started, will pursue their endless journey without interference.

Now another case. The two ends of the same tube are fastened. If preferred, a violin string may be used. What happens now if a wave is created at one end of the rubber tube or cord? The wave begins its journey as in the previous example, but it is soon reflected by the other end of the tube. We now have two waves: one created by oscillation, the other by reflection; they travel in opposite directions and interfere with each other. It would not be difficult to trace the interference of the two waves and discover the one wave resulting from their super-position; it is called the *standing wave*. The two words "standing" and "wave" seem to contradict each other; their combination is, neverthe-less, justified by the result of the superposition of the two waves.

The simplest example of a standing wave is the motion of a cord with the two ends fixed, an up-and-down motion, as shown in our drawing

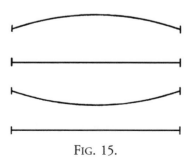

FIG. 15.

318

(figure 15). This motion is the result of one wave lying on the other when the two are traveling in opposite directions. The characteristic feature of this motion is: only the two end points are at rest. They are called *nodes*. The wave stands, so to speak, between the two nodes, all points of the cord reaching simultaneously the maxima and minima of their deviation.

But this is only the simplest kind of a standing wave. There are others. For example, a standing wave can have three nodes, one at each end and one in the center. In this case three points are always at rest. A glance at figures 16 and 17 shows that here the wave-length is half as

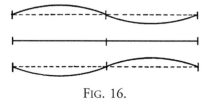

FIG. 16.

great as the one with two nodes. Similarly, standing waves can have four, five, and more nodes. The wave-length in each case will depend on the number of nodes. This number can only be an integer and can

FIG. 17.

change only by jumps. The sentence, "the number of nodes in a standing wave is 3.576," is pure nonsense. Thus the wave-length can only change discontinuously. Here, in this most classical problem, we recognize the familiar features of the quantum theory. The standing wave produced by a violin player is, in fact, still more complicated, being a mixture of very many waves with two, three, four, five, and more nodes and, therefore, a mixture of several wave-lengths. Physics can analyze such a mixture into the simple standing waves from which it is composed. Or, using our previous terminology, we could say that the oscillating string has its spectrum, just as an element emitting radiation. And, in the same way as for the spectrum of an element, only certain wave-lengths are allowed, all others being prohibited.

We have thus discovered some similarity between the oscillating cord and the atom emitting radiation. Strange as this analogy may seem, let us draw further conclusions from it and try to proceed with the comparison, once having chosen it. The atoms of every element are composed of elementary particles, the heavier constituting the nucleus, and the lighter the electrons. Such a system of particles behaves like a small acoustical instrument in which standing waves are produced.

Yet the standing wave is the result of interference between two or, generally, even more moving waves. If there is some truth in our analogy, a still simpler arrangement than that of the atom should correspond to a spreading wave. What is the simplest arrangement? In our material world, nothing can be simpler than an electron, an elementary particle, on which no forces are acting, that is, an electron at rest or in uniform motion. We could guess a further link in the chain of our analogy: electron moving uniformly → waves of a definite length. This was de Broglie's new and courageous idea.

It was previously shown that there are phenomena in which light reveals its wave-like character and others in which light reveals its corpuscular character. After becoming used to the idea that light is a wave, we found, to our astonishment, that in some cases, for instance in the photoelectric effect, it behaves like a shower of photons. Now we have just the opposite state of affairs for electrons. We accustomed ourselves to the idea that electrons are particles, elementary quanta of electricity and matter. Their charge and mass were investigated. If there is any truth in de Broglie's idea, then there must be some phenomena in which matter reveals its wave-like character. At first, this conclusion, reached by following the acoustical analogy, seems strange and incomprehensible. How can a moving corpuscle have anything to do with a wave? But this is not the first time we have faced a difficulty of this kind in physics. We met the same problem in the domain of light phenomena.

Fundamental ideas play the most essential role in forming a physical theory. Books on physics are full of complicated mathematical formulae. But thought and ideas, not formulae, are the beginning of every physical theory. The ideas must later take the mathematical form of a quantitative theory, to make possible the comparison with experiment. This can be explained by the example of the problem with which we are now dealing. The principal guess is that the uniformly moving electron will behave, in some phenomena, like a wave. Assume that an electron or a shower of electrons, provided they all have the same velocity, is moving uniformly. The mass, charge, and velocity of each individual electron is known. If we wish to associate in some way a wave concept with a uniformly moving electron or electrons, our next question must be: what is the wave-length? This is a quantitative question and a more or less quantitative theory must be built up to answer it. This is indeed a simple matter. The mathematical simplicity of de Broglie's work, providing an answer to this question, is most astonishing. At the time his work was done, the mathematical technique of other physical theories was very subtle and complicated, comparatively speaking. The mathematics dealing with the problem of waves of matter is extremely simple and elementary but the fundamental ideas are deep and far-reaching.

Previously, in the case of light waves and photons, it was shown that every statement formulated in the wave language can be translated into the language of photons or light corpuscles. The same is true for electronic waves. For uniformly moving electrons, the corpuscular language is already known. But every statement expressed in the corpuscular language can be translated into the wave language, just as in the case of photons. Two clews laid down the rules of translation. The analogy between light waves and electronic waves or photons and electrons is one clew. We try to use the same method of translation for matter as for light. The special relativity theory furnished the other clew. The laws of nature must be invariant with respect to the Lorentz and not to the classical transformation. These two clews together determine the wave-length corresponding to a moving electron. It follows from the theory that an electron moving with a velocity of, say, 10,000 miles per second, has a wave-length which can be easily calculated, and which turns out to lie in the same region as the X-ray wave-lengths. Thus we conclude further that if the wave character of matter can be detected, it should be done experimentally in an analogous way to that of X rays.

Imagine an electron beam moving uniformly with a given velocity, or, to use the wave terminology, a homogeneous electronic wave, and assume that it falls on a very thin crystal, playing the part of a diffraction grating. The distances between the diffracting obstacles in the crystal are so small that diffraction for X rays can be produced. One might expect a similar effect for electronic waves with the same order of wave-length. A photographic plate would register this diffraction of electronic waves passing through the thin layer of crystal. Indeed, the experiment produces what is undoubtedly one of the great achievements of the theory: the phenomenon of diffraction for electronic waves. The similarity between the diffraction of an electronic wave and that of an X ray is particularly marked as seen from a comparison of the patterns in Plate III. We know that such pictures enable us to determine the wave-lengths of X rays. The same holds good for electronic waves. The diffraction pattern gives the length of a wave of matter and the perfect quantitative agreement between theory and experiment confirms the chain of our argument splendidly.

Our previous difficulties are broadened and deepened by this result. This can be made clear by an example similar to the one given for a light wave. An electron shot at a very small hole will bend like a light wave. Light and dark rings appear on the photographic plate. There may be some hope of explaining this phenomenon by the interaction between the electron and the rim, though such an explanation does not seem to be very promising. But what about the two pinholes? Stripes appear instead of rings. How is it possible that the presence of the other hole completely changes the effect? The electron is indivisible and can, it would seem, pass through only one of the two holes. How could an electron passing through a hole possibly know that another hole has been made some distance away?

We asked before: what is light? Is it a shower of corpuscles or a wave? We now ask: what is matter, what is an electron? Is it a particle or a wave? The electron behaves like a particle when moving in an external electric or magnetic field. It behaves like a wave when diffracted by a crystal. With the elementary quanta of matter we came across the same difficulty that we met with in the light quanta. One of the most fundamental questions raised by recent advance in science is how to reconcile the two contradictory views of matter and wave. It is one of those fundamental difficulties which, once formulated, must lead, in the long run, to scientific progress. Physics has tried to solve this problem. The future must decide whether the solution suggested by modern physics is enduring or temporary.

## PROBABILITY WAVES

If, according to classical mechanics, we know the position and velocity of a given material point and also what external forces are acting, we can predict, from the mechanical laws, the whole of its future path. The sentence: "The material point has such-and-such position and velocity at such-and-such an instant," has a definite meaning in classical mechanics. If this statement were to lose its sense, our argument about foretelling the future path would fail.

In the early nineteenth century, scientists wanted to reduce all physics to simple forces acting on material particles that have definite positions and velocities at any instant. Let us recall how we described motion when discussing mechanics at the beginning of our journey through the realm of physical problems. We drew points along a definite path showing the exact positions of the body at certain instants and then tangent vectors showing the direction and magnitude of the velocities. This was both simple and convincing. But it cannot be repeated for our elementary quanta of matter, that is electrons, or for quanta of energy, that is photons. We cannot picture the journey of a photon or electron in the way we imagined motion in classical mechanics. The example of the two pinholes shows this clearly. Electron and photon seem to pass through the two holes. It is thus impossible to explain the effect by picturing the path of an electron or a photon in the old classical way.

We must, of course, assume the presence of elementary actions, such as the passing of electrons or photons through the holes. The existence of elementary quanta of matter and energy cannot be doubted. But the elementary laws certainly cannot be formulated by specifying positions and velocities at any instant in the simple manner of classical mechanics.

Let us, therefore, try something different. Let us continually repeat the same elementary processes. One after the other, the electrons are sent in the direction of the pinholes. The word "electron" is used here for the sake of definiteness; our argument is also valid for photons.

The same experiment is repeated over and over again in exactly the same way; the electrons all have the same velocity and move in the direction of the two pinholes. It need hardly be mentioned that this is an idealized experiment which cannot be carried out in reality but may well be imagined. We cannot shoot out single photons or electrons at given instants, like bullets from a gun.

The outcome of repeated experiments must again be dark and light rings for one hole and dark and light stripes for two. But there is one essential difference. In the case of one individual electron, the experimental result was incomprehensible. It is more easily understood when the experiment is repeated many times. We can now say: light stripes appear where many electrons fall. The stripes become darker at the place where fewer electrons are falling. A completely dark spot means that there are no electrons. We are not, of course, allowed to assume that all the electrons pass through one of the holes. If this were so it could not make the slightest difference whether or not the other is covered. But we already know that covering the second hole does make a difference. Since one particle is indivisible we cannot imagine that it passes through both the holes. The fact that the experiment was repeated many times points to another way out. Some of the electrons may pass through the first hole and others through the second. We do not know why individual electrons choose particular holes, but the net result of repeated experiments must be that both pinholes participate in transmitting the electrons from the source to the screen. If we state only what happens to the crowd of elecrons when the experiment is repeated, not bothering about the behavior of individual particles, the difference between the ringed and the striped pictures becomes comprehensible. By the discussion of a sequence of experiments a new idea was born, that of a crowd with the individuals behaving in an unpredictable way. We cannot foretell the course of one single electron, but we can predict that, in the net result, the light and dark stripes will appear on the screen.

Let us leave quantum physics for the moment.

We have seen in classical physics that if we know the position and velocity of a material point at a certain instant and the forces acting upon it, we can predict its future path. We also saw how the mechanical point of view was applied to the kinetic theory of matter. But in this theory a new idea arose from our reasoning. It will be helpful in understanding later arguments to grasp this idea thoroughly.

There is a vessel containing gas. In attempting to trace the motion of every particle one would have to commence by finding the initial states, that is, the initial positions and velocities of all the particles. Even if this were possible, it would take more than a human lifetime to set down the result on paper, owing to the enormous number of particles which would have to be considered. If one then tried to employ the known methods of classical mechanics for calculating the final positions of the particles, the difficulties would be insurmountable. In principle, it is possible to use the method applied for the motion of planets, but in practice this is useless and must give way to the *method of statistics*. This method dispenses with any exact knowledge of initial states. We know less about the system at any given moment and are thus less able to say anything about its past or future. We become indifferent to the fate of the individual gas particles. Our problem is of a different nature. For example: we do not ask, "What is the speed of every particle at this moment?" But we may ask: "How many particles have a speed between 1000 and 1100 feet per second?" We care nothing for individuals. What we seek to determine are average values typifying the whole aggregation. It is clear that there can be some point in a statistical method of reasoning only when the system consists of a large number of individuals.

By applying the statistical method we cannot foretell the behavior of an individual in a crowd. We can only foretell the chance, *the probability*, that it will behave in some particular manner. If our statistical laws tell us that one-third of the particles have a speed between 1000 and 1100 feet per second, it means that by repeating our observations for many particles, we shall really obtain this average, or in other words, that the probability of finding a particle within this limit is equal to one-third.

Similarly, to know the birth rate of a great community does not mean knowing whether any particular family is blessed with a child. It means a knowledge of statistical results in which the contributing personalities play no role.

By observing the registration plates of a great many cars we can soon discover that one-third of their numbers are divisible by three. But we cannot foretell whether the car which will pass in the next moment will have this property. Statistical laws can be applied only to big aggregations, but not to their individual members.

We can now return to our quantum problem.

The laws of quantum physics are of a statistical character. This means: they concern not one single system but an aggregation of identical systems; they cannot be verified by measurement of one individual, but only by a series of repeated measurements.

Radioactive disintegration is one of the many events for which quantum physics tries to formulate laws governing the spontaneous transmutation from one element to another. We know, for example, that in 1600 years half of one gram of radium will disintegrate, and half will remain. We can foretell approximately how many atoms will disintegrate during the next half-hour, but we cannot say, even in our theoretical descriptions, why just these particular atoms are doomed. According to our present knowledge, we have no power to designate the individual atoms condemned to disintegration. The fate of an atom does not depend on its age. There is not the slightest trace of a law governing their individual behavior. Only statistical laws can be formulated, laws governing large aggregations of atoms.

Take another example. The luminous gas of some element placed before a spectroscope shows lines of definite wave-length. The appearance of a discontinuous set of definite wave-lengths is characteristic of the atomic phenomena in which the existence of elementary quanta is revealed. But there is still another aspect of this problem. Some of the spectrum lines are very distinct, others are fainter. A distinct line means that a comparatively large number of photons belonging to this particular wave-length are emitted; a faint line means that a comparatively small number of photons belonging to this wave-length are emitted. Theory again gives us statements of a statistical nature only. Every line corresponds to a transition from higher to lower energy level. Theory tells us only about the probability of each of these possible transitions, but nothing about the actual transition of an individual atom. The theory works splendidly because all these phenomena involve large aggregations and not single individuals.

It seems that the new quantum physics resembles somewhat the kinetic theory of matter, since both are of a statistical nature and both refer to great aggregations. But this is not so! In this analogy an understanding not only of the similarities but also of the differences is most important. The similarity between the kinetic theory of matter and quantum physics lies chiefly in their statistical character. But what are the differences?

If we wish to know how many men and women over the age of twenty live in a city, we must get every citizen to fill out a form under the headings: "male," "female," and "age." Provided every answer is correct, we can obtain, by counting and segregating them, a result of a statistical nature. The individual names and addresses on the forms are of no account. Our statistical view is gained by the knowledge of individual cases. Similarly, in the kinetic theory of matter, we have statistical laws governing the aggregation, gained on the basis of individual laws.

But in quantum physics the state of affairs is entirely different. Here the statistical laws are given immediately. The individual laws are discarded. In the example of a photon or an electron and two pin-holes we have seen that we cannot describe the possible motion of elementary particles in space and time as we did in classical physics. Quantum physics abandons individual laws of elementary particles and states *directly* the statistical laws governing aggregations. It is impossible, on the basis of quantum physics, to describe positions and velocities of an elementary particle or to predict its future path as in classical physics. Quantum physics deals only with aggregations, and its laws are for crowds and not for individuals.

It is hard necessity and not speculation or a desire for novelty which forces us to change the old classical view. The difficulties of applying the old view have been outlined for one instance only, that of diffraction phenomena. But many others, equally convincing, could be quoted. Changes of view are continually forced upon us by our attempts to understand reality. But it always remains for the future to decide whether we chose the only possible way out and whether or not a better solution of our difficulties could have been found.

We have had to forsake the description of individual cases as objective happenings in space and time; we have had to introduce laws of a statistical nature. These are the chief characteristics of modern quantum physics.

Previously, when introducing new physical realities, such as the electromagnetic and gravitational field, we tried to indicate in general terms the characteristic features of the equations through which the ideas have been mathematically formulated. We shall now do the same with quantum physics, referring only very briefly to the work of Bohr, De Broglie, Schrödinger, Heisenberg, Dirac and Born.

Let us consider the case of one electron. The electron may be under the influence of an arbitrary foreign electromagnetic field, or free from all external influences. It may move, for instance, in the field of an atomic nucleus or it may diffract on a crystal. Quantum physics teaches us how to formulate the mathematical equations for any of these problems.

We have already recognized the similarity between an oscillating cord, the membrane of a drum, a wind instrument, or any other acoustical instrument on the one hand, and a radiating atom on the other. There is also some similarity between the mathematical equations governing the acoustical problem and those governing the problem of quantum physics. But again the physical interpretation of the quantities determined in these two cases is quite different. The physical quantities describing the oscillating cord and the radiating atom have quite a different meaning, despite some formal likeness in the equations. In the case of the cord, we ask about the deviation of an arbitrary point from its normal position at an arbitrary moment. Knowing the form of the oscillating cord at a given instant, we know everything we wish. The deviation from the normal can thus be calculated for any other moment from the mathematical equations for the oscillating cord. The fact that some definite deviation from the normal position corresponds to every point of the cord is expressed more rigorously as follows: for any instant, the deviation from the normal value is a *function* of the co-ordinates of the cord. All points of the cord form a one-dimensional continuum, and the deviation is a function defined in this one-dimensional continuum, to be calculated from the equations of the oscillating cord.

Analogously, in the case of an electron a certain function is determined for any point in space and for any moment. We shall call this function the *probability wave*. In our analogy the probability wave corresponds to the deviation from the normal position in the acoustical problem. The probability wave is, at a given instant, a function of a three-dimensional continuum, whereas, in the case of the cord the deviation was, at a given moment, a function of the one-dimensional continuum. The probability wave forms the catalogue of our knowledge of the quantum system under consideration and will enable us to answer all sensible statistical questions concerning this system. It does not tell us the position and velocity of the electron at any moment because such a question has no sense in quantum physics. But it will tell us the probability of meeting the electron on a particular spot, or where we have the greatest chance of meeting an electron. The result does not refer to one, but to many repeated measurements. Thus the equations of quantum physics determine the probability wave just as Maxwell's equations determine the electromagnetic field and the gravitational equations determine the gravitational field. The laws of quantum physics are again structure laws. But the meaning of physical concepts determined by these equations of quantum physics is much more abstract than in the case of electromagnetic and gravitational fields; they provide only the mathematical means of answering questions of a statistical nature.

So far we have considered the electron in some external field. If it were not the electron, the smallest possible charge, but some respectable charge containing billions of electrons, we could disregard the whole quantum theory and treat the problem according to our old pre-quantum physics. Speaking of currents in a wire, of charged conductors, of electromagnetic waves, we can apply our old simple physics contained in Maxwell's equations. But we cannot do this when speaking of the photoelectric effect, intensity of spectral lines, radioactivity, diffraction of electronic waves and many other phenomena in which the quantum character of matter and energy is revealed. We must then, so to speak, go one floor higher. Whereas in classical physics we spoke of positions and velocities of one particle, we must now consider probability waves, in a three-dimensional continuum corresponding to this one-particle problem.

Quantum physics gives its own prescription for treating a problem if we have previously been taught how to treat an analogous problem from the point of view of classical physics.

For one elementary particle, electron or photon, we have probability waves in a three-dimensional continuum, characterizing the statistical behavior of the system if the experiments are often repeated. But what about the case of not one but two interacting particles, for instance, two electrons, electron and photon, or electron and nucleus? We cannot treat them separately and describe each of them through a probability wave in three dimensions, just because of their mutual interaction. Indeed, it is not very difficult to guess how to describe in quantum physics a system composed of two interacting particles. We have to descend one floor, to return for a moment to classical physics. The position of two material points in space, at any moment, is characterized by six numbers, three for each of the points. All possible positions of the two material points form a six-dimensional continuum and not a three-dimensional one as in the case of one point. If we now again ascend one floor, to quantum physics, we shall have probability waves in a six-dimensional continuum and not in a three-dimensional continuum as in the case of one particle. Similarly, for three, four, and more particles the probability waves will be functions in a continuum of nine, twelve, and more dimensions.

This shows clearly that the probability waves are more abstract than the electromagnetic and gravitational field existing and spreading in our three-dimensional space. The continuum of many dimensions forms the background for the probability waves, and only for one particle does the number of dimensions equal that of physical space. The only physical significance of the probability wave is that it enables us to answer sensible statistical questions in the case of many particles as well as of one. Thus, for instance, for one electron we could ask about the probability of meeting an electron in some particular spot. For two particles our question could be: what is the probability of meeting the two particles at two definite spots at a given instant?

Our first step away from classical physics was abandoning the description of individual cases as objective events in space and time. We were forced to apply the statistical method provided by the probability waves. Once having chosen this way, we are obliged to go further toward abstraction. Probability waves in many dimensions corresponding to the many-particle problem must be introduced.

Let us, for the sake of briefness, call everything except quantum physics, classical physics. Classical and quantum physics differ radically. Classical physics aims at a description of objects existing in space, and the formulation of laws governing their changes in time. But the phenomena revealing the particle and wave nature of matter and radiation, the apparently statistical character of elementary events such as radioactive disintegration, diffraction, emission of spectral lines, and

SELECTIONS FROM *THE EVOLUTION OF PHYSICS*

many others, forced us to give up this view. Quantum physics does not aim at the description of individual objects in space and their changes in time. There is no place in quantum physics for statements such as: "This object is so-and-so, has this-and-this property." Instead we have statements of this kind: "There is such-and-such a probability that the individual object is so-and-so and has this-and-this property." There is no place in quantum physics for laws governing the changes in time of the individual object. Instead, we have laws governing the changes in time of the probability. Only this fundamental change, brought into physics by the quantum theory, made possible an adequate explanation of the apparently discontinuous and statistical character of events in the realm of phenomena in which the elementary quanta of matter and radiation reveal their existence.

Yet new, still more difficult problems arise which have not been definitely settled as yet. We shall mention only some of these unsolved problems. Science is not and will never be a closed book. Every important advance brings new questions. Every development reveals, in the long run, new and deeper difficulties.

We already know that in the simple case of one or many particles we can rise from the classical to the quantum description, from the objective description of events in space and time to probability waves. But we remember the all-important field concept in classical physics. How can we describe interaction between elementary quanta of matter and field? If a probability wave in thirty dimensions is needed for the quantum description of ten particles, then a probability wave with an infinite number of dimensions would be needed for the quantum description of a field. The transition from the classical field concept to the corresponding problem of probability waves in quantum physics is a very difficult step. Ascending one floor is here no easy task and all attempts so far made to solve the problem must be regarded as unsatisfactory. There is also one other fundamental problem. In all our arguments about the transition from classical physics to quantum physics we used the old prerelativistic description in which space and time are treated differently. If, however, we try to begin from the classical description as proposed by the relativity theory, then our ascent to the quantum problem seems much more complicated. This is another problem tackled by modern physics, but still far from a complete and satisfactory solution. There is still a further difficulty in forming a consistent physics for heavy particles, constituting the nuclei. In spite of the many experimental data and the many attempts to throw light on the nuclear problem, we are still in the dark about some of the most fundamental questions in this domain.

There is no doubt that quantum physics explained a very rich variety of facts, achieving, for the most part, splendid agreement between theory and observation. The new quantum physics removes us still further from the old mechanical view, and a retreat to the former position seems, more than ever, unlikely. But there is also no doubt that quantum physics must still be based on the two concepts: matter and field. It is, in this sense, a dualistic theory and does not bring our old problem of reducing everything to the field concept even one step nearer realization.

Will the further development be along the line chosen in quantum physics, or is it more likely that new revolutionary ideas will be introduced into physics? Will the road of advance again make a sharp turn, as it has so often done in the past?

During the last few years all the difficulties of quantum physics have been concentrated around a few principal points. Physics awaits their solution impatiently. But there is no way of foreseeing when and where the clarification of these difficulties will be brought about.

## PHYSICS AND REALITY

What are the general conclusions which can be drawn from the development of physics indicated here in a broad outline representing only the most fundamental ideas?

Science is not just a collection of laws, a catalogue of unrelated facts. It is a creation of the human mind, with its freely invented ideas and concepts. Physical theories try to form a picture of reality and to establish its connection with the wide world of sense impressions. Thus the only justification for our mental structures is whether and in what way our theories form such a link.

We have seen new realities created by the advance of physics. But this chain of creation can be traced back far beyond the starting point of physics. One of the most primitive concepts is that of an object. The concepts of a tree, a horse, any material body, are creations gained on the basis of experience, though the impressions from which they arise are primitive in comparison with the world of physical phenomena. A cat teasing a mouse also creates, by thought, its own primitive reality. The fact that the cat reacts in a similar way toward any mouse it meets shows that it forms concepts and theories which are its guide through its own world of sense impressions.

"Three trees" is something different from "two trees." Again "two trees" is different from "two stones." The concepts of the pure numbers 2, 3, 4 . . ., freed from the objects from which they arose, are creations of the thinking mind which describe the reality of our world.

The psychological subjective feeling of time enables us to order our impressions, to state that one event precedes another. But to connect every instant of time with a number, by the use of a clock, to regard time as a one-dimensional continuum, is already an invention. So also are the concepts of Euclidean and non-Euclidean geometry, and our space understood as a three-dimensional continuum.

Physics really began with the invention of mass, force, and an inertial system. These concepts are all free inventions. They led to the formulation of the mechanical point of view. For the physicist of the early nineteenth century, the reality of our outer world consisted of particles with simple forces acting between them and depending only on the distance. He tried to retain as long as possible his belief that he would succeed in explaining all events in nature by these fundamental concepts of reality. The difficulties connected with the deflection of the magnetic needle, the difficulties connected with the structure of the ether, induced us to create a more subtle reality. The important invention of the electromagnetic field appears. A courageous scientific imagination was needed to realize fully that not the behavior of bodies, but the behavior of something between them, that is, the field, may be essential for ordering and understanding events.

Later developments both destroyed old concepts and created new ones. Absolute time and the inertial co-ordinate system were abandoned by the relativity theory. The background for all events was no longer the one-dimensional time and the three-dimensional space continuum, but the four-dimensional time-space continuum, another free invention, with new transformation properties. The inertial co-ordinate system was no longer needed. Every co-ordinate system is equally suited for the description of events in nature.

The quantum theory again created new and essential features of our reality. Discontinuity replaced continuity. Instead of laws governing individuals, probability laws appeared.

The reality created by modern physics is, indeed, far removed from the reality of the early days. But the aim of every physical theory still remains the same.

With the help of physical theories we try to find our way through the maze of observed facts, to order and understand the world of our sense impressions. We want the observed facts to follow logically from our concept of reality. Without the belief that it is possible to grasp the reality with our theoretical constructions, without the belief in the inner harmony of our world, there could be no science. This belief is and always will remain the fundamental motive for all scientific creation. Throughout all our efforts, in every dramatic struggle between old and new views, we recognize the eternal longing for understanding, the ever-firm belief in the harmony of our world, continually strengthened by the increasing obstacles to comprehension.

# WE SUMMARIZE

*Again the rich variety of facts in the realm of atomic phenomena forces us to invent new physical concepts. Matter has a granular structure; it is composed of elementary particles, the elementary quanta of matter. Thus, the electric charge has a granular structure and—most important from the point of view of the quantum theory—so has energy. Photons are the energy quanta of which light is composed.*

*Is light a wave or a shower of photons? Is a beam of electrons a shower of elementary particles or a wave? These fundamental questions are forced upon physics by experiment. In seeking to answer them we have to abandon the description of atomic events as happenings in space and time, we have to retreat still further from the old mechanical view. Quantum physics formulates laws governing crowds and not individuals. Not properties but probabilities are described, not laws disclosing the future of systems are formulated, but laws governing the changes in time of the probabilities and relating to great congregations of individuals.*

# Autobiographical Notes

A lbert Einstein famously said of himself, "Do not worry about your difficulties in mathematics. I can assure you mine are still greater." Though modest about his own abilities and often caricatured as a poor student (though in reality, simply a willful one), Einstein showed an unusually intense curiosity about the natural world, and a drive to learn as much of the scientific and mathematical cannon as possible. In his "Autobiographical Notes," Einstein presents his own very unusual scientific history, unusual if for no other reason than that it is replete with equations.

This work, perhaps more than any other in the volume, gives us insight into why Einstein is the icon that he is. In describing his own education, Einstein gives us a guided tour of the state of science in his youth. By gradually describing both his contributions and those of others in relativity and quantum mechanics, we begin to see how much the world of physics was revolutionized over his lifetime.

While only twelve, Einstein first read Euclid's "Elements," which he called the "Holy Little Geometry Book." He was awed by the idea that with a few simple principles, one could derive proofs pertaining to the real universe. He spent the rest of his life in pursuit of these proofs, though he was occasionally confounded when his intuition contradicted what could be or was observed. For example, Euclid's theory of geometry formed the foundation of our understanding of the physical universe. By starting with an assumption that physics is the same for all observers and that time flows at a constant rate, Sir Isaac Newton's theory of mechanics could be directly inferred from Euclid.

Despite his admiration for their work, Einstein would ultimately be responsible for overturning the foundations of both Euclidean geometry as the coordinate system for our universe, and Newtonian mechanics as the basis for physics. The dogma for much of the nineteenth century was that Newton's laws of motions were the fundamental foundation upon which all future discoveries would be made. Newton's picture, simply, was that all of the forces in the universe were produced by particles, and that all of physics could be cast in terms of the interactions.

By the time Einstein was born, cracks were already beginning to appear in the edifice of the particle nature of the physics. In 1864, James Clerk Maxwell developed a theory of electrodynamics. Einstein derived considerable inspiration from Maxwell in two important ways. First, Maxwell's equations showed that an electromagnetic wave (light) propagates at a constant speed, regardless of the speed of the source. This was the important foundation for Einstein's theory of special relativity.

Second, Maxwell's equations formed a field theory. It was the electrical and magnetic field that defined how charged particles behaved, not the charged particles interacting with each other. This may seem a subtle distinction, but it is an important one. Ultimately, this concept of the field would form the foundation not only of electromagnetism, but also for advances in unifying the fundamental forces of nature.

Einstein concludes his notes with a discussion of general relativity, his theory of gravity. Part of the elegance of this theory stems from the fact that to those versed in the mathematics it looks almost identical to Maxwell's theory of electrodynamics. This fact was not lost on Einstein. Indeed, it was one of Einstein's perennial disappointments that he was unable to unify electromagnetism and gravity into a single unified theory. This remains one of the great unsolved problems of modern theoretical physics.

# AUTOBIOGRAPHICAL NOTES*

Here I sit in order to write, at the age of 67, something like my own obituary. I am doing this not merely because Dr. Schilpp has persuaded me to do it; but because I do, in fact, believe that it is a good thing to show those who are striving alongside of us, how one's own striving and searching appears to one in retrospect. After some reflection, I felt how insufficient any such attempt is bound to be. For, however brief and limited one's working life may be, and however predominant may be the ways of error, the exposition of that which is worthy of communication does nonetheless not come easy—today's person of 67 is by no means the same as was the one of 50, of 30, or of 20. Every reminiscence is colored by today's being what it is, and therefore by a deceptive point of view. This consideration could very well deter. Nevertheless much can be lifted out of one's own experience which is not open to another consciousness.

Even when I was a fairly precocious young man the nothingness of the hopes and strivings which chases most men restlessly through life came to my consciousness with considerable vitality. Moreover, I soon discovered the cruelty of that chase, which in those years was much more carefully covered up by hypocrisy and glittering words than is the case today. By the mere existence of his stomach everyone was condemned to participate in that chase. Moreover, it was possible to satisfy the stomach by such participation, but not man in so far as he is a thinking and feeling being. As the first way out there was religion, which is implanted into every child by way of the traditional education-machine. Thus I came—despite the fact that I was the son of entirely irreligious ( Jewish) parents—to a deep religiosity, which, however, found an abrupt ending at the age of 12. Through the reading of popular scientific books I soon reached the conviction that much in the stories of the Bible could not be true. The consequence was a positively fanatic [orgy of] freethinking coupled with the

* Reprinted by permission of Open Court Publishing Company, a division of Carus Publishing Company, Peru, IL, from *A. Einstein: Autobiographical Notes* translated and edited by Paul Arthur Schilpp, first published in *Albert Einstein: Philosopher-Scientist* in The Library of Living Philosophers Series Volume VII, (c) 1949, 1951, 1970, 1979 by The Library of Living Philosophers, Inc., and the Estate of Albert Einstein.

impression that youth is intentionally being deceived by the state through lies; it was a crushing impression. Suspicion against every kind of authority grew out of this experience, a skeptical attitude towards the convictions which were alive in any specific social environment—an attitude which has never again left me, even though later on, because of a better insight into the causal connections, it lost some of its original poignancy.

It is quite clear to me that the religious paradise of youth, which was thus lost, was a first attempt to free myself from the chains of the "merely-personal," from an existence which is dominated by wishes, hopes and primitive feelings. Out yonder there was this huge world, which exists independently of us human beings and which stands before us like a great, eternal riddle, at least partially accessible to our inspection and thinking. The contemplation of this world beckoned like a liberation, and I soon noticed that many a man whom I had learned to esteem and to admire had found inner freedom and security in devoted occupation with it. The mental grasp of this extra-personal world within the frame of the given possibilities swam as highest aim half consciously and half unconsciously before my mind's eye. Similarly motivated men of the present and of the past, as well as the insights which they had achieved, were the friends which could not be lost. The road to this paradise was not as comfortable and alluring as the road to the religious paradise; but it has proved itself as trustworthy, and I have never regretted having chosen it.

What I have here said is true only within a certain sense, just as a drawing consisting of a few strokes can do justice to a complicated object, full of perplexing details, only in a very limited sense. If an individual enjoys well-ordered thoughts, it is quite possible that this side of his nature may grow more pronounced at the cost of other sides and thus may determine his mentality in increasing degree. In this case it is well possible that such an individual in retrospect sees a uniformly systematic development, whereas the actual experience takes place in kaleidoscopic particular situations. The manifoldness of the external situations and the narrowness of the momentary content of consciousness bring about a sort of atomizing of the life of every human being. In a man of my type the turning-point of the development lies in the fact that gradually the major interest disengages itself to a far-reaching degree from the momentary and the merely personal and turns towards the striving for a mental grasp of things. Looked at from this point of view the above schematic remarks contain as much truth as can be uttered in such brevity.

What, precisely, is "thinking"? When, at the reception of sense-impressions, memory-pictures emerge, this is not yet "thinking." And when such pictures form series, each member of which calls forth another, this too is not yet "thinking." When, however, a certain picture turns up in many such series, then—precisely through such return—it becomes an ordering element for such series, in that it connects series which in themselves are unconnected. Such an element becomes an instrument, a concept. I think that the transition from free association or "dreaming" to thinking is characterized by the more or less dominating rôle which the "concept" plays in it. It is by no means necessary that a concept must be connected with a sensorily cognizable and reproducible sign (word); but when this is the case thinking becomes by means of that fact communicable.

With what right—the reader will ask—does this man operate so carelessly and primitively with ideas in such a problematic realm without making even the least effort to prove anything? My defense: all our thinking is of this nature of a free play with concepts; the justification for this play lies in the measure of survey over the experience of the senses which we are able to achieve with its aid. The concept of "truth" can not yet be applied to such a structure; to my thinking this concept can come in question only when a far-reaching agreement (*convention*) concerning the elements and rules of the game is already at hand.

For me it is not dubious that our thinking goes on for the most part without use of signs (words) and beyond that to a considerable degree unconsciously. For how, otherwise, should it happen that sometimes we "wonder" quite spontaneously about some experience? This "wondering" seems to occur when an experience comes into conflict with a world of concepts which is already sufficiently fixed in us. Whenever such a conflict is experienced hard and intensively it reacts back upon our thought world in a decisive way. The development of this thought world is in a certain sense a continuous flight from "wonder."

A wonder of such nature I experienced as a child of 4 or 5 years, when my father showed me a compass. That this needle behaved in such a determined way did not at all fit into the nature of events, which could find a place in the unconscious world of concepts (effect connected with direct "touch"). I can still remember—or at least believe I can remember—that this experience made a deep and lasting impression upon me. Something deeply hidden had to be behind things. What man sees before him from infancy causes no reaction of this kind; he is not surprised over the falling of bodies, concerning wind and rain, nor concerning the moon or about the fact that the moon does not fall down, nor concerning the differences between living and non-living matter.

At the age of 12 I experienced a second wonder of a totally different nature: in a little book dealing with Euclidian plane geometry, which came into my hands at the beginning of a schoolyear. Here were assertions, as for example the intersection of the three altitudes of a triangle in one point, which—though by no means evident—could nevertheless be proved with such certainty that any doubt appeared to be out of the question. This lucidity and certainty made an indescribable impression upon me. That the axiom had to be accepted unproved did not disturb me. In any case it was quite sufficient for me if I could peg proofs upon propositions the validity of which did not seem to me to be dubious. For example I remember that an uncle told me the Pythagorean theorem before the holy geometry booklet had come into my hands. After much effort I succeeded in "proving" this theorem on the basis of the similarity of triangles; in doing so it seemed to me "evident" that the relations of the sides of the right-angled triangles would have to be completely determined by one of the acute angles. Only something which did not in similar fashion seem to be "evident" appeared to me to be in need of any proof at all. Also, the objects with which geometry deals seemed to be of no different type than the objects of sensory perception, "which can be seen and touched." This primitive idea, which probably also lies at the bottom of the well known Kantian problematic concerning the possibility of "synthetic judgments *a priori*," rests obviously upon the fact that the relation of geometrical concepts to objects of direct experience (rigid rod, finite interval, etc.) was unconsciously present.

If thus it appeared that it was possible to get certain knowledge of the objects of experience by means of pure thinking, this "wonder" rested upon an error. Nevertheless, for anyone who experiences it for the first time, it is marvellous enough that man is capable at all to reach such a degree of certainty and purity in pure thinking as the Greeks showed us for the first time to be possible in geometry.

Now that I have allowed myself to be carried away sufficiently to interrupt my scantily begun obituary, I shall not hesitate to state here in a few sentences my epistemological credo, although in what precedes something has already incidentally been said about this. This credo actually evolved only much later and very slowly and does not correspond with the point of view I held in younger years.

I see on the one side the totality of sense-experiences, and, on the other, the totality of the concepts and propositions which are laid down in books. The relations between the concepts and propositions among themselves and each other are of a logical nature, and the business of logical thinking is strictly limited to the achievement of the connection between concepts and propositions among each other according to firmly laid down rules, which are the concern of logic. The concepts and propositions get "meaning," viz., "content," only through their connection with sense-experiences. The connection of the latter with the former is purely intuitive, not itself of a logical nature. The degree of certainty with which this relation, viz., intuitive connection, can be undertaken, and nothing else, differentiates empty phantasy from scientific "truth." The system of concepts is a creation of man together with the rules of syntax, which constitute the structure of the conceptual systems. Although the conceptual systems are logically entirely arbitrary, they are bound by the aim to permit the most nearly possible certain (intuitive) and complete co-ordination with the totality of sense-experiences; secondly they aim at greatest possible sparsity of their logically independent elements (basic concepts and axioms), i.e., undefined concepts and underived [postulated] propositions.

A proposition is correct if, within a logical system, it is deduced according to the accepted logical rules. A system has truth-content according to the certainty and completeness of its co-ordination-possibility to the totality of experience. A correct proposition borrows its "truth" from the truth-content of the system to which it belongs.

A remark to the historical development. Hume saw clearly that certain concepts, as for example that of causality, cannot be deduced from the material of experience by logical methods. Kant, thoroughly convinced of the indispensability of certain concepts, took them—just as they are selected—to be the necessary premises of every kind of thinking and differentiated them from concepts of empirical origin. I am convinced, however, that this differentiation is erroneous, i.e., that it does not do justice to the problem in a natural way. All concepts, even those which are closest to experience, are from the point of view of logic freely chosen conventions, just as is the case with the concept of causality, with which this problematic concerned itself in the first instance.

And now back to the obituary. At the age of 12–16 I familiarized myself with the elements of mathematics together with the principles of differential and integral calculus. In doing so I had the good fortune of hitting on books which were not too particular in their logical rigour, but which made up for this by permitting the main thoughts to stand out clearly and synoptically. This occupation was, on the whole, truly fascinating; climaxes were reached whose impression could easily compete with that of elementary geometry—the basic idea of analytical geometry, the infinite series, the concepts of differential and integral. I also had the good fortune of getting to know the essential results and methods of the entire field of the natural sciences in an excellent popular exposition, which limited itself almost throughout to qualitative aspects (Bernstein's *People's Books on Natural Science*, a work of 5 or 6 volumes), a work which I read with breathless attention. I had also already studied some theoretical physics when, at the age of 17, I entered the Polytechnic Institute of Zürich as a student of mathematics and physics.

There I had excellent teachers (for example, Hurwitz, Minkowski), so that I really could have gotten a sound mathematical education. However, I worked most of the time in the physical laboratory, fascinated by the direct contact with experience. The balance of the time I used in the main in order to study at home the works of Kirchhoff, Helmholtz, Hertz, etc. The fact that I neglected mathematics to a certain extent had its cause not merely in my stronger interest in the natural sciences than in mathematics but also in the following strange experience. I saw that mathematics was split up into numerous specialities, each of which could easily absorb the short lifetime granted to us. Consequently I saw myself in the position of Buridan's ass which was unable to decide upon any specific bundle of hay. This was obviously due to the fact that my intuition was not strong enough in the field of mathematics in order to differentiate clearly the fundamentally important, that which is really basic, from the rest of the more or less dispensable erudition. Beyond this, however, my interest in the knowledge of nature was also unqualifiedly stronger; and it was not clear to me as a student that the approach to a more profound knowledge of the basic principles of physics is tied up with the most intricate mathematical methods. This dawned upon me only gradually after years of independent scientific work. True enough, physics also was divided into separate fields, each of which was capable of devouring a short lifetime of work without having satisfied the hunger for deeper knowledge. The mass of insufficiently connected experimental data was overwhelming here also. In this field, however, I soon learned to scent out that which was able to lead to fundamentals and to turn aside from everything else, from the multitude of things which clutter up the mind and divert it from the essential. The hitch in this was, of course, the fact that one had to cram all this stuff into one's mind for the examinations, whether one liked it or not. This coercion had such a deterring effect [upon me] that, after I had passed the final examination, I found the consideration of any scientific problems distasteful to me for an entire year. In justice I must add, moreover, that in Switzerland we had to suffer far less under such coercion, which smothers every truly scientific impulse, than is the case in many another locality. There were altogether only two examinations; aside from these, one could just about do as one pleased. This was especially the case if one had a friend, as did I, who attended the lectures regularly and who worked over their content conscientiously. This gave one freedom in the choice of pursuits until a few months before the examination, a freedom which I enjoyed to a great extent and have gladly taken into the bargain the bad conscience connected with it as by far the lesser evil. It is, in fact, nothing short of a miracle that the modern methods of instruction have not yet entirely strangled the holy curiosity of inquiry; for this delicate little plant, aside from stimulation, stands mainly in need of freedom; without this it goes to wreck and ruin without fail. It is a very grave mistake to think that the enjoyment of seeing and searching can be promoted by means of coercion and a sense of duty. To the contrary, I believe that it would be possible to rob even a healthy beast of prey of its voraciousness, if it were possible, with the aid of a whip, to force the beast to devour continuously, even when not hungry, especially if the food, handed out under such coercion, were to be selected accordingly. - - -

Now to the field of physics as it presented itself at that time. In spite of all the fruitfulness in particulars, dogmatic rigidity prevailed in matters of principles: In the beginning (if there was such a thing) God created Newton's laws of motion together with the necessary masses and forces. This is all; everything beyond this follows from the development of appropriate mathematical methods by means of deduction. What the nineteenth century achieved on the strength of this basis, especially through the application of the partial differential equations, was bound to arouse the admiration of every receptive person. Newton was probably first to reveal, in his theory of sound-transmission, the efficacy of partial differential equations. Euler had already created the foundation of hydrodynamics. But the more precise development of the mechanics of discrete masses, as the basis of all physics, was the achievement of the 19th century. What made the greatest impression upon the student, however, was less the technical construction of mechanics or the solution of complicated problems than the achievements of mechanics in areas which apparently had nothing to do with mechanics: the mechanical theory of light, which conceived of light as the wave-motion of a quasi-rigid elastic ether, and above all the kinetic theory of gases:—the independence of the specific heat of monatomic gases of the atomic weight, the derivation of the equation of state of a gas and its relation to the specific heat, the kinetic theory of the dissociation of gases, and above all the quantitative connection of viscosity, heat-conduction and diffusion of gases, which also furnished the absolute magnitude of the atom. These results supported at the same time mechanics as the foundation of physics and of the atomic hypothesis, which latter was already firmly anchored in chemistry. However, in chemistry only the ratios of the atomic masses played any rôle, not their absolute magnitudes, so that atomic theory could be viewed more as a visualizing symbol than as knowledge concerning the factual construction of matter. Apart from this it was also of profound interest that the statistical theory of classical mechanics was able to deduce the basic laws of thermodynamics, something which was in essence already accomplished by Boltzmann.

We must not be surprised, therefore, that, so to speak, all physicists of the last century saw in classical mechanics a firm and final foundation for all physics, yes, indeed, for all natural science, and that they never grew tired in their attempts to base Maxwell's theory of electro-magnetism, which, in the meantime, was slowly beginning to win out, upon mechanics as well. Even Maxwell and H. Hertz, who in retrospect appear as those who demolished the faith in mechanics as the final basis of all physical thinking, in their conscious thinking adhered throughout to mechanics as the secured basis of physics. It was Ernst Mach who, in his *History of Mechanics*, shook this dogmatic faith; this book exercised a profound influence upon me in this regard while I was a student. I see Mach's greatness in his incorruptible skepticism and independence; in my younger years, however, Mach's epistemological position also influenced me very greatly, a position which today appears to me to be essentially untenable. For he did not place in the correct light the essentially constructive and speculative nature of thought and more especially of scientific thought; in consequence of which he condemned theory on precisely those points where its constructive-speculative character unconcealably comes to light, as for example in the kinetic atomic theory.

Before I enter upon a critique of mechanics as the foundation of physics, something of a broadly general nature will first have to be said concerning the points of view according to which it is possible to criticize physical theories at all. The first point of view is obvious: the theory must not contradict empirical facts. However evident this demand may in the first place appear, its application turns out to be quite delicate. For it is often, perhaps even always, possible to adhere to a general theoretical foundation by securing the adaptation of the theory to the facts by means of artificial additional assumptions. In any case, however, this first point of view is concerned with the confirmation of the theoretical foundation by the available empirical facts.

The second point of view is not concerned with the relation to the material of observation but with the premises of the theory itself, with what may briefly but vaguely be characterized as the "naturalness" or "logical simplicity" of the premises (of the basic concepts and of the relations between these which are taken as a basis). This point of view, an exact formulation of which meets with great difficulties, has played an important rôle in the selection and evaluation of theories since time immemorial. The problem here is not simply one of a kind of enumeration of the logically independent premises (if anything like this were at all unequivocally possible), but that of a kind of reciprocal weighing of incommensurable qualities. Furthermore, among theories of equally "simple" foundation that one is to be taken as superior which most sharply delimits the qualities of systems in the abstract (i.e., contains the most definite claims). Of the "realm" of theories I need not speak here, inasmuch as we are confining ourselves to such theories whose object is the *totality* of all physical appearances. The second point of view may briefly be characterized as concerning itself with the "inner perfection" of the theory, whereas the first point of view refers to the "external confirmation." The following I reckon as also belonging to the "inner perfection" of a theory: We prize a theory more highly if, from the logical standpoint, it is not the result of an arbitrary choice among theories which, among themselves, are of equal value and analogously constructed.

The meager precision of the assertions contained in the last two paragraphs I shall not attempt to excuse by lack of sufficient printing space at my disposal, but confess herewith that I am not, without more ado [immediately], and perhaps not at all, capable to replace these hints by more precise definitions. I believe, however, that a sharper formulation would be possible. In any case it turns out that among the "augurs" there usually is agreement in judging the "inner perfection" of the theories and even more so concerning the "degree" of "external confirmation."

And now to the critique of mechanics as the basis of physics.

From the first point of view (confirmation by experiment) the incorporation of wave-optics into the mechanical picture of the world was bound to arouse serious misgivings. If light was to be interpreted as undulatory motion in an elastic body (ether), this had to be a medium which permeates everything; because of the transversality of the lightwaves in the main similar to a solid body, yet incompressible, so that longitudinal waves did not exist. This ether had to lead a ghostly existence alongside the rest of matter, inasmuch as it seemed to offer no resistance whatever to the motion of "ponderable" bodies. In order to explain the refraction-indices of transparent bodies as well as the processes of emission and absorption of radiation, one would have had to assume complicated reciprocal actions between the two types of matter, something which was not even seriously tried, let alone achieved.

Furthermore, the electromagnetic forces necessitated the introduction of electric masses, which, although they had no noticeable inertia, yet interacted with each other, and whose interaction was, moreover, in contrast to the force of gravitation, of a polar type.

The factor which finally succeeded, after long hesitation, to bring the physicists slowly around to give up the faith in the possibility that all of physics could be founded upon Newton's mechanics, was the electrodynamics of Faraday and Maxwell. For this theory and its confirmation by Hertz's experiments showed that there are electromagnetic phenomena which by their very nature are detached from every ponderable matter—namely the waves in empty space which consist of electromagnetic "fields." If mechanics was to be maintained as the foundation of physics, Maxwell's equations had to be interpreted mechanically. This was zealously but fruitlessly attempted, while the equations were proving themselves fruitful in mounting degree. One got used to operating with these fields as independent substances without finding it necessary to give one's self an account of their mechanical nature; thus mechanics as the basis of physics was being abandoned, almost unnoticeably, because its adaptability to the facts presented itself finally as hopeless. Since then there exist two types of conceptual elements, on the one hand, material points with forces at a distance between them, and, on the other hand, the continuous field. It presents an intermediate state in physics without a uniform basis for the entirety, which—although unsatisfactory—is far from having been superseded. - - -

Now for a few remarks to the critique of mechanics as the foundation of physics from the second, the "interior," point of view. In today's state of science, i.e., after the departure from the mechanical foundation, such critique has only an interest in method left. But such a critique is well suited to show the type of argumentation which, in the choice of theories in the future will have to play an all the greater rôle the more the basic concepts and axioms distance themselves from what is directly observable, so that the confrontation of the implications of theory by the facts becomes constantly more difficult and more drawn out. First in line to be mentioned is Mach's argument, which, however, had already been clearly recognized by Newton (bucket experiment). From the standpoint of purely geometrical description all "rigid" co-ordinate systems are among themselves logically equivalent. The equations of mechanics (for example this is already true of the law of inertia) claim validity only when referred to a specific class of such systems, i.e., the "inertial systems." In this the co-ordinate system as bodily object is without any significance. It is necessary, therefore, in order to justify the necessity of the specific choice, to look for something which lies outside of the objects (masses, distances) with which the theory is concerned. For this reason "absolute space" as originally determinative was quite explicitly introduced by Newton as the omnipresent active participant in all mechanical events; by "absolute" he obviously means uninfluenced by the masses and by their motion. What makes this state of affairs appear particularly offensive is the fact that there are supposed to be infinitely many inertial systems, relative to each other in uniform translation, which are supposed to be distinguished among all other rigid systems.

Mach conjectures that in a truly rational theory inertia would have to depend upon the interaction of the masses, precisely as was true for Newton's other forces, a conception which for a long time I considered as in principle the correct one. It presupposes implicitly, however, that the basic theory should be of the general type of Newton's mechanics: masses and their interaction as the original concepts. The attempt at such a solution does not fit into a consistent field theory, as will be immediately recognized.

How sound, however, Mach's critique is in essence can be seen particularly clearly from the following analogy. Let us imagine people construct a mechanics, who know only a very small part of the earth's surface and who also can not see any stars. They will be inclined to ascribe special physical attributes to the vertical dimension of space (direction of the acceleration of falling bodies) and, on the ground of such a conceptual basis, will offer reasons that the earth is in most places horizontal. They might not permit themselves to be influenced by the argument that as concerns the geometrical properties space is isotrope and that it is therefore supposed to be unsatisfactory to postulate basic physical laws, according to which there is supposed to be a preferential direction; they will probably be inclined (analogously to Newton) to assert the absoluteness of the vertical, as proved by experience as something with which one simply would have to come to terms. The preference given to the vertical over all other spatial directions is precisely analogous to the preference given to inertial systems over other rigid co-ordination systems.

Now to [a consideration of] other arguments which also concern themselves with the inner simplicity, i.e., naturalness, of mechanics. If one puts up with the concepts of space (including geometry) and time without critical doubts, then there exists no reason to object to the idea of action-at-a-distance, even though such a concept is unsuited to the ideas which one forms on the basis of the raw experience of daily life. However, there is another consideration which causes mechanics, taken as the basis of physics, to appear as primitive. Essentially there exist two laws

(1) the law of motion

(2) the expression for force or potential energy.

The law of motion is precise, although empty, as long as the expression for the forces is not given. In postulating the latter, however, there exists great latitude for arbitrary [choice], especially if one omits the demand, which is not very natural in any case, that the forces depend only on the co-ordinates (and, for example, not on their differential quotients with respect to time). Within the framework of theory alone it is entirely arbitrary that the forces of gravitation (and electricity), which come from one point are governed by the potential function (1/r). Additional remark: it has long been known that this function is the central-symmetrical solution of the simplest (rotation-invariant) differential equation $\Delta\varphi = 0$; it would therefore have been a suggestive idea to regard this as a sign that this function is to be regarded as determined by a law of space, a procedure by which the arbitrariness in the choice of the law of energy would have been removed. This is really the first insight which suggests a turning away from the theory of distant forces, a development which—prepared by Faraday, Maxwell and Hertz—really begins only later on under the external pressure of experimental data.

I would also like to mention, as one internal asymmetry of this theory, that the inert mass occuring in the law of motion also appears in the expression for the gravitational force, but not in the expression for the other forces. Finally I would like to point to the fact that the division of energy into two essentially different parts, kinetic and potential energy, must be felt as unnatural; H. Hertz felt this as so disturbing that, in his very last work, he attempted to free mechanics from the concept of potential energy (i.e., from the concept of force). - - -

Enough of this, Newton, forgive me; you found the only way which, in your age, was just about possible for a man of highest thought- and creative power. The concepts, which you created, are even today still guiding our thinking in physics, although we now know that they will have to be replaced by others farther removed from the sphere of immediate experience, if we aim at a profounder understanding of relationships.

"Is this supposed to be an obituary?" the astonished reader will likely ask. I would like to reply: essentially yes. For the essential in the being of a man of my type lies precisely in *what* he thinks and *how* he thinks, not in what he does or suffers. Consequently, the obituary can limit itself in the main to the communicating of thoughts which have played a considerable rôle in my endeavors.—A theory is the more impressive the greater the simplicity of its premises is, the more different kinds of things it relates, and the more extended is its area of applicability. Therefore the deep impression which classical thermodynamics made upon me. It is the only physical theory of universal content concerning which I am convinced that, within the framework of the applicability of its basic concepts, it will never be overthrown (for the special attention of those who are skeptics on principle).

The most fascinating subject at the time that I was a student was Maxwell's theory. What made this theory appear revolutionary was the transition from forces at a distance to fields as fundamental variables. The incorporation of optics into the theory of electromagnetism, with its relation of the speed of light to the electric and magnetic absolute system of units as well as the relation of the refraction coëfficient to the dielectric constant, the qualitative relation between the reflection coëfficient and the metallic conductivity of the body—it was like a revelation. Aside from the transition to field-theory, i.e., the expression of the elementary laws through differential equations, Maxwell needed only one single hypothetical step—the introduction of the electrical displacement current in the vacuum and in the dielectrica and its magnetic effect, an innovation which was almost prescribed by the formal properties of the differential equations. In this connection I cannot suppress the remark that the pair Faraday-Maxwell has a most remarkable inner similarity with the pair Galileo-Newton—the former of each pair grasping the relations intuitively, and the second one formulating those relations exactly and applying them quantitatively.

What rendered the insight into the essence of electromagnetic theory so much more difficult at that time was the following peculiar situation. Electric or magnetic "field intensities" and "displacements" were treated as equally elementary variables, empty space as a special instance of a dielectric body. *Matter* appeared as the bearer of the field, not *space*. By this it was implied that the carrier of the field could have velocity, and this was naturally to apply to the "vacuum" (ether) also. Hertz's electrodynamics of moving bodies rests entirely upon this fundamental attitude.

It was the great merit of H. A. Lorentz that he brought about a change here in a convincing fashion. In principle a field exists, according to him, only in empty space. Matter—considered as atoms—is the only seat of electric charges; between the material particles there is empty space, the seat of the electromagnetic field, which is created by the position and velocity of the point charges which are located on the material particles. Dielectricity, conductivity, etc., are determined exclusively by the type of mechanical tie connecting the particles, of which the bodies consist. The particle-charges create the field, which, on the other hand, exerts forces upon the charges of the particles, thus determining the motion of the latter according to Newton's law of motion. If one compares this with Newton's system, the change consists in this: action at a distance is replaced by the field, which thus also describes the radiation. Gravitation is usually not taken into account because of its relative smallness; its consideration, however, was always possible by means of the enrichment of the structure of the field, i.e., expansion of Maxwell's law of the field. The physicist of the present generation regards the point of view achieved by Lorentz as the only possible one; at that time, however, it was a surprising and audacious step, without which the later development would not have been possible.

If one views this phase of the development of theory critically, one is struck by the dualism which lies in the fact that the material point in Newton's sense and the field as continuum are used as elementary concepts side by side. Kinetic energy and field-energy appear as essentially different things. This appears all the more unsatisfactory inasmuch as, according to Maxwell's theory, the magnetic field of a moving electric charge represents inertia. Why not then *total* inertia? Then only field-energy would be left, and the particle would be merely an area of special density of field-energy. In that case one could hope to deduce the concept of the mass-point together with the equations of the motion of the particles from the field equations—the disturbing dualism would have been removed.

H. A. Lorentz knew this very well. However, Maxwell's equations did not permit the derivations of the equilibrium of the electricity which constitutes a particle. Only other, nonlinear field equations could possibly accomplish such a thing. But no method existed by which this kind of field equations could be discovered without deteriorating into adventurous arbitrariness. In any case one could believe that it would be possible by and by to find a new and secure foundation for all of physics upon the path which had been so successfully begun by Faraday and Maxwell. - - -

Accordingly, the revolution begun by the introduction of the field was by no means finished. Then it happened that, around the turn of the century, independently of what we have just been discussing, a second fundamental crisis set in, the seriousness of which was suddenly recognized due to Max Planck's investigations into heat radiation (1900). The history of this event is all the more remarkable because, at least in its first phase, it was not in any way influenced by any surprising discoveries of an experimental nature.

On thermodynamic grounds Kirchhoff had concluded that the energy density and the spectral composition of radiation in a *Hohlraum*, surrounded by impenetrable walls of the temperature $T$, would be independent of the nature of the walls. That is to say, the nonchromatic density of radiation $\varrho$ is a universal function of the frequency $\nu$ and of the absolute temperature $T$. Thus arose the interesting problem of determining this function $\varrho(\nu, T)$ What could theoretically be ascertained about this function? According to Maxwell's theory the radiation had to exert a pressure on the walls, determined by the total energy density. From this Boltzmann concluded by means of pure thermodynamics, that the entire energy density of the radiation ($\int \varrho \, d\nu$) is proportional to $T^4$. In this way he found a theoretical justification of a law which had previously been discovered empirically by Stefan, i.e., in this way he connected this empirical law with the basis of Maxwell's theory. Thereafter, by way of an ingenious thermodynamic consideration, which also made use of Maxwell's theory, W. Wien found that the universal function $\varrho$ of the two variables $\nu$ and $T$ would have to be of the form

$$\rho \approx \nu^3 f\left(\frac{\nu}{T}\right),$$

whereby $f(\nu/T)$ is a universal function of one variable $\nu/T$ only. It was clear that the theoretical determination of this universal function $f$ was of fundamental importance—this was precisely the task which confronted Planck. Careful measurements had led to a very precise empirical determination of the function $f$. Relying on those empirical measurements, he succeeded in the first place in finding a statement which rendered the measurements very well indeed:

$$\rho = \frac{8\pi h\nu^3}{c^3} \frac{1}{exp(h\nu/kT) - 1}$$

whereby $h$ and $k$ are two universal constants, the first of which led to quantum theory. Because of the denominator this formula looks a bit queer. Was it possible to derive it theoretically? Planck actually did find a derivation, the imperfections of which remained at first hidden, which latter fact was most fortunate for the development of physics. If this formula was correct, it permitted, with the aid of Maxwell's theory, the calculation of the average energy $E$ of a quasi-monochromatic oscillator within the field of radiation:

$$E = \frac{h\nu}{exp(h\nu/kT) - 1}$$

Planck preferred to attempt calculating this latter magnitude theoretically. In this effort, thermodynamics, for the time being, proved no longer helpful, and neither did Maxwell's theory. The following circumstance was unusually encouraging in this formula. For high temperatures (with a fixed $\nu$) it yielded the expression

$$E = kT.$$

This is the same expression as the kinetic theory of gases yields for the average energy of a mass-point which is capable of oscillating elastically in one dimension. For in kinetic gas theory one gets.

$$E = (R/N)T,$$

whereby $R$ means the constant of the equation of state of a gas and $N$ the number of molecules per mol, from which constant one can compute the absolute size of the atom. Putting these two expressions equal to each other one gets

$$N = R/k.$$

The one constant of Planck's formula consequently furnishes exactly the correct size of the atom. The numerical value agreed satisfactorily with the determinations of $N$ by means of kinetic gas theory, even though these latter were not very accurate.

This was a great success, which Planck clearly recognized. But the matter has a serious drawback, which Planck fortunately overlooked at first. For the same considerations demand in fact that the relation $E = kT$ would also have to be valid for low temperatures. In that case, however, it would be all over with Planck's formula and with the constant $h$. From the existing theory, therefore, the correct conclusion would have been: the average kinetic energy of the oscillator is either given incorrectly by the theory of gases, which would imply a refutation of [statistical] mechanics; or else the average energy of the oscillator follows incorrectly from Maxwell's theory, which would imply a refutation of the latter. Under such circumstances it is most probable that both theories are correct only at the limits, but are otherwise false; this is indeed the situation, as we shall see in what follows. If Planck had drawn this conclusion, he probably would not have made his great discovery, because the foundation would have been withdrawn from his deductive reasoning.

Now back to Planck's reasoning. On the basis of the kinetic theory of gases Boltzmann had discovered that, aside from a constant factor, entropy is equivalent to the logarithm of the "probability" of the state under consideration. Through this insight he recognized the nature of courses of events which, in the sense of thermodynamics, are "irreversible." Seen from the molecular-mechanical point of view, however, all courses of events are reversible. If one calls a molecular-theoretically defined state a microscopically described one, or, more briefly, micro-state, and a state described in terms of thermodynamics a macro-state, then an immensely large number ($Z$) of states belong to a macroscopic condition. $Z$ then is a measure of the probability of a chosen macro-state. This idea appears to be of outstanding importance also because of the fact that its usefulness is not limited to microscopic description on the basis of mechanics. Planck recognized this and applied the Boltzmann principle to a system which consists of very many resonators of the same frequency $\nu$. The macroscopic situation is given through the total energy of the oscillation of all resonators, a micro-condition through determination of the (instantaneous) energy of each individual resonator. In order then to be able to express the number of the micro-states belonging to a macro-state by means of a finite number, he [Planck] divided the total energy into a large but finite number of identical energy-elements ε and asked: in how many ways can these energy-elements be divided among the resonators. The logarithm of this number, then, furnishes the entropy and thus (via thermodynamics) the temperature of the system. Planck got his radiation-formula if he chose his energy-elements ε of the magnitude $\varepsilon = h\nu$. The decisive element in doing this lies in the fact that the result depends on taking for ε a definite finite value, i.e., that one does not go to the limit $\varepsilon = 0$. This form of reasoning does not make obvious the fact that it contradicts the mechanical and electrodynamic basis, upon which the derivation otherwise depends. Actually, however, the derivation presupposes implicitly that energy can be absorbed and emitted by the individual resonator only in "quanta" of magnitude $h\nu$, i.e., that the energy of a mechanical structure capable of oscillations as well as the energy of radiation can be transferred only in such quanta—in contradiction to the laws of mechanics and electrodynamics. The contradiction with dynamics was here fundamental; whereas, the contradiction with electrodynamics could be less fundamental. For the expression for the density of radiation-energy, although it is *compatible* with Maxwell's equations, is not a necessary consequence of these equations. That this expression furnishes important average-values is shown by the fact that the Stefan-Boltzmann law and Wien's law, which are based on it, are in agreement with experience.

All of this was quite clear to me shortly after the appearance of Planck's fundamental work; so that, without having a substitute for classical mechanics, I could nevertheless see to what kind of consequences this law of temperature-radiation leads for the photo-electric effect and for other related phenomena of the transformation of radiation-energy, as well as for the specific heat of (especially) solid bodies. All my attempts, however, to adapt the theoretical foundation of physics to this [new type of] knowledge failed completely. It was as if the ground had been pulled out from under one, with no firm foundation to be seen anywhere, upon which one could have built. That this insecure and contradictory foundation was sufficient to enable a man of Bohr's unique instinct and tact to discover the major laws of the spectral lines and of the electron-shells of the atoms together with their significance for chemistry appeared to me like a miracle—and appears to me as a miracle even today. This is the highest form of musicality in the sphere of thought.

My own interest in those years was less concerned with the detailed consequences of Planck's results, however important these might be. My major question was: What general conclusions can be drawn from the radiation-formula concerning the structure of radiation and even more generally concerning the electro-magnetic foundation of physics? Before I take this up, I must briefly mention a number of investigations which relate to the Brownian motion and related objects (fluctuation-phenomena) and which in essence rest upon classical molecular mechanics. Not acquainted with the earlier investigations of Boltzmann and Gibbs, which had appeared earlier and actually exhausted the subject, I developed the statistical mechanics and the molecular-kinetic theory of thermodynamics which was based on the former. My major aim in this was to find facts which would guarantee as much as possible the existence of atoms of definite finite size. In the midst of this I discovered that, according to atomistic theory, there would have to be a movement of suspended microscopic particles open to observation, without knowing that observations concerning the Brownian motion were already long familiar. The simplest derivation rested upon the following consideration. If the molecular-kinetic theory is essentially correct, a suspension of visible particles must possess the same kind of osmotic pressure fulfilling the laws of gases as a solution of molecules. This osmotic pressure depends upon the actual magnitude of the molecules, i.e., upon the number of molecules in a gram-equivalent. If the density of the suspension is inhomogeneous, the osmotic pressure is inhomogeneous, too, and gives rise to a compensating diffusion, which can be calculated from the well known mobility of the particles. This diffusion can, on the other hand, also be considered as the result of the random displacement—unknown in magnitude originally—of the suspended particles due to thermal agitation. By comparing the amounts obtained for the diffusion current from both types of reasoning one reaches quantitatively the statistical law for those displacements, i.e., the law of the Brownian motion. The agreement of these considerations with experience together with Planck's determination of the true molecular size from the law of radiation (for high temperatures) convinced the sceptics, who were quite numerous at that time (Ostwald, Mach) of the reality of atoms. The antipathy of these scholars towards atomic theory can indubitably be traced back to their positivistic philosophical attitude. This is an interesting example of the fact that even scholars of audacious spirit and fine instinct can be obstructed in the interpretation of facts by philosophical prejudices. The prejudice—which has by no means died out in the meantime—consists in the faith that facts by themselves can and should yield scientific knowledge without free conceptual construction. Such a misconception is possible only because one does not easily become aware of the free choice of such concepts, which, through verification and long usage, appear to be immediately connected with the empirical material.

The success of the theory of the Brownian motion showed again conclusively that classical mechanics always offered trustworthy results whenever it was applied to motions in which the higher time derivatives of velocity are negligibly small. Upon this recognition a relatively direct method can be based which permits us to learn something concerning the constitution of radiation from Planck's formula. One may conclude in fact that, in a space filled with radiation, a (vertically to its plane) freely moving, quasi monochromatically reflecting mirror would have to go through a kind of Brownian movement, the average kinetic energy of which equals $\frac{1}{2}(R/N)T$ ($R =$ constant of the gas-equation for one gram-molecule, $N$ equals the number of the molecules per mol, $T =$ absolute temperature). If radiation were not subject to local fluctuations, the mirror would gradually come to rest, because, due to its motion, it reflects more radiation on its front than on its reverse side. However, the mirror must experience certain random fluctuations of the pressure exerted upon it due to the fact that the wave-packets, constituting the radiation, interfere with one another. These can be computed from Maxwell's theory. This calculation, then, shows that these pressure variations (especially in the case of small radiation-densities) are by no means sufficient to impart to the mirror the average kinetic energy $\frac{1}{2}(R/N)T$. In order to get this result one has to assume rather that there exists a second type of pressure variations, which can not be derived from Maxwell's theory, which corresponds to the assumption that radiation energy consists of indivisible point-like localized quanta of the energy $h\nu$ (and of momentum $(h\nu/c)$, (c = velocity of light)), which are reflected undivided. This way of looking at the problem showed in a drastic and direct way that a type of immediate reality has to be ascribed to Planck's quanta, that radiation must, therefore, possess a kind of molecular structure in energy, which of course contradicts Maxwell's theory. Considerations concerning radiation which are based directly on Boltzmann's entropy-probability-relation (probability taken equal to statistical temporal frequency) also lead to the same result. This double nature of radiation (and of material corpuscles) is a major property of reality, which has been interpreted by quantum-mechanics in an ingenious and amazingly successful fashion. This interpretation, which is looked upon as essentially final by almost all contemporary physicists, appears to me as only a temporary way out; a few remarks to this [point] will follow later. - - -

Reflections of this type made it clear to me as long ago as shortly after 1900, i.e., shortly after Planck's trailblazing work, that neither mechanics nor electrodynamics could (except in limiting cases) claim exact validity. By and by I despaired of the possibility of discovering the true laws by means of constructive efforts based on known facts. The longer and the more despairingly I tried, the more I came to the conviction that only the discovery of a universal formal principle could lead us to assured results. The example I saw before me was thermodynamics. The general principle was there given in the theorem: the laws of nature are such that it is impossible to construct a *perpetuum mobile* (of the first and second kind). How, then, could such a universal principle be found? After ten years of reflection such a principle resulted from a paradox upon which I had already hit at the age of sixteen: If I pursue a beam of light with the velocity $c$ (velocity of light in a vacuum), I should observe such a beam of light as a spatially oscillatory electromagnetic field at rest. However, there seems to be no such thing, whether on the basis of experience or according to Maxwell's equations. From the very beginning it appeared to me intuitively clear that, judged from the stand-point of such an observer, everything would have to happen according to the same laws as for an observer who, relative to the earth, was at rest. For how, otherwise, should the first observer know, i.e., be able to determine, that he is in a state of fast uniform motion?

One sees that in this paradox the germ of the special relativity theory is already contained. Today everyone knows, of course, that all attempts to clarify this paradox satisfactorily were condemned to failure as long as the axiom of the absolute character of time, viz., of simultaneity, unrecognizedly was anchored in the unconscious. Clearly to recognize this axiom and its arbitrary character really implies already the solution of the problem. The type of critical reasoning which was required for the discovery of this central point was decisively furthered, in my case, especially by the reading of David Hume's and Ernst Mach's philosophical writings.

One had to understand clearly what the spatial co-ordinates and the temporal duration of events meant in physics. The physical interpretation of the spatial co-ordinates presupposed a fixed body of reference, which, moreover, had to be in a more or less definite state of motion (inertial system). In a given inertial system the co-ordinates meant the results of certain measurements with rigid (stationary) rods. (One should always be conscious of the fact that the presupposition of the existence in principle of rigid rods is a presupposition suggested by approximate experience, but which is, in principle, arbitrary.) With such an interpretation of the spatial co-ordinates the question of the validity of Euclidean geometry becomes a problem of physics.

If, then, one tries to interpret the time of an event analogously, one needs a means for the measurement of the difference in time (in itself determined periodic process realized by a system of sufficiently small spatial extension). A clock at rest relative to the system of inertia defines a local time. The local times of all space points taken together are the "time," which belongs to the selected system of inertia, if a means is given to "set" these clocks relative to each other. One sees that *a priori* it is not at all necessary that the "times" thus defined in different inertial systems agree with one another. One would have noticed this long ago, if, for the practical experience of everyday life light did not appear (because of the high value of $c$), as the means for the statement of absolute simultaneity.

The presupposition of the existence (in principle) of (ideal, viz., perfect) measuring rods and clocks is not independent of each other; since a lightsignal, which is reflected back and forth between the ends of a rigid rod, constitutes an ideal clock, provided that the postulate of the constancy of the light-velocity in vacuum does not lead to contradictions.

The above paradox may then be formulated as follows. According to the rules of connection, used in classical physics, of the spatial co-ordinates and of the time of events in the transition from one inertial system to another the two assumptions of

(1) the constancy of the light velocity

(2) the independence of the laws (thus specially also of the law of the constancy of the light velocity) of the choice of the inertial system (principle of special relativity)

are mutually incompatible (despite the fact that both taken separately are based on experience).

The insight which is fundamental for the special theory of relativity is this: The assumptions (1) and (2) are compatible if relations of a new type ("Lorentz-transformation") are postulated for the conversion of co-ordinates and the times of events. With the given physical interpretation of co-ordinates and time, this is by no means merely a conventional step, but implies certain hypotheses concerning the actual behavior of moving measuring-rods and clocks, which can be experimentally validated or disproved.

The universal principle of the special theory of relativity is contained in the postulate: The laws of physics are invariant with respect to the

Lorentz-transformations (for the transition from one inertial system to any other arbitrarily chosen system of inertia). This is a restricting principle for natural laws, comparable to the restricting principle of the non-existence of the *perpetuum mobile* which underlies thermodynamics.

First a remark concerning the relation of the theory to "four-dimensional space." It is a wide-spread error that the special theory of relativity is supposed to have, to a certain extent, first discovered, or at any rate, newly introduced, the four-dimensionality of the physical continuum. This, of course, is not the case. Classical mechanics, too, is based on the four-dimensional continuum of space and time. But in the four-dimensional continuum of classical physics the subspaces with constant time value have an absolute reality, independent of the choice of the reference system. Because of this [fact], the four-dimensional continuum falls naturally into a three-dimensional and a one-dimensional (time), so that the four-dimensional point of view does not force itself upon one as *necessary*. The special theory of relativity, on the other hand, creates a formal dependence between the way in which the spatial co-ordinates, on the one hand, and the temporal coordinates, on the other, have to enter into the natural laws.

Minkowski's important contribution to the theory lies in the following: Before Minkowski's investigation it was necessary to carry out a Lorentz-transformation on a law in order to test its invariance under such transformations; he, on the other hand, succeeded in introducing a formalism such that the mathematical form of the law itself guarantees its invariance under Lorentz-transformations. By creating a four-dimensional tensor-calculus he achieved the same thing for the four-dimensional space which the ordinary vector-calculus achieves for the three spatial dimensions. He also showed that the Lorentz-transformation (apart from a different algebraic sign due to the special character of time) is nothing but a rotation of the coordinate system in the four-dimensional space.

First, a remark concerning the theory as it is characterized above. One is struck [by the fact] that the theory (except for the four-dimensional space) introduces two kinds of physical things, i.e., (1) measuring rods and clocks, (2) all other things, e.g., the electro-magnetic field, the material point, etc. This, in a certain sense, is inconsistent; strictly speaking measuring rods and clocks would have to be represented as solutions of the basic equations (objects consisting of moving atomic configurations), not, as it were, as theoretically self-sufficient entities. However, the procedure justifies itself because it was clear from the very beginning that the postulates of the theory are not strong enough to deduce from them sufficiently complete equations for physical events sufficiently free from arbitrariness, in order to base upon such a foundation a theory of measuring rods and clocks. If one did not wish to forego a physical interpretation of the co-ordinates in general (something which, in itself, would be possible), it was better to permit such inconsistency—with the obligation, however, of eliminating it at a later stage of the theory. But one must not legalize the mentioned sin so far as to imagine that intervals are physical entities of a special type, intrinsically different from other physical variables ("reducing physics to geometry," etc.).

We now shall inquire into the insights of definite nature which physics owes to the special theory of relativity.

(1) There is no such thing as simultaneity of distant events; consequently there is also no such thing as immediate action at a distance in the sense of Newtonian mechanics. Although the introduction of actions at a distance, which propogate with the speed of light, remains thinkable, according to this theory, it appears unnatural; for in such a theory there could be no such thing as a reasonable statement of the principle of conservation of energy. It therefore appears unavoidable that physical reality must be described in terms of continuous functions in space. The material point, therefore, can hardly be conceived any more as the basic concept of the theory.

(2) The principles of the conservation of momentum and of the conservation of energy are fused into one single principle. The inert mass of a closed system is identical with its energy, thus eliminating mass as an independent concept.

Remark. The speed of light $c$ is one of the quantities which occurs as "universal constant" in physical equations. If, however, one introduces as unit of time instead of the second the time in which light travels 1 cm, $c$ no longer occurs in the equations. In this sense one could say that the constant $c$ is only an *apparently* universal constant.

It is obvious and generally accepted that one could eliminate two more universal constants from physics by introducing, instead of the gram and the centimeter, properly chosen "natural" units (for example, mass and radius of the electron).

If one considers this done, then only "dimension-less" constants could occur in the basic equations of physics. Concerning such I would like to state a theorem which at present can not be based upon anything more than upon a faith in the simplicity, i.e., intelligibility, of nature: there are no *arbitrary* constants of this kind; that is to say, nature is so constituted that it is possible logically to lay down such strongly determined laws that within these laws only rationally completely determined constants occur (not constants, therefore, whose numerical value could be changed without destroying the theory). - - -

The special theory of relativity owes its origin to Maxwell's equations of the electromagnetic field. Inversely the latter can be grasped formally in satisfactory fashion only by way of the special theory of relativity. Maxwell's equations are the simplest Lorentz-invariant field equations which can be postulated for an anti-symmetric tensor derived from a vector field. This in itself would be satisfactory, if we did not know from quantum phenomena that Maxwell's theory does not do justice to the energetic properties of radiation. But how Maxwell's theory would have to be modified in a natural fashion, for this even the special theory of relativity offers no adequate foothold. Also to Mach's question: "how does it come about that inertial systems are physically distinguished above all other co-ordinate systems?" this theory offers no answer.

That the special theory of relativity is only the first step of a necessary development became completely clear to me only in my efforts to represent gravitation in the framework of this theory. In classical mechanics, interpreted in terms of the field, the potential of gravitation appears as a *scalar* field (the simplest theoretical possibility of a field with a single component). Such a scalar theory of the gravitational field can easily be made invariant under the group of Lorentz-transformations. The following program appears natural, therefore: The total physical field consists of a scalar field (gravitation) and a vector field (electromagnetic field); later insights may eventually make necessary the introduction of still more complicated types of fields; but to begin with one did not need to bother about this.

The possibility of the realization of this program was, however, dubious from the very first, because the theory had to combine the following things:

(1) From the general considerations of special relativity theory it was clear that the *inert* mass of a physical system increases with the total energy (therefore, e.g., with the kinetic energy).

(2) From very accurate experiments (specially from the torsion balance experiments of Eötvös) it was empirically known with very high accuracy that the gravitational mass of a body is exactly equal to its *inert* mass.

It followed from (1) and (2) that the *weight* of a system depends in a precisely known manner on its total energy. If the theory did not accomplish this or could not do it naturally, it was to be rejected. The condition is most naturally expressed as follows: the acceleration of a system falling freely in a given gravitational field is independent of the nature of the falling system (specially therefore also of its energy content).

It then appeared that, in the framework of the program sketched, this elementary state of affairs could not at all or at any rate not in any natural fashion, be represented in a satisfactory way. This convinced me that, within the frame of the special theory of relativity, there is no room for a satisfactory theory of gravitation.

Now it came to me: The fact of the equality of inert and heavy mass, i.e., the fact of the independence of the gravitational acceleration of the nature of the falling substance, may be expressed as follows: In a gravitational field (of small spatial extension) things behave as they do in a space free of gravitation, if one introduces in it, in place of an "inertial system," a reference system which is accelerated relative to an inertial system.

If then one conceives of the behavior of a body, in reference to the latter reference system, as caused by a "real" (not merely apparent) gravitational field, it is possible to regard this reference system as an "inertial system" with as much justification as the original reference system.

So, if one regards as possible, gravitational fields of arbitrary extension which are not initially restricted by spatial limitations, the concept of the "inertial system" becomes completely empty. The concept, "acceleration relative to space," then loses every meaning and with it the principle of inertia together with the entire paradox of Mach.

The fact of the equality of inert and heavy mass thus leads quite naturally to the recognition that the basic demand of the special theory of relativity (invariance of the laws under Lorentz-transformations) is too narrow, i.e., that an invariance of the laws must be postulated also relative to *non-linear* transformations of the co-ordinates in the four-dimensional continuum.

This happened in 1908. Why were another seven years required for the construction of the general theory of relativity? The main reason lies in the fact that it is not so easy to free oneself from the idea that co-ordinates must have an immediate metrical meaning. The transformation took place in approximately the following fashion.

We start with an empty, field-free space, as it occurs—related to an inertial system—in the sense of the special theory of relativity, as the simplest of all imaginable physical situations. If we now think of a non-inertial system introduced by assuming that the new system is uniformly accelerated against the inertial system (in a three-dimensional description) in one direction (conveniently defined), then there exists with reference to this system a static parallel gravitational field. The reference system may thereby be chosen as rigid, of Euclidian type, in three-dimensional metric relations. But the time, in which the field appears as static, is *not* measured by *equally constituted* stationary clocks. From this special example one can already recognize that the immediate metric significance of the co-ordinates is lost if one admits non-linear transformations of co-ordinates at all. To do the latter is, however, *obligatory* if one wants to do justice to the equality of gravitational and inert mass by means of the basis of the theory, and if one wants to overcome Mach's paradox as concerns the inertial systems.

If, then, one must give up the attempt to give the co-ordinates an immediate metric meaning (differences of co-ordinates = measurable lengths, viz., times), one will not be able to avoid treating as equivalent all co-ordinate systems, which can be created by the continuous transformations of the co-ordinates.

The general theory of relativity, accordingly, proceeds from the following principle: Natural laws are to be expressed by equations which are covariant under the group of continuous co-ordinate transformations. This group replaces the group of the Lorentz-transformations of the special theory of relativity, which forms a sub-group of the former.

This demand by itself is of course not sufficient to serve as point of departure for the derivation of the basic concepts of physics. In the first instance one may even contest [the idea] that the demand by itself contains a real restriction for the physical laws; for it will always be possible thus to reformulate a law, postulated at first only for certain co-ordinate systems, such that the new formulation becomes formally universally co-variant. Beyond this it is clear from the beginning that an infinitely large number of field-laws can be formulated which have this property of covariance. The eminent heuristic significance of the general principles of relativity lies in the fact that it leads us to the search for those systems of equations which are *in their general covariant* formulation the *simplest ones possible*; among these we shall have to look for the field equations of physical space. Fields which can be transformed into each other by such transformations describe the same real situation.

The major question for anyone doing research in this field is this: Of which mathematical type are the variables (functions of the co-ordinates) which permit the expression of the physical properties of space ("structure")? Only after that: Which equations are satisfied by those variables?

The answer to these questions is today by no means certain. The path chosen by the first formulation of the general theory of relativity can be characterized as follows. Even though we do not know by what type field-variables (structure) physical space is to be characterized, we do know with certainty a special case: that of the "field-free" space in the special theory of relativity. Such a space is characterized by the fact that for a properly chosen co-ordinate system the expression

$$ds^2 = dx_1^2 + dx_2^2 + dx_3^2 - dx_4^2 \quad \cdot \quad \cdot \quad \cdot \quad (1)$$

belonging to two neighboring points, represents a measurable quantity (square of distance), and thus has a real physical meaning. Referred to an arbitrary system this quantity is expressed as follows:

$$ds^2 = g_{ik} dx_i dx_k \quad \cdot \quad \cdot \quad \cdot \quad \cdot \quad \cdot \quad (2)$$

whereby the indices run from 1 to 4. The $g_{ik}$ form a (real) symmetrical tensor. If, after carrying out a transformation on field (1), the first derivatives of the $g_{ik}$ with respect to the co-ordinates do not vanish, there exists a gravitational field with reference to this system of co-ordinates in the sense of the above consideration, a gravitational field, moreover, of a very special type. Thanks to Riemann's investigation of $n$-dimensional metrical spaces this special field can be invariantly characterized:

(1) Riemann's curvature-tensor $R_{iklm}$, formed from the coefficients of the metric (2) vanishes.

(2) The orbit of a mass-point in reference to the inertial system (relative to which (1) is valid) is a straight line, therefore an extremal (geodetic). The latter, however, is already a characterization of the law of motion based on (2).

The *universal* law of physical space must now be a generalization of the law just characterized. I now assume that there are two steps of generalization:

(a) pure gravitational field

(b) general field (in which quantities corresponding some-how to the electromagnetic field occur, too).

The instance (a) was characterized by the fact that the field can still be represented by a Riemann-metric (a), i.e., by a symmetric tensor, whereby, however, there is no representation in the form (1) (except in infinitesimal regions). This means that in the case (a) the Riemann-tensor does not vanish. It is clear, however, that in this case a field-law must be valid, which is a generalization (loosening) of this law. If this law also is to be of the second order of differentiation and linear in the second derivations, then only the equation, to be obtained by a single contraction

$$0 = R_{kl} = g^{im} R_{iklm}$$

came under consideration as field-equation in the case of (a). It appears natural, moreover, to assume that also in the case of (a) the geodetic line is still to be taken as representing the law of motion of the material point.

It seemed hopeless to me at that time to venture the attempt of representing the total field (b) and to ascertain field-laws for it. I preferred, therefore, to set up a preliminary formal frame for the representation of the entire physical reality; this was necessary in order to be able to investigate, at least preliminarily, the usefulness of the basic idea of general relativity. This was done as follows.

In Newton's theory one can write the field-law of gravitation thus:

$$\Delta\varphi = 0$$

($\varphi$ = gravitation-potential) at points, where the density of matter, $\varrho$, vanishes. In general one may write (Poisson equation)

$$\Delta\varphi = 4\pi k\varrho \cdot \quad (\varrho = \text{mass-density}).$$

In the case of the relativistic theory of the gravitational field $R_{ik}$ takes the place of $\Delta\varphi$. On the right side we shall then have to place a tensor also in place of $\varrho$. Since we know from the special theory of relativity that the (inert) mass equals energy, we shall have to put on the right side the tensor of energy-density—more precisely the entire energy-density, insofar as it does not belong to the pure gravitational field. In this way one gets the field-equations

$$R_{ik} - \tfrac{1}{2} g_{ik} R = -k T_{ik}.$$

The second member on the left side is added because of formal reasons; for the left side is written in such a way that its divergence disappears identically in the sense of the absolute differential calculus. The right side is a formal condensation of all things whose comprehension in the sense of a field-theory is still problematic. Not for a moment, of course, did I doubt that this formulation was merely a makeshift in order to give the general principle of relativity a preliminary closed expression. For it was essentially not anything *more* than a theory of the gravitational field, which was somewhat artificially isolated from a total field of as yet unknown structure.

If anything in the theory as sketched—apart from the demand of the invariance of the equations under the group of the continuous co-ordinate-transformations—can possibly make the claim to final significance, then it is the theory of the limiting case of the pure gravitational field and its relation to the metric structure of space. For this reason, in what immediately follows we shall speak only of the equations of the pure gravitational field.

The peculiarity of these equations lies, on the one hand, in their complicated construction, especially their non-linear character, as regards the field-variables and their derivatives, and, on the other hand, in the almost compelling necessity with which the transformation-group determines this complicated field-law. If one had stopped with the special theory of relativity, i.e., with the invariance under the Lorentz-group, then the field-law $R_{ik} = 0$ would remain invariant also within the frame of this narrower group. But, from the point of view of the narrower group there would at first exist no reason for representing gravitation by so complicated a structure as is represented by the symmetric tensor $g_{ik}$. If, nonetheless, one would find sufficient reasons for it, there would then arise an immense number of field-laws out of quantities $g_{ik}$, all of which are co-variant under Lorentz-transformations (not, however, under the general group). However, even if, of all the conceivable Lorentz-invariant laws, one had accidentally guessed precisely yet the law which belongs to the wider group, one would still not yet be on the plane of insight achieved by the general principle of relativity. For, from the standpoint of the Lorentz-group two solutions would incorrectly have to be viewed as physically different from each other, if they can be transformed into each other by a non-linear transformation of co-ordinates, i.e., if they are, from the point of view of the wider field, only different representations of the same field.

One more general remark concerning field-structure and the group. It is clear that in general one will judge a theory to be the more nearly perfect the simpler a "structure" it postulates and the broader the group is concerning which the field-equations are invariant. One sees now that these two demands get in each other's way. For example: according to the special theory of relativity (Lorentz-Group) one can set up a covariant law for simplest structure imaginable (a scalar field), whereas in the general theory of relativity (wider group of the continuous transformations of co-ordinates) there is an invariant field-law only for the more complicated structure of the symmetric tensor. We have already given *physical* reasons for the fact that in physics invariance under the wider group has to be demanded:[1] from a purely mathematical standpoint I can see no necessity for sacrificing the simpler structure to the generality of the group.

The group of the general relativity is the first one which demands that the simplest invariant law be no longer linear or homogeneous in the field-variables and in their differential quotients. This is of fundamental importance for the following reason. If the field-law is linear (and homogeneous), then the sum of two solutions is again a solution; as, for example: in Maxwell's field-equations for the vacuum. In such a theory it is impossible to deduce from the field equations alone an interaction between bodies, which can be described separately by means of solutions of the system. For this reason all theories up to now required, in addition to the field equations, special equations for the motion of material bodies under the influence of the fields. In the relativistic theory of gravitation, it is true, the law of motion (geodetic line) was originally postulated independently in addition to the field-law equations. Afterwards, however, it became apparent that the law of motion need

---

[1] To remain with the narrower group and at the same time to base the relativity theory of gravitation upon the more complicated (tensor-) structure implies a naïve inconsequence. Sin remains sin, even if it is committed by otherwise ever so respectable men.

not (and must not) be assumed independently, but that it is already implicitly contained within the law of the gravitational field.

The essence of this genuinely complicated situation can be visualized as follows: A single material point at rest will be represented by a gravitational field which is everywhere finite and regular, except at the position where the material point is located: there the field has a singularity. If, however, one computes by means of the integration of the field-equations the field which belongs to two material points at rest, then this field has, in addition to the singularities at the positions of the material points, a line consisting of singular points, which connects the two points. However, it is possible to stipulate a motion of the material points in such a way that the gravitational field which is determined by them does not become singular anywhere at all except at the material points. These are precisely those motions which are described in first approximation by Newton's laws. One may say, therefore: The masses move in such fashion that the solution of the field-equation is nowhere singular except in the mass points. This attribute of the gravitational equations is intimately connected with their non-linearity, and this is a consequence of the wider group of transformations.

Now it would of course be possible to object: If singularities are permitted at the positions of the material points, what justification is there for forbidding the occurrence of singularities in the rest of space? This objection would be justified if the equations of gravitation were to be considered as equations of the total field. [Since this is not the case], however, one will have to say that the field of a material particle may the less be viewed as a *pure gravitational field* the closer one comes to the position of the particle. If one had the field-equation of the total field, one would be compelled to demand that the particles themselves would *everywhere* be describable as singularity-free solutions of the completed field-equations. Only then would the general theory of relativity be a *complete* theory.

Before I enter upon the question of the completion of the general theory of relativity, I must take a stand with reference to the most successful physical theory of our period, viz., the statistical quantum theory which, about twenty-five years ago, took on a consistent logical form (Schrödinger, Heisenberg, Dirac, Born). This is the only theory at present which permits a unitary grasp of experiences concerning the quantum character of micro-mechanical events. This theory, on the one hand, and the theory of relativity on the other, are both considered correct in a certain sense, although their combination has resisted all efforts up to now. This is probably the reason why among contemporary theoretical physicists there exist entirely differing opinions concerning the question as to how the theoretical foundation of the physics of the future will appear. Will it be a field theory; will it be in essence a statistical theory? I shall briefly indicate my own thoughts on this point.

Physics is an attempt conceptually to grasp reality as it is thought independently of its being observed. In this sense one speaks of "physical reality." In pre-quantum physics there was no doubt as to how this was to be understood. In Newton's theory reality was determined by a material point in space and time; in Maxwell's theory, by the field in space and time. In quantum mechanics it is not so easily seen. If one asks: does a $\psi$-function of the quantum theory represent a real factual situation in the same sense in which this is the case of a material system of points or of an electromagnetic field, one hesitates to reply with a simple "yes" or "no"; why? What the $\psi$-function (at a definite time) asserts, is this: What is the probability for finding a definite physical magnitude $q$ (or $p$) in a definitely given interval, if I measure it at time $t$? The probability is here to be viewed as an empirically determinable, and therefore certainly as a "real" quantity which I may determine if I create the same $\psi$-function very often and perform a $q$-measurement each time. But what about the single measured value of $q$? Did the respective individual system have this $q$-value even before the measurement? To this question there is no definite answer within the framework of the [existing] theory, since the measurement is a process which implies a finite disturbance of the system from the outside; it would therefore be thinkable that the system obtains a definite numerical value for $q$ (or $p$), i.e., the measured numerical value, only through the measurement itself. For the further discussion I shall assume two physicists, $A$ and $B$, who represent a different conception with reference to the real situation as described by the $\psi$-function.

$A$. The individual system (before the measurement) has a definite value of $q$ (i.e., $p$) for all variables of the system, and more specifically, *that* value which is determined by a measurement of this variable. Proceeding from this conception, he will state: The $\psi$-function is no exhaustive description of the real situation of the system but an incomplete description; it expresses only what we know on the basis of former measurements concerning the system.

$B$. The individual system (before the measurement) has no definite value of $q$ (i.e., $p$). The value of the measurement only arises in cooperation with the unique probability which is given to it in view of the $\psi$-function only through the act of measurement itself. Proceeding from this conception, he will (or, at least, he may) state: the $\psi$-function is an exhaustive description of the real situation of the system.

We now present to these two physicists the following instance: There is to be a system which at the time $t$ of our observation consists of two partial systems $S_1$ and $S_2$, which at this time are spatially separated and (in the sense of the classical physics) are without significant reciprocity. The total system is to be completely described through a known $\psi$-function $\psi_{12}$ in the sense of quantum mechanics. All quantum theoreticians now agree upon the following: If I make a complete measurement of $S_1$, I get from the results of the measurement and from $\psi_{12}$ an entirely definite $\psi$-function $\psi_2$ of the system $S_2$. The character of $\psi_2$ then depends upon *what kind* of measurement I undertake on $S_1$.

Now it appears to me that one may speak of the real factual situation of the partial system $S_2$. Of this real factual situation, we know to begin with, before the measurement of $S_1$, even less than we know of a system described by the $\psi$-function. But on one supposition we should, in my opinion, absolutely hold fast: the real factual situation of the system $S_2$ is independent of what is done with the system $S_1$, which is spatially separated from the former. According to the type of measurement which I make of $S_1$, I get, however, a very different $\psi_2$ for the second partial system ($\Psi_2$, $\Psi_2^1$, . . .). Now, however, the real situation of $S_2$ must be independent of what happens to $S_1$. For the same real situation of $S_2$ it is possible therefore to find, according to one's choice, different types of $\psi$-function. (One can escape from this conclusion only by either assuming that the measurement of $S_1$ ((telepathically)) changes the real situation of $S_2$ or by denying independent real situations as such to things which are spatially separated from each other. Both alternatives appear to me entirely unacceptable.)

If now the physicists, $A$ and $B$, accept this consideration as valid, then $B$ will have to give up his position that the $\Psi$-function constitutes a complete description of a real factual situation. For in this case it would be impossible that two different types of $\psi$-functions could be co-ordinated with the identical factual situation of $S_2$.

The statistical character of the present theory would then have to be a necessary consequence of the incompleteness of the description of the systems in quantum mechanics, and there would no longer exist any ground for the supposition that a future basis of physics must be based upon statistics. - - -

It is my opinion that the contemporary quantum theory by means of certain definitely laid down basic concepts, which on the whole have been taken over from classical mechanics, constitutes an optimum formulation of the connections. I believe, however, that this theory offers no useful point of departure for future development. This is the point at which my expectation departs most widely from that of contemporary physicists. They are convinced that it is impossible to account for the essential aspects of quantum phenomena (apparently discontinuous and temporally not determined changes of the situation of a system, and at the same time corpuscular and undulatory qualities of the elementary bodies of energy) by means of a theory which describes the real state of things [objects] by continuous functions of space for which differential equations are valid. They are also of the opinion that in this way one can not understand the atomic structure of matter and of radiation. They rather expect that systems of differential equations, which could come under consideration for such a theory, in any case would have no solutions which would be regular (free from singularity) everywhere in four-dimensional space. Above everything else, however, they believe that the apparently discontinuous character of elementary events can be described only by means of an essentially statistical theory, in which the discontinuous changes of the systems are taken into account by way of the continuous changes of the probabilities of the possible states.

All of these remarks seem to me to be quite impressive. However, the question which is really determinative appears to me to be as follows: What can be attempted with some hope of success in view of the present situation of physical theory? At this point it is the experiences with the theory of gravitation which determine my expectations. These equations give, from my point of view, more warrant for the expectation to assert something *precise* than all other equations of physics. One may, for example, call on Maxwell's equations of empty space by way of comparison. These are formulations which coincide with the experiences of infinitely weak electro-magnetic fields. This empirical origin already determines their linear form; it has, however, already been emphasized above that the true laws can not be linear. Such linear laws fulfill the super-position-principle for their solutions, but contain no assertions concerning the interaction of elementary bodies. The true laws can not be linear nor can they be derived from such. I have learned something else from the theory of gravitation: No ever so inclusive collection of empirical facts can ever lead to the setting up of such complicated equations. A theory can be tested by experience, but there is no way from experience to the setting up of a theory. Equations of such complexity as are the equations of the gravitational field can be found only through the discovery of a logically simple mathematical condition which determines the equations completely or [at least] almost completely. Once one has those sufficiently strong formal conditions, one requires only little knowledge of facts for the setting up of a theory; in the case of the equations of gravitation it is the four-dimensionality and the symmetric tensor as expression for the structure of space which, together with the invariance concerning the continuous transformation-group, determine the equations almost completely.

Our problem is that of finding the field equations for the total field. The desired structure must be a generalization of the symmetric tensor. The group must not be any narrower than that of the continuous transformations of co-ordinates. If one introduces a richer structure, then the group will no longer determine the equations as strongly as in the case of the symmetrical tensor as structure. Therefore it would be most beautiful, if one were to succeed in expanding the group once more, analogous to the step which led from special relativity to general relativity. More specifically I have attempted to draw upon the group of the complex transformations of the co-ordinates. All such endeavors were unsuccessful. I also gave up an open or concealed raising of the number of dimensions of space, an endeavor which was originally undertaken by Kaluza and which, with its projective variant, even today has its adherents. We shall limit ourselves to the four-dimensional space and to the group of the continuous real transformations of co-ordinates. After many years of fruitless searching I consider the solution sketched in what follows as the logically most satisfactory.

In place of the symmetrical $g_{ik}$ ($g_{ik} = g_{ki}$), the non-symmetrical tensor $g_{ik}$ is introduced. This magnitude is constituted by a symmetric part $s_{ik}$ and by a real or purely imaginary anti-symmetric $a_{ik}$, thus:

$$g_{ik} = s_{ik} + a_{ik}.$$

Viewed from the standpoint of the group the combination of $s$ and $a$ is arbitrary, because the tensors $s$ and $a$ individually have tensor-character. It turns out, however, that these $g_{ik}$ (viewed as a whole) play a quite analogous rôle in the construction of the new theory as the symmetric $g_{ik}$ in the theory of the pure gravitational field.

This generalization of the space structure seems natural also from the standpoint of our physical knowledge, because we know that the electro-magnetic field has to do with an anti-symmetric tensor.

For the theory of gravitation it is furthermore essential that from the symmetric $g_{ik}$ it is possible to form the scalar density $\sqrt{|g_{ik}|}$ as well as the contravariant tensor $g^{ik}$ according to the definition

$$g_{ik}g^{il} = \delta_k^l \quad (\delta_k^l = \text{Kronecker-Tensor}).$$

These concepts can be defined in precisely corresponding manner for the non-symmetric $g_{ik}$, also for tensor-densities.

In the theory of gravitation it is further essential that for a given symmetrical $g_{ik}$-field a field $\Gamma_{ik}^l$ can be defined, which is symmetric in the lower indices and which, considered geometrically, governs the parallel displacement of a vector. Analogously for the non-symmetric $g_{ik}$ a non-symmetric $\Gamma_{ik}^l$ can be defined, according to the formula

$$g_{ik,l} - g_{sk}\Gamma_{il}^s - g_{is}\Gamma^s = 0, \ldots \tag{A}$$

which coincides with the respective relation of the symmetrical $g$, only that it is, of course, necessary to pay attention here to the position of the lower indices in the $g$ and $\Gamma$.

Just as in the theory of a symmetrical $g_{ik}$, it is possible to form a curvature $R^i{}_{klm}$ out of the $\Gamma$ and a contracted curvature $R_{kl}$. Finally, with the use of a variation principle, together with (A), it is possible to find compatible field-equations:

$$g^{\underline{ik}} = \tfrac{1}{2}(g^{ik} - g^{ki})\sqrt{-|g_{ik}|)} \tag{$B_1$}$$

$$\Gamma_{\underline{is}}^{a} = 0 \left( \Gamma_{is}^s = \frac{1}{2}(\Gamma_{is}^s - \Gamma_{si}^s) \right) \tag{$B_2$}$$

$$R_{\underline{ik}} = 0 \tag{$C_1$}$$

$$R_{\underline{kl},m} + R_{\underline{lm},k} + R_{\underline{mk},l} = 0 \tag{$C_2$}$$

Each of the two equations $(B_1)$, $(B_2)$ is a consequence of the other, if $(A)$ is satisfied. $R_{kl}$ means the symmetric, $R_{kl}$ the anti-symmetric part of $R_{kl}$.

If the anti-symmetric part of $g_{ik}$ vanishes, these formulas reduce to $(A)$ and $(C_1)$—the case of the pure gravitational field.

I believe that these equations constitute the most natural generalization of the equations of gravitation.[2] The proof of their physical usefulness is a tremendously difficult task, inasmuch as mere approximations will not suffice. The question is:

[2]The theory here proposed, according to my view, represents a fair probability of being found valid, if the way to an exhaustive description of physical reality on the basis of the continuum turns out to be possible at all.

What are the everywhere regular solutions of these equations? - - -

This exposition has fulfilled its purpose if it shows the reader how the efforts of a life hang together and why they have led to expectations os a definite form.

INSTITUTE FOR ADVANCED STUDY
PRINCETON, NEW JERSEY

# Selections from
# *Out of My Later Years*

This collection of essays was written during the last twenty years of Einstein's life, after he had made his greatest contributions to science and had achieved international celebrity as the preeminent thinker of the time. In a change from his earlier works, Einstein no longer sought to explain the basic workings of his greatest theory—relativity—instead laying out a broader historical perspective on the development of physics. In 1936, when Einstein wrote the longest and most detailed of these essays, "Physics and Reality," the scientific world was undergoing a series of revolutions based on new understanding of both Einstein's theory of relativity, and quantum mechanics.

Although Einstein was instrumental in the development of quantum theory with his 1905 paper on the Photoelectric Effect, very few of his popular writings focus on it. Unlike relativity, which provided a deterministic explanation of physical phenomena, quantum mechanics is fundamentally probabilistic, which Einstein had great difficulty accepting. Consider what quantum theory says: a particle can exist in two states simultaneously, and will only be forced to make a particular (and random) choice when the system is observed. Such systems are so incompatible with the macroscopic world that Einstein posited that if we were able to investigate microscopic phenomena on the smallest scales, we would be able to find deterministic relations.

He also took issue with the fact that quantum mechanics requires an absolute time and space, concepts that were ruled out by his own theory of relativity. Einstein, Podolsky, and Rosen argued one year earlier that the two theories created a paradox.

Two subatomic particles that were created in a high-energy experiment would be entangled with one another and thus the measurement of one would "force" the other, even far away, into a particular quantum state. This idea seemed to suggest that because the effect would occur instantaneously, a signal between the two was traveling faster than light, and relativity precludes faster-than-light travel. The modern interpretation is that the Einstein-Podolsky-Rosen paradox can be resolved by the fact that no information is flowing from one particle to the other.

It is clear from his writings that Einstein was well aware that he was in the midst of a revolution—one that he had, in large part, helped to bring about. His concerns about the philosophical problems with relativity and quantum mechanics ultimately resolved themselves through the development of relativistic quantum mechanics, quantum field theory, and may ultimately form the foundation for string theory, which in turn may satisfy Einstein's dream of unifying the forces of physics.

# THE THEORY OF RELATIVITY

From Albert Einstein: *Out of my Later Years*, Philosophical Library, New York 1950.

Under the title "Relativity: Essence of the Theory of Relativity" originally published in *The American People's Encyclopedia* XVI, Chicago 1949.

Mathematics deals exclusively with the relations of concepts to each other without consideration of their relation to experience. Physics too deals with mathematical concepts; however, these concepts attain physical content only by the clear determination of their relation to the objects of experience. This in particular is the case for the concepts of motion, space, time.

The theory of relativity is that physical theory which is based on a consistent physical interpretation of these three concepts. The name "theory of relativity" is connected with the fact that motion from the point of view of possible experience always appears as the *relative* motion of one object with respect to another (e.g., of a car with respect to the ground, or the earth with respect to the sun and the fixed stars). Motion is never observable as "motion with respect to space" or, as it has been expressed, as "absolute motion." The "principle of relativity" in its widest sense is contained in the statement: The totality of physical phenomena is of such a character that it gives no basis for the introduction of the concept of "absolute motion"; or shorter but less precise: There is no absolute motion.

It might seem that our insight would gain little from such a negative statement. In reality, however, it is a strong restriction for the (conceivable) laws of nature. In this sense there exists an analogy between the theory of relativity and thermodynamics. The latter too is based on a negative statement: "There exists no perpetuum mobile."

The development of the theory of relativity proceeded in two steps, "special theory of relativity" and "general theory of relativity." The latter presumes the validity of the former as a limiting case and is its consistent continuation.

# A. SPECIAL THEORY OF RELATIVITY
## Physical Interpretation of Space and Time
### in Classical Mechanics

Geometry, from a physical standpoint, is the totality of laws according to which rigid bodies mutually at rest can be placed with respect to each other (e.g., a triangle consists of three rods whose ends touch permanently). It is assumed that with such an interpretation the Euclidean laws are valid. "Space" in this interpretation is in principle an infinite rigid body (or skeleton) to which the position of all other bodies is related (body of reference). Analytic geometry (Descartes) uses as the body of reference, which represents space, three mutually perpendicular rigid rods on which the "coordinates" $(x, y, z)$ of space points are measured in the known manner as perpendicular projections (with the aid of a rigid unit-measure).

Physics deals with "events" in space and time. To each event belongs, besides its place coordinates $x, y, z$, a time value $t$. The latter was considered measurable by a clock (ideal periodic process) of negligible spatial extent. This clock $C$ is to be considered at rest at one point of the coordinate system, e.g., at the coordinate origin $(x = y = z = O)$. The time of an event taking place at a point $P(x, y, z)$ is then defined as the time shown on the clock $C$ simultaneously with the event. Here the concept "simultaneous" was assumed as physically meaningful without special definition. This is a lack of exactness which seems harmless only since with the help of light (whose velocity is practically infinite from the point of view of daily experience) the simultaneity of spatially distant events can apparently be decided immediately. The special theory of relativity removes this lack of precision by defining simultaneity physically with the use of light signals. The time $t$ of the event in $P$ is the reading of the clock $C$ at the time of arrival of a light signal emitted from the event, corrected with respect to the time needed for the light signal to travel the distance. This correction presumes (postulates) that the velocity of light is constant.

This definition reduces the concept of simultaneity of spatially distant events to that of the simultaneity of events happening at the same place (coincidence), namely the arrival of the light signal at $C$ and the reading of $C$.

Classical mechanics is based on Galileo's principle: A body is in rectilinear and uniform motion as long as other bodies do not act on it. This statement cannot be valid for arbitrary moving systems of coordinates. It can claim validity only for so-called "inertial systems." Inertial systems are in rectilinear and uniform motion with respect to each other. In classical physics laws claim validity only with respect to all inertial systems (special principle of relativity).

It is now easy to understand the dilemma which has led to the special theory of relativity. Experience and theory have gradually led to the conviction that light in empty space always travels with the same velocity $c$ independent of its color and the state of motion of the source of light (principle of the constancy of the velocity of light— in the following referred to as "L-principle"). Now elementary intuitive considerations seem to show that the same light ray *cannot* move with respect to all inertial systems with the same velocity $c$. The L-principle seems to contradict the special principle of relativity.

It turns out, however, that this contradiction is only an apparent one which is based essentially on the prejudice about the absolute character of time or rather of the simultaneity of distant events. We just saw that $x, y, z$ and $t$ of an event can, for the moment, be defined only with respect to a certain chosen system of coordinates (inertial system). The transformation of the $x, y, z, t$ of events which has to be carried out with the passage from one inertial system to another (coordinate transformation), is a problem which cannot be solved without special physical assumptions. However, the following postulate is exactly sufficient for a solution: *The L-principle holds for all inertial systems* (application of the special principle of relativity to the L-principle). The transformations thus defined, which are linear in $x, y, z, t$, are called Lorentz transformations. Lorentz transformations are formally characterized by the demand that the expression

$$dx^2 + dy^2 + dz^2 - c^2 dt^2,$$

which is formed from the coordinate differences $dx$, $dy$, $dz$, $dt$ of two infinitely close events, be invariant (i.e., that through the transformation it goes over into the *same* expression formed from the coordinate differences in the new system).

With the help of the Lorentz transformations the special principle of relativity can be expressed thus: The laws of nature are invariant with respect to Lorentz-transformations (i.e., a law of nature does not change its form if one introduces into it a new inertial system with the help of a Lorentz-transformation on $x$, $y$, $z$, $t$).

The special theory of relativity has led to a clear understanding of the physical concepts of space and time and in connection with this to a recognition of the behavior of moving measuring rods and clocks. It has in principle removed the concept of absolute simultaneity and thereby also that of instantaneous action at a distance in the sense of Newton. It has shown how the law of motion must be modified in dealing with motions that are not negligibly small as compared with the velocity of light. It has led to a formal clarification of Maxwell's equations of the electromagnetic field; in particular it has led to an understanding of the essential oneness of the electric and the magnetic field. It has unified the laws of conservation of momentum and of energy into one single law and has demonstrated the equivalence of mass and energy. From a formal point of view one may characterize the achievement of the special theory of relativity thus: it has shown generally the role which the universal constant $c$ (velocity of light) plays in the laws of nature and has demonstrated that there exists a close connection between the form in which time on the one hand and the spatial coordinates on the other hand enter into the laws of nature.

## B. GENERAL THEORY OF RELATIVITY

The special theory of relativity retained the basis of classical mechanics in one fundamental point, namely the statement: The laws of nature are valid only with respect to inertial systems. The "permissible" transformations for the coordinates (i.e., those which leave the form of the laws unchanged) are *exclusively* the (linear) Lorentz-transformations. Is this restriction really founded in physical facts? The following argument convincingly denies it.

Principle of equivalence. A body has an inertial mass (resistance to acceleration) and a heavy mass (which determines the weight of the body in a given gravitational field, e.g., that at the surface of the earth). These two quantities, so different according to their definition, are according to experience measured by one and the same number. There must be a deeper reason for this. The fact can also be described thus: In a gravitational field different masses receive the same acceleration. Finally, it can also be expressed thus: Bodies in a gravitational field behave as in the absence of a gravitational field if, in the latter case, the system of reference used is a uniformly accelerated coordinate system (instead of an inertial system).

There seems, therefore, to be no reason to ban the following interpretation of the latter case. One considers the system as being "at rest" and considers the "apparent" gravitational field which exists with respect to it as a "real" one. This gravitational field "generated" by the acceleration of the coordinate system would of course be of unlimited extent in such a way that it could not be caused by gravitational masses in a finite region; however, if we are looking for a field-like theory, this fact need not deter us. With this interpretation the inertial system loses its meaning and one has an "explanation" for the equality of heavy and inertial mass (the same property of matter appears as weight or as inertia depending on the mode of description).

Considered formally, the admission of a coordinate system which is accelerated with respect to the original "inertial" coordinates means the admission of non-linear coordinate transformations, hence a mighty enlargement of the idea of invariance, i.e., the principle of relativity.

First, a penetrating discussion, using the results of the special theory of relativity, shows that with such a generalization the coordinates can no longer be interpreted directly as the results of measurements.

Only the coordinate difference together with the field quantities which describe the gravitational field determine measurable distances between events. After one has found oneself forced to admit non-linear coordinate transformations as transformations between equivalent coordinate systems, the simplest demand appears to admit all continuous coordinate transformations (which form a group), i.e., to admit arbitrary curvilinear coordinate systems in which the fields are described by regular functions (general principle of relativity).

Now it is not difficult to understand why the general principle of relativity (*on the basis of the equivalence principle*) has led to a theory of gravitation. There is a special kind of space whose physical structure (field) we can presume as precisely known on the basis of the special theory of relativity. This is empty space without electromagnetic field and without matter. It is completely determined by its "metric" property: Let $dx_0$, $dy_0$, $dz_0$, $dt_0$ be the coordinate differences of two infinitesimally near points (events); then

$$(1) \qquad ds^2 = dx_0{}^2 + dy_0{}^2 + dz_0{}^2 - c^2 dt_0{}^2$$

is a measurable quantity which is independent of the special choice of the inertial system. If one introduces in this space the new coordinates $x_1, x_2, x_3, x_4$ through a general transformation of coordinates, then the quantity $ds^2$ for the same pair of points has an expression of the form

$$(2) \qquad ds^2 = \sum g_{ik} dx^i dx^k \text{ (summed for } i \text{ and } k \text{ from 1 to 4)}$$

where $g_{ik} = g_{kl}$. The $g_{ik}$ which form a "symmetric tensor" and are continuous functions of $x_1, \ldots x_4$ then describe according to the "principle of equivalence" a gravitational field of a special kind (namely one which can be retransformed to the form (1)). From Riemann's investigations on metric spaces the mathematical properties of this $g_{ik}$ field can be given exactly ("Riemann-condition"). However, what we are looking for are the equations satisfied by "general" gravitational fields. It is natural to assume that they too can be described as tensor-fields of the type $g_{ik}$, which in general do *not* admit a transformation to the form (1), i.e., which do not satisfy the "Riemann condition," but weaker conditions, which, just as the

Riemann condition, are independent of the choice of coordinates (i.e., are generally invariant). A simple formal consideration leads to weaker conditions which are closely connected with the Riemann condition. These conditions are the very equations of the pure gravitational field (on the outside of matter and at the absence of an electromagnetic field).

These equations yield Newton's equations of gravitational mechanics as an approximate law and in addition certain small effects which have been confirmed by observation (deflection of light by the gravitational field of a star, influence of the gravitational on the frequency of emitted light, slow rotation of the elliptic circuits of planets—perihelion motion of the planet Mercury). They further yield an explanation for the expanding motion of galactic systems, which is manifested by the red-shift of the light omitted from these systems.

The general theory of relativity is as yet incomplete insofar as it has been able to apply the general principle of relativity satisfactorily only to gravitational fields, but not to the total field. We do not yet know with certainty, by what mathematical mechanism the total field in space is to be described and what the general invariant laws are to which this total field is subject. One thing, however, seems certain: namely, that the general principle of relativity will prove a necessary and effective tool for the solution of the problem of the total field.

# $E = Mc^2$

From Albert Einstein: *Out of my Later Years*, Philosophical Library,
New York 1950.

First published in *Science Illustrated*, first issue, April 1946.

In order to understand the law of the equivalence of mass and energy, we must go back to two conservation or "balance" principles which, independent of each other, held a high place in pre-relativity physics. These were the principle of the conservation of energy and the principle of the conservation of mass. The first of these, advanced by Leibnitz as long ago as the seventeenth century, was developed in the nineteenth century essentially as a corollary of a principle of mechanics.

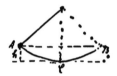

Drawing from Dr. Einstein's manuscript.

Consider, for example, a pendulum whose mass swings back and forth between the points $A$ and $B$. At these points the mass $m$ is higher by the amount $h$ than it is at $C$, the lowest point of the path (see drawing). At $C$, on the other hand, the lifting height has disappeared and instead of it the mass has a velocity $v$. It is as though the lifting height could be converted entirely into velocity, and vice versa. The exact relation would be expressed as mgh $= \frac{m}{2} v^2$, with $g$ representing the acceleration of gravity. What is interesting here is that this relation is independent of both the length of the pendulum and the form of the path through which the mass moves.

The significance is that something remains constant throughout the process, and that something is energy. At $A$ and at $B$ it is an energy of position, or "potential" energy; at $C$ it is an energy of motion, or "kinetic" energy. If this concept is correct, then the sum $mgh + m\frac{v^2}{2}$

must have the same value for any position of the pendulum, if $h$ is understood to represent the height above $C$, and $v$ the velocity at that point in the pendulum's path. And such is found to be actually the case. The generalization of this principle gives us the law of the conservation of mechanical energy. But what happens when friction stops the pendulum?

The answer to that was found in the study of heat phenomena. This study, based on the assumption that heat is an indestructible substance which flows from a warmer to a colder object, seemed to give us a principle of the "conservation of heat." On the other hand, from time immemorial it has been known that heat could be produced by friction, as in the fire-making drills of the Indians. The physicists were for long unable to account for this kind of heat "production." Their difficulties were overcome only when it was successfully established that, for any given amount of heat produced by friction, an exactly proportional amount of energy had to be expended. Thus did we arrive at a principle of the "equivalence of work and heat." With our pendulum, for example, mechanical energy is gradually converted by friction into heat.

In such fashion the principles of the conservation of mechanical and thermal energies were merged into one. The physicists were thereupon persuaded that the conservation principle could be further extended to take in chemical and electromagnetic processes—in short, could be applied to all fields. It appeared that in our physical system there was a sum total of energies that remained constant through all changes that might occur.

Now for the principle of the conservation. Mass is defined by the resistance that a body opposes to its acceleration (inert mass). It is also measured by the weight of the body (heavy mass). That these two radically different definitions lead to the same value for the mass of a body is, in itself, an astonishing fact. According to the principle—namely, that masses remain unchanged under any physical or chemical changes—the mass appeared to be the essential (because unvarying) quality of matter. Heating, melting, vaporization, or combining into chemical compounds would not change the total mass.

Physicists accepted this principle up to a few decades ago. But it proved inadequate in the face of the special theory of relativity. It was therefore merged with the energy principle just as, about 60 years before, the principle of the conservation of mechanical energy had been combined with the principle of the conservation of heat. We might say that the principle of the conservation of energy, having previously swallowed up that of the conservation of heat, now proceeded to swallow that of the conservation of mass—and holds the field alone.

It is customary to express the equivalence of mass and energy (though somewhat inexactly) by the formula $E = mc^2$, in which $c$ represents the velocity of light, about 186,000 miles per second. $E$ is the energy that is contained in a stationary body; $m$ is its mass. The energy that belongs to the mass $m$ is equal to this mass, multiplied by the square of the enormous speed of light—which is to say, a vast amount of energy for every unit of mass.

But if every gram of material contains this tremendous energy, why did it go so long unnoticed? The answer is simple enough: so long as none of the energy is given off externally, it cannot be observed. It is as though a man who is fabulously rich should never spend or give away a cent; no one could tell how rich he was.

Now we can reverse the relation that an increase of $E$ in the amount of energy must be accompanied by an increase of $\dfrac{E}{c^2}$ in the mass. I can easily supply energy to the mass—for instance, if I heat it by 10 degrees. So why not measure the mass increase, or weight increase, connected with this change? The trouble here is that in the mass increase the enormous factor $c^2$ occurs in the denominator of the fraction. In such a case the increase is too small to be measured directly; even with the most sensitive balance.

For a mass increase to be measurable, the change of energy per mass unit must be enormously large. We know of only one sphere in which such amounts of energy per mass unit are released: namely, radioactive disintegration. Schematically, the process goes like this: An atom of the mass $M$ splits into two atoms of the mass $M'$ and $M''$, which separate with tremendous kinetic energy. If we imagine these two masses as brought to rest—that is, if we take this energy of motion from them—then, considered together, they are essentially poorer in energy than was the original atom. According to the equivalence principle, the mass sum $M' + M''$ of the disintegration products must also be somewhat smaller than the original mass $M$ of the disintegrating atom—in contradiction to the old principle of the conservation of mass. The relative difference of the two is on the order of $\frac{1}{10}$ of one percent.

Now, we cannot actually weigh the atoms individually. However, there are indirect methods for measuring their weights exactly. We can likewise determine the kinetic energies that are transferred to the disintegration products $M'$ and $M''$. Thus it has become possible to test and confirm the equivalence formula. Also, the law permits us to calculate in advance, from precisely determined atom weights, just how much energy will be released with any atom disintegration we have in mind. The law says nothing, of course, as to whether—or how—the disintegration reaction can be brought about.

What takes place can be illustrated with the help of our rich man. The atom $M$ is a rich miser who, during his life, gives away no money (*energy*). But in his will he bequeaths his fortune to his sons $M'$ and $M''$, on condition that they give to the community a small amount, less than one thousandth of the whole estate (*energy or mass*). The sons together have somewhat less than the father had (*the mass sum $M' + M''$ is somewhat smaller than the mass M of the radioactive atom*). But the part given to the community, though relatively small, is still so enormously large (*considered as kinetic energy*) that it brings with it a great threat of evil. Averting that threat has become the most urgent problem of our time.

# WHAT IS THE THEORY
# OF RELATIVITY?

From Albert Einstein: *Out of my Later Years,* Philosophical Library,
New York 1950.

Written at the request of *Times,* London. First published under the tittle
"My Theory" in *Times,* Nov 28, 1919.

I gladly accede to the request of your colleague to write something for
*The Times* on relativity. After the lamentable breakdown of the old
active intercourse between men of learning, I welcome this opportu-
nity of expressing my feelings of joy and gratitude towards the
astronomers and physicists of England. It is thoroughly in keeping
with the great and proud traditions of scientific work in your country that eminent scientists should have spent much time and trouble,
and your scientific institutions have spared no expense, to test the
implications of a theory which was perfected and published during
the War in the land of your enemies. Even though the investigation
of the influence of the gravitational field of the sun on light rays is a
purely objective matter, I cannot forbear to express my personal thanks
to my English colleagues for their work; for without it I could hardly
have lived to see the most important implication of my theory tested.

We can distinguish various kinds of theories in physics. Most of
them are constructive. They attempt to build up a picture of the more
complex phenomena out of the materials of a relatively simple formal
scheme from which they start out. Thus the kinetic theory of gases
seeks to reduce mechanical, thermal and diffusional processes to movements of molecules—i.e., to build them up out of the hypothesis of
molecular motion. When we say that we have succeeded in understanding a group of natural processes, we invariably mean that a constructive theory has been found which covers the processes in question.

Along with this most important class of theories there exists a second, which I will call "principle-theories." These employ the analytic,

not the synthetic, method. The elements which form their basis and starting-point are not hypothetically constructed but empirically discovered ones, general characteristics of natural processes, principles that give rise to mathematically formulated criteria which the separate processes or the theoretical representations of them have to satisfy. Thus the science of thermodynamics seeks by analytical means to deduce necessary connections, which separate events have to satisfy, from the universally experienced fact that perpetual motion is impossible.

The advantages of the constructive theory are completeness, adaptability and clearness, those of the principle theory are logical perfection and security of the foundations.

The theory of relativity belongs to the latter class. In order to grasp its nature, one needs first of all to become acquainted with the principles on which it is based. Before I go into these, however, I must observe that the theory of relativity resembles a building consisting of two separate stories, the special theory and the general theory. The special theory, on which the general theory rests, applies to all physical phenomena with the exception of gravitation; the general theory provides the law of gravitation and its relations to the other forces of nature.

It has, of course, been known since the days of the ancient Greeks that in order to describe the movement of a body, a second body is needed to which the movement of the first is referred. The movement of a vehicle is considered in reference to the earth's surface, that of a planet to the totality of the visible fixed stars. In physics the body to which events are spatially referred is called the co-ordinate system. The laws of the mechanics of Galileo and Newton, for instance, can only be formulated with the aid of a coordinate system.

The state of motion of the co-ordinate system may not, however, be arbitrarily chosen, if the laws of mechanics are to be valid (it must be free from rotation and acceleration). A co-ordinate system which is admitted in mechanics is called an "inertial system." The state of motion of an inertial system is according to mechanics not one that is determined uniquely by nature. On the contrary, the following definition holds good:—a co-ordinate system that is moved uniformly and in a straight line relatively to an inertial system is likewise an inertial system. By the "special principle of relativity" is meant the generalization of this definition to include any natural event whatever: thus, every universal law of nature which is valid in relation to a co-ordinate system $C$, must also be valid, as it stands, in relation to a co-ordinate system $C'$, which is in uniform translatory motion relatively to $C$.

The second principle, on which the special theory of relativity rests, is the "principle of the constant velocity of light in vacuo." This principle asserts that light in vacuo always has a definite velocity of propagation (independent of the state of motion of the observer or of the source of the light). The confidence which physicists place in this principle springs from the successes achieved by the electro-dynamics of Clerk Maxwell and Lorentz.

Both the above-mentioned principles are powerfully supported by experience, but appear not to be logically reconcilable. The special theory of relativity finally succeeded in reconciling them logically by a modification of kinematics—i.e., of the doctrine of the laws relating to space and time (from the point of view of physics). It became clear that to speak of the simultaneity of two events had no meaning except in relation to a given co-ordinate system, and that the shape of measuring devices and the speed at which clocks move depend on their state of motion with respect to the co-ordinate system.

But the old physics, including the laws of motion of Galileo and Newton, did not fit in with the suggested relativist kinematics. From the latter, general mathematical conditions issued, to which natural laws had to conform, if the above-mentioned two principles were really to apply. To these, physics had to be adapted. In particular, scientists arrived at a new law of motion for (rapidly moving) mass points, which was admirably confirmed in the case of electrically charged particles. The most important upshot of the special theory of relativity concerned the inert mass of corporeal systems. It turned out that the inertia of a system necessarily depends on its energy-content, and this led straight to the notion that inert mass is simply latent energy. The principle of the conservation of mass lost its independence and became fused with that of the conservation of energy.

The special theory of relativity, which was simply a systematic development of the electro-dynamics of Clerk Maxwell and Lorentz, pointed beyond itself, however. Should the independence of physical laws of the state of motion of the co-ordinate system be restricted to the uniform translatory motion of co-ordinate systems in respect to each other? What has nature to do with our coordinate systems and their state of motion? If it is necessary for the purpose of describing nature, to make use of a co-ordinate system arbitrarily introduced by us, then the choice of its state of motion ought to be subject to no restriction; the laws ought to be entirely independent of this choice (general principle of relativity).

The establishment of this general principle of relativity is made easier by a fact of experience that has long been known, namely that the weight and the inertia of a body are controlled by the same constant. (Equality of inertial and gravitational mass.) Imagine a co-ordinate system which is rotating uniformly with respect to an inertial system in the Newtonian manner. The centrifugal forces which manifest themselves in relation to this system must, according to Newton's teaching, be regarded as effects of inertia. But these centrifugal forces are, exactly like the forces of gravity, proportional to the masses of the bodies. Ought it not to be possible in this case to regard the co-ordinate system as stationary and the centrifugal forces as gravitational forces? This seems the obvious view, but classical mechanics forbid it.

This hasty consideration suggests that a general theory of relativity must supply the laws of gravitation, and the consistent following up of the idea has justified our hopes.

But the path was thornier than one might suppose, because it demanded the abandonment of Euclidean geometry. This is to say, the laws according to which fixed bodies may be arranged in space, do not completely accord with the spatial laws attributed to bodies by Euclidean geometry. This is what we mean when we talk of the "curvature of space." The fundamental concepts of the "straight line," the "plan," etc., thereby lose their precise significance in physics.

In the general theory of relativity the doctrine of space and time, or kinematics, no longer figures as a fundamental independent of the rest of physics. The geometrical behavior of bodies and the motion of clocks rather depend on gravitational fields, which in their turn are produced by matter.

The new theory of gravitation diverges considerably, as regards principles, from Newton's theory. But its practical results agree so nearly with those of Newton's theory that it is difficult to find criteria for distinguishing them which are accessible to experience. Such have been discovered so far:—

1. In the revolution of the ellipses of the planetary orbits round the sun (confirmed in the case of Mercury).

2. In the curving of light rays by the action of gravitational fields (confirmed by the English photographs of eclipses).

3. In a displacement of the spectral lines towards the red end of the spectrum in the case of light transmitted to us from stars of considerable magnitude (unconfirmed so far).*

The chief attraction of the theory lies in its logical completeness. If a single one of the conclusions drawn from it proves wrong, it must be given up; to modify it without destroying the whole structure seems to be impossible.

Let no one suppose, however, that the mighty work of Newton can really be superseded by this or any other theory. His great and lucid ideas will retain their unique significance for all time as the foundation of our whole modern conceptual structure in the sphere of natural philosophy.

NOTE: Some of the statements in your paper concerning my life and person owe their origin to the lively imagination of the writer. Here is yet another application of the principle of relativity for the delectation of the reader:—Today I am described in Germany as a "German savant," and in England as a "Swiss Jew." Should it ever be my fate to be represented as a *bête noire*, I should, on the contrary, become a "Swiss Jew" for the Germans and a "German savant" for the English.

---

*Editor's Note: This criterion has also been confirmed in the meantime.

# PHYSICS AND REALITY

From Albert Einstein: *Out of my Later Years,* Philosophical Library,
New York 1950.

Originally published in the *Journal of the Franklin Institute,* vol. 221,
March 1936.

## 1. GENERAL CONSIDERATION CONCERNING THE METHOD OF SCIENCE

It has often been said, and certainly not without justification, that the man of science is a poor philosopher. Why then should it not be the right thing for the physicist to let the philosopher do the philosophizing? Such might indeed be the right thing at a time when the physicist believes he has at his disposal a rigid system of fundamental concepts and fundamental laws which are so well established that waves of doubt can not reach them; but, it can not be right at a time when the very foundations of physics itself have become problematic as they are now. At a time like the present, when experience forces us to seek a newer and more solid foundation, the physicist cannot simply surrender to the philosopher the critical contemplation of the theoretical foundations; for, he himself knows best, and feels more surely where the shoe pinches. In looking for a new foundation, he must try to make clear in his own mind just how far the concepts which he uses are justified, and are necessities.

The whole of science is nothing more than a refinement of everyday thinking. It is for this reason that the critical thinking of the physicist cannot possibly be restricted to the examination of the concepts of his own specific field. He cannot proceed without considering critically a much more difficult problem, the problem of analyzing the nature of everyday thinking.

On the stage of our subconscious mind appear in colorful succession sense experiences, memory pictures of them, representations and feelings. In contrast to psychology, physics treats directly only of sense experiences and of the "understanding" of their connection. But even the concept of the "real external world" of everyday thinking rests exclusively on sense impressions.

Now we must first remark that the differentiation between sense impressions and representations is not possible; or, at least it is not possible with absolute certainty. With the discussion of this problem, which affects also the notion of reality, we will not concern ourselves but we shall take the existence of sense experiences as given, that is to say as psychic experiences of special kind.

I believe that the first step in the setting of a "real external world" is the formation of the concept of bodily objects and of bodily objects of various kinds. Out of the multitude of our sense experiences we take, mentally and arbitrarily, certain repeatedly occurring complexes of sense impression (partly in conjunction with sense impressions which are interpreted as signs for sense experiences of others), and we attribute to them a meaning—the meaning of the bodily object. Considered logically this concept is not identical with the totality of sense impressions referred to; but it is an arbitrary creation of the human (or animal) mind. On the other hand, the concept owes its meaning and its justification exclusively to the totality of the sense impressions which we associate with it.

The second step is to be found in the fact that, in our thinking (which determines our expectation), we attribute to this concept of the bodily object a significance, which is to a high degree independent of the sense impression which originally gives rise to it. This is what we mean when we attribute to the bodily object "a real existence." The justification of such a setting rests exclusively on the fact that, by means of such concepts and mental relations between them, we are able to orient ourselves in the labyrinth of sense impressions. These notions and relations, although free statements of our thoughts, appear to us as stronger and more unalterable than the individual sense experience itself, the character of which as anything other than the result of an illusion or hallucination is never completely guaranteed. On the other hand, these concepts and relations, and indeed the setting of real objects and, generally speaking, the existence of "the real world," have justification only in so far as they are connected with sense impressions between which they form a mental connection.

The very fact that the totality of our sense experiences is such that by means of thinking (operations with concepts, and the creation and use of definite functional relations between them, and the coordination of sense experiences to these concepts) it can be put in order, this fact is one which leaves us in awe, but which we shall never understand. One may say "the eternal mystery of the world is its comprehensibility." It is one of the great realizations of Immanuel Kant that the setting up of a real external world would be senseless without this comprehensibility.

In speaking here concerning "comprehensibility," the expression is used in its most modest sense. It implies: the production of some sort of order among sense impressions, this order being produced by the creation of general concepts, relations between these concepts, and by relations between the concepts and sense experience, these relations being determined in any possible manner. It is in this sense that the world of our sense experiences is comprehensible. The fact that it is comprehensible is a miracle.

In my opinion, nothing can be said concerning the manner in which the concepts are to be made and connected, and how we are to coordinate them to the experiences. In guiding us in the creation of such an order of sense experiences, success in the result is alone the determining factor. All that is necessary is *the statement* of a set of rules, since without such rules the acquisition of knowledge in the desired sense would be impossible. One may compare these rules with the rules of a game in which, while the rules themselves are arbitrary, it is their rigidity alone which makes the game possible. However, the fixation will never be final. It will have validity only for a special field of application (i.e. there are no final categories in the sense of Kant).

The connection of the elementary concepts of everyday thinking with complexes of sense experiences can only be comprehended intuitively and it is unadaptable to scientifically logical fixation. The totality of these connections—none of which is expressible in notional terms—is the only thing which differentiates the great building which is science from a logical but empty scheme of concepts. By means of these connections, the purely notional theorems of science become statements about complexes of sense experiences.

We shall call "primary concepts" such concepts as are directly and intuitively connected with typical complexes of sense experiences. All other notions are—from the physical point of view—possessed of meaning, only in so far as they are connected, by theorems, with the primary notions. These theorems are partially definitions of the concepts (and of the statements derived logically from them) and partially theorems not derivable from the definitions, which express at least indirect relations between the "primary concepts," and in this way between sense experiences. Theorems of the latter kind are "statements about reality" or laws of nature, i.e. theorems which have to show their usefulness when applied to sense experiences comprehended by primary concepts. The question as to which of the theorems shall be considered as definitions and which as natural laws will depend largely upon the chosen representation. It really becomes absolutely necessary, to make this differentiation only when one examines the degree to which the whole system of concepts considered is not empty from the physical point of view.

## STRATIFICATION OF THE SCIENTIFIC SYSTEM

The aim of science is, on the one hand, a comprehension, as *complete* as possible, of the connection between the sense experiences in their totality, and, on the other hand, the accomplishment of this aim by *the use of a minimum of primary concepts and relations*. (Seeking, as far as possible, logical unity in the world picture, i.e. paucity in logical elements.)

Science concerns the totality of the primary concepts, i.e. concepts directly connected with sense experiences, and theorems connecting them. In its first stage of development, science does not contain anything else. Our everyday thinking is satisfied on the whole with this level. Such a state of affairs cannot, however, satisfy a spirit which is really scientifically minded; because, the totality of concepts and relations obtained in this manner is utterly lacking in logical unity. In order to supplement this deficiency, one invents a system poorer in concepts and relations, a system retaining the primary concepts and relations of the "first layer" as logically derived concepts and relations. This new "secondary system" pays for its higher logical unity by having, as its own elementary concepts (concepts of the second layer), only those which are no longer directly connected with complexes of sense experiences. Further striving for logical unity brings us to a tertiary system, still poorer in concepts and relations, for the deduction of the concepts and relations of the secondary (and so indirectly of the primary) layer. Thus the story goes on until we have arrived at a system of the greatest conceivable unity, and of the greatest poverty of concepts of the logical foundations, which are still compatible with the observation made by our senses. We do not know whether or not this ambition will ever result in a definite system. If one is asked for his opinion, he is inclined to answer no. While wrestling with the problems, however, one will never give up the hope that this greatest of all aims can really be attained to a very high degree.

An adherent to the theory of abstraction or induction might call our layers "degrees of abstraction"; but, I do not consider it justifiable to veil the logical independence of the concept from the sense experiences. The relation is not analogous to that of soup to beef but rather of wardrobe number to overcoat.

The layers are furthermore not clearly separated. It is not even absolutely clear which concepts belong to the primary layer. As a matter of fact, we are dealing with freely formed concepts, which, with a certainty sufficient for practical use, are intuitively connected with complexes of sense experiences in such a manner that, in any given case of experience, there is no uncertainty as to the applicability or non-applicability of the statement. The essential thing is the aim to represent the multitude of concepts and theorems, close to experience, as theorems, logically deduced and belonging to a basis, as narrow as possible, of fundamental concepts and fundamental relations which themselves can be chosen freely (axioms). The liberty of choice, however, is of a special kind; it is not in any way similar to the liberty of a writer of fiction. Rather, it is similar to that of a man engaged in solving a well designed word puzzle. He may, it is true, propose any word as the solution; but, there is only one word which really solves the puzzle in all its forms. It is an outcome of faith that nature—as she is perceptible to our five senses—takes the character of such a well formulated puzzle. The successes reaped up to now by science do, it is true, give a certain encouragement for this faith.

The multitude of layers discussed above corresponds to the several stages of progress which have resulted from the struggle for unity in the course of development. As regards the final aim, intermediary layers are only of temporary nature. They must eventually disappear as irrelevant. We have to deal, however, with the science of today, in which these strata represent problematic partial successes which support one another but which also threaten one another, because today's systems of concepts contain deep seated incongruities which we shall meet later on.

It will be the aim of the following lines to demonstrate what paths the constructive human mind has entered, in order to arrive at a basis of physics which is logically as uniform as possible.

## 2. MECHANICS AND THE ATTEMPTS TO BASE ALL PHYSICS UPON IT

An important property of our sense experiences, and, more generally, of all of our experience, is its time-like order. This kind of order leads to the mental conception of a subjective time, an ordinating scheme for our experience. The subjective time leads then through the concept of the bodily object and of space, to the concept of objective time, as we shall see later on.

Ahead of the notion of objective time there is, however, the concept of space; and, ahead of the latter we find the concept of the bodily object. The latter is directly connected with complexes of sense experiences. It has been pointed out that one property which is characteristic of the notion "bodily object" is the property which provides that we coordinate to it an existence, independent of (subjective) time, and independent of the fact that it is perceived by our senses. We do this in spite of the fact that we perceive temporal alterations in it.

Poincaré has justly emphasized the fact that we distinguish two kinds of alterations of the bodily object, "changes of state" and "changes of position." The latter, he remarked, are alterations which we can reverse by arbitrary motions of our bodies.

That there are bodily objects to which we have to ascribe, within a certain sphere of perception, no alteration of state, but only alterations of position, is a fact of fundamental importance for the formation of the concept of space (in a certain degree even for the justification of the notion of the bodily object itself). Let us call such an object "practically rigid."

If, as the object of our perception, we consider simultaneously (i.e. as a single unit) two practically rigid bodies, then there exist for this ensemble such alterations as can *not* possibly be considered as changes of position of the whole, notwithstanding the fact that this is the case for each one of the two constituents. This leads to the notion of "change of relative position" of the two objects; and, in this way, also to the notion of "relative position" of the two objects. It is found moreover that among the relative positions, there is one of a specific kind which we designate as "Contact."* Permanent contact of two bodies in three or more "points" means that they are united as a quasi rigid compound body. It is permissible to say that the second body forms then a (quasi rigid) continuation on the first body and may, in its turn, be continued quasi rigidly. The possibility of the quasi rigid continuation of a body is unlimited. The real essence of the conceivable quasi rigid continuation of a body $B_0$ is the infinite "space" determined by it.

In my opinion, the fact that every bodily object situated in any arbitrary manner can be put into contact with the quasi rigid continuation of a predetermined and chosen body $B_0$ (body of relation), this fact is the empirical basis of our conception of space. In pre-scientific thinking, the solid earth's crust plays the role of $B_0$ and its continuation. The

---

*It is in the nature of things that we are able to talk about these objects only by means of concepts of our own creation, concepts which themselves are not subject to definition. It is essential, however, that we make use only of such concepts concerning whose coordination to our experience we feel no doubt.

very name geometry indicates that the concept of space is psychologically connected with the earth as an assigned body.

The bold notion of "space" which preceded all scientific geometry transformed our mental concept of the relations of positions of bodily objects into the notion of the position of these bodily objects in "space." This, of itself, represents a great formal simplification. Through this concept of space one reaches, moreover, an attitude in which any description of position is admittedly a description of contact; the statement that a point of a bodily object is located at a point $P$ of space means that the object touches the point $P$ of the standard body of reference $B_0$ (supposed appropriately continued) at the point considered.

In the geometry of the Greeks, space plays only a qualitative role, since the position of bodies in relation to space is considered as given, it is true, but is not described by means of numbers. Descartes was the first to introduce this method. In his language, the whole content of Euclidian geometry can axiomatically be founded upon the following statements: (1) Two specified points of a rigid body determine a distance. (2) We may coordinate triplets of numbers $X_1, X_2, X_3$, to points of space in such a manner that for every distance $P' - P''$ under consideration, the coordinates of whose end points are $X_1', X_2', X_3'; X_1'', X_2'', X_3''$, the expression

$$S^2 = (X_1'' - X_1')^2 + (X_2'' - X_2')^2 + (X_3'' - X_3')^2$$

is independent of the position of the body, and of the positions of any and all other bodies.

The (positive) number $S$ means the length of the stretch, or the distance between the two points $P'$ and $P''$ of space (which are coincident with the points $P'$ and $P''$ of the stretch).

The formulation is chosen, intentionally, in such a way that it expresses clearly, not only the logical and axiomatic, but also the empirical content of Euclidian geometry. The purely logical (axiomatic) representation of Euclidian geometry has, it is true, the advantage of greater simplicity and clarity. It pays for this, however, by renouncing representation of the connection between the notional construction and the sense experience upon which connection, alone, the significance of geometry for physics rests. The fatal error that the necessity of thinking, preceding all experience, was at the basis of Euclidian geometry and the concept of space belonging to it, this fatal error arose from the fact that the empirical basis, on which the axiomatic construction of Euclidian geometry rests, had fallen into oblivion.

In so far as one can speak of the existence of rigid bodies in nature, Euclidian geometry is a physical science, the usefulness of which must be shown by application to sense experiences. It relates to the totality of laws which must hold for the relative positions of rigid bodies independently of time. As one may see, the physical notion of space also, as originally used in physics, is tied to the existence of rigid bodies.

From the physicist's point of view, the central importance of Euclidian geometry rests in the fact that its laws are independent of the specific nature of the bodies whose relative positions it discusses. Its formal simplicity is characterized by the properties of homogeneity and isotropy (and the existence of similar entities).

The concept of space is, it is true, useful, but not indispensable for geometry proper, i.e. for the formulation of rules about the relative positions of rigid bodies. In opposition to this, the concept of objective time, without which the formulation of the fundamentals of classical mechanics is impossible, is linked with the concept of the special continuum.

The introduction of objective time involves two statements which are independent of each other.

(1) The introduction of the objective local time by connecting the temporal sequence of experiences with the indications of a "clock," i.e. of a closed system with periodical occurrence.

(2) The introduction of the notion of objective time for the happenings in the whole space, by which notion alone the idea of local time is enlarged to the idea of time in physics.

Note concerning (1). As I see it, it does not mean a "petitio principii" if one puts the concept of periodical occurrence ahead of the concept of time, while one is concerned with the clarification of the origin and of the empirical content of the concept of time. Such a conception corresponds exactly to the precedence of the concept of the rigid (or quasi rigid) body in the interpretation of the concept of space.

Further discussion of (2). The illusion which prevailed prior to the enunciation of the theory of relativity—that, from the point of view of experience the meaning of simultaneity in relation to happenings distant in space and consequently that the meaning of time in physics is a priori clear—this illusion had its origin in the fact that in our everyday experience, we can neglect the time of propagation of light. We are accustomed on this account to fail to differentiate between "simultaneously seen" and "simultaneously happening"; and, as a result the difference between time and local time fades away.

The lack of definiteness which, from the point of view of empirical importance, adheres to the notion of time in classical mechanics was veiled by the axiomatic representation of space and time as things given independently of our senses. Such a use of notions—independent of the empirical basis, to which they owe their existence—does not necessarily damage science. One may however easily be led into the error of believing that these notions, whose origin is forgotten, are necessary and unalterable accompaniments to our thinking, and this error may constitute a serious danger to the progress of science.

It was fortunate for the development of mechanics and hence also for the development of physics in general, that the lack of definiteness in the concept of objective time remained obscured from the earlier philosophers as regards its empirical interpretation. Full of confidence in the real meaning of the space-time construction they developed the foundations of mechanics which we shall characterize, schematically, as follows:

(*a*) Concept of a material point: a bodily object which—as regards its position and motion—can be described with sufficient exactness as a point with coordinates $X_1, X_2, X_3$. Description of its motion (in relation to the "space" $B_0$) by giving $X_1, X_2, X_3$, as functions of the time.

(*b*) Law of inertia: the disappearance of the components of acceleration for the material point which is sufficiently far away from all other points.

(*c*) Law of motion (for the material point): Force = mass × acceleration.

(*d*) Laws of force (actions and reactions between material points).

In this (*b*) is nothing more than an important special case of (*c*). A real theory exists only when the laws of force are given. The forces must in the first place only obey the law of equality of action and reaction in order that a system of points—permanently connected to each other—may behave like *one* material point.

These fundamental laws, together with Newton's law for gravitational force, form the basis of the mechanics of celestial bodies. In this mechanics of Newton, and in contrast to the above conceptions of space derived from rigid bodies, the space $B_0$ enters in a form which contains a new idea; it is not for every $B_0$ that validity is required (for a given law of force) by (*b*) and (*c*), but only for a $B_0$ in the appropriate condition of motion (inertial system). On account of this fact, the coordinate space acquired an independent physical property which is not contained in the purely geometrical notion of space, a circumstance which gave Newton considerable food for thought (pail-experiment)*

Classical mechanics is only a general scheme; it becomes a theory only by explicit indication of the force laws (*d*) as was done so very successfully by Newton for celestial mechanics. From the point of view of the aim of the greatest logical simplicity of the foundations, this theoretical method is deficient in so far as the laws of force cannot be obtained by logical and formal considerations, so that their choice is *a priori* to a large extent arbitrary. Also Newton's gravitation law of force is distinguished from other conceivable laws of force exclusively by its *success*.

In spite of the fact that, today, we know positively that classical mechanics fails as a foundation dominating all physics, it still occupies

---

*This defect of the theory could only be eliminated by such a formulation of mechanics as would command validity for all $B_0$. This is one of the steps which lead to the general theory of relativity. A second defect, also eliminated only by the introduction of the general theory of relativity, lies in the fact that there is no reason given by mechanics itself for the equality of the gravitational and inertial mass of the material point.

the center of all of our thinking in physics. The reason for this lies in the fact that, regardless of important progress reached since the time of Newton, we have not yet arrived at a new foundation of physics concerning which we may be certain that the whole complexity of investigated phenomena, and of partial theoretical systems of a successful kind, could be deduced logically from it. In the following lines I shall try to describe briefly how the matter stands.

First we try to get clearly in our minds how far the system of classical mechanics has shown itself adequate to serve as a basis for the whole of physics. Since we are dealing here only with the foundations of physics and with its development, we need not concern ourselves with the purely *formal* progresses of mechanics (equation of Lagrange, canonical equations, etc.). *One* remark, however, appears indispensable. The notion "material point" is fundamental for mechanics. If now we seek the mechanics of a bodily object which itself can *not* be treated as a material point—and strictly speaking every object "perceptible to our senses" is of this category—then the question arises: How shall we imagine the object to be built up out of material points, and what forces must we assume as acting between them? The formulation of this question is indispensable, if mechanics is to pretend to describe the object *completely.*

It is natural to the tendency of mechanics to assume these material points, and the laws of forces acting between them, as invariable, since time alterations would lie outside of the scope of mechanical explanation. From this we can see that classical mechanics must lead us to an atomistic construction of matter. We now realize, with special clarity, how much in error are those theorists who believe that theory comes inductively from experience. Even the great Newton could not free himself from this error ("Hypotheses non fingo").*

In order to save itself from becoming hopelessly lost in this line of thought (atomistic), science proceeded first in the following manner. The mechanics of a system is determined if its potential energy is given as a function of its configuration. Now, if the acting forces

---

* "I make no hypotheses."

are of such a kind as to guarantee maintenance of certain qualities of order of the system's configuration, then the configuration may be described with sufficient accuracy by a relatively small number of configuration variables $q_r$; the potential energy is considered only insofar as it is dependent upon *these* variables (for instance, description of the configuration of a practically rigid body by six variables).

A second method of application of mechanics, which avoids the consideration of a subdivision of matter down to "real" material points, is the mechanics of so-called continuous media. This mechanics is characterized by the fiction that the density of matter and speed of matter is dependent in a continuous manner upon coordinates and time, and that the part of the interactions not explicitly given can be considered as surface forces (pressure forces) which again are continuous functions of location. Herein we find the hydrodynamic theory, and the theory of elasticity of solid bodies. These theories avoid the explicit introduction of material points by fictions which, in the light of the foundation of classical mechanics, can only have an approximate significance.

In addition to their great *practical* significance, these categories of science have—by enlargement of the mathematical world of ideas—created those formal auxiliary instruments (partial differential equations) which have been necessary for the subsequent attempts at formulating the total scheme of physics in a manner which is new as compared with that of Newton.

These two modes of application of mechanics belong to the so-called "phenomenological" physics. It is characteristic of this kind of physics that it makes as much use as possible of concepts which are close to experience but which, for this reason, have to give up, to a large degree, unity in the foundations. Heat, electricity and light are described by special variables of state and constants of matter other than the mechanical state, and to determine all of these variables in their relative dependence was a rather empirical task. Many contemporaries of Maxwell saw in such a manner of presentation the ultimate aim of physics, which they thought could be obtained purely inductively from experience on account of the relative closeness of the concepts used to the experience. From the point of view of theories of knowledge St. Mill and E. Mach took their stand approximately on this ground.

According to my belief, the greatest achievement of Newton's mechanics lies in the fact that its consistent application has led beyond this phenomenological representation, particularly in the field of heat phenomena. This occurred in the kinetic theory of gases and, in a general way, in statistical mechanics. The former connected the equation of state of the ideal gases, viscosity, diffusion and heat conductivity of gases and radiometric phenomena of gases, and gave the logical connection of phenomena which, from the point of view of direct experience, had nothing whatever to do with one another. The latter gave a mechanical interpretation of the thermodynamic ideas and laws as well as the discovery of the limit of applicability of the notions and laws to the classical theory of heat. This kinetic theory which surpassed, by far, the phenomenological physics as regards the logical unity of its foundations, produced moreover definite values for the true magnitudes of atoms and molecules which resulted from several independent methods and were thus placed beyond the realm of reasonable doubt. These decisive progresses were paid for by the coordination of atomistic entities to the material points, the constructively speculative character of which entities being obvious. Nobody could hope ever to "perceive directly" an atom. Laws concerning variables connected more directly with experimental facts (for example: temperature, pressure, speed) were deduced from the fundamental ideas by means of complicated calculations. In this manner physics (at least part of it), originally more phenomenologically constructed, was reduced, by being founded upon Newton's mechanics for atoms and molecules, to a basis further removed from direct experiment, but more uniform in character.

## 3. THE FIELD CONCEPT

In explaining optical and electrical phenomena Newton's mechanics has been far less successful than it had been in the fields cited above. It is true that Newton tried to reduce light to the motion of material points in his corpuscular theory of light. Later on, however, as the phenomena of polarization, diffraction and interference of light forced upon his theory more and more unnatural modifications, Huyghens' undulatory theory of light, prevailed. Probably this theory owes its origin essentially to the phenomena of crystallographic optics and to the theory of sound, which was then already elaborated to a certain degree. It must be admitted that Huyghens' theory also was based in the first instance upon classical mechanics; but, the all-penetrating ether had to be assumed as the carrier of the waves and the structure of the ether, formed from material points, could not be explained by any known phenomenon. One could never get a clear picture of the interior forces governing the ether, nor of the forces acting between the ether and the "ponderable" matter. The foundations of this theory remained, therefore, eternally in the dark. The true basis was a partial differential equation, the reduction of which to mechanical elements remained always problematic.

For the theoretical conception of electric and magnetic phenomena one introduced, again, masses of a special kind, and between these masses one assumed the existence of forces acting at a distance, similar to Newton's gravitational forces. This special kind of matter, however, appeared to be lacking in the fundamental property of inertia; and, the forces acting between these masses and the ponderable matter remained obscure. To these difficulties there had to be added the polar character of these kinds of matter which did not fit into the scheme of classical mechanics. The basis of the theory became still more unsatisfactory when electro-dynamic phenomena became known, notwithstanding the fact that these phenomena brought the physicist to the explanation of magnetic phenomena through electrodynamic phenomena and, in this way, made the assumption of magnetic masses superfluous. This progress had, indeed, to be paid for by increasing the complexity of the forces of interaction which had to be assumed as existing between electrical masses in motion.

The escape from this unsatisfactory situation by the electric field theory of Faraday and Maxwell represents probably the most profound transformation which has been experienced by the foundations of physics since Newton's time. Again, it has been a step in the direction of constructive speculation which has increased the distance between the foundation of the theory and what can be experienced by means of our five senses. The existence of the field manifests itself, indeed, only when electrically charged bodies are introduced into it. The differential equations of Maxwell connect the special and temporal differential coefficients of the electric and magnetic fields. The electric masses are nothing more than places of non-disappearing divergency of the electric field. Light waves appear as undulatory electromagnetic field processes in space.

To be sure, Maxwell still tried to interpret his field theory mechanically by means of mechanical ether models. But these attempts receded gradually to the background following the representation— purged of any unnecessary additions—by Heinrich Hertz, so that, in this theory the field finally took the fundamental position which had been occupied in Newton's mechanics by the material points. At first, however, this applies only for electromagnetic fields in empty space.

In its initial stage the theory was yet quite unsatisfactory for the interior of matter, because there, two electric vectors had to be introduced, which were connected by relations dependent on the nature of the medium, these relations being inaccessible to any theoretical analysis. An analogous situation arose in connection with the magnetic field, as well as in the relation between electric current density and the field.

Here H. A. Lorentz found an escape which showed, at the same time, the way to an electrodynamic theory of bodies in motion, a theory which was more or less free of arbitrary assumption. His theory was built on the following fundamental hypothesis:

Everywhere (including the interior of ponderable bodies) the seat of the field is the empty space. The participation of matter in electromagnetic phenomena has its origin only in the fact that the elementary particles of matter carry unalterable electric charges, and, on this account are subject on the one hand to the actions of ponderomotive forces and on the other hand possess the property of generating a field.

The elementary particles obey Newton's law of motion for the material point.

This is the basis on which H. A. Lorentz obtained his synthesis of Newton's mechanics and Maxwell's field theory. The weakness of this theory lies in the fact that it tried to determine the phenomena by a combination of partial differential equations (Maxwell's field equations for empty space) and total differential equations (equations of motion of points), which procedure was obviously unnatural. The unsatisfactory part of the theory showed up externally by the necessity of assuming finite dimensions for the particles in order to prevent the electromagnetic field existing at their surfaces from becoming infinitely great. The theory failed moreover to give any explanation concerning the tremendous forces which hold the electric charges on the individual particles. H. A. Lorentz accepted these weaknesses of his theory, which were well known to him, in order to explain the phenomena correctly at least as regards their general lines.

Furthermore, there was one consideration which reached beyond the frame of Lorentz's theory. In the environment of an electrically charged body there is a magnetic field which furnishes an (apparent) contribution to its inertia. Should it not be possible to explain the *total* inertia of the particles electromagnetically? It is clear that this problem could be worked out satisfactorily only if the particles could be interpreted as regular solutions of the electromagnetic partial differential equations. The Maxwell equations in their original form do not, however, allow such a description of particles, because their corresponding solutions contain a singularity. Theoretical physicists have tried for a long time, therefore, to reach the goal by a modification of Maxwell's equations. These attempts have, however, not been crowned with success. Thus it happened that the goal of erecting a pure electromagnetic field theory of matter remained unattained for the time being, although in principle no objection could be raised against the possibility of reaching such a goal. The thing which deterred one in any further attempt in this direction was the lack of any systematic method leading to the solution. What appears certain to me, however, is that, in the foundations of any consistent field theory, there shall not be, in addition to the concept of field, any concept concerning particles. The whole theory must be based solely on partial differential equations and their singularity-free solutions.

# 4. THE THEORY OF RELATIVITY

There is no inductive method which could lead to the fundamental concepts of physics. Failure to understand this fact constituted the basic philosophical error of so many investigators of the nineteenth century. It was probably the reason why the molecular theory, and Maxwell's theory were able to establish themselves only at a relatively late date. Logical thinking is necessarily deductive; it is based upon hypothetical concepts and axioms. How can we hope to choose the latter in such a manner as to justify us in expecting success as a consequence?

The most satisfactory situation is evidently to be found in cases where the new fundamental hypotheses are suggested by the world of experience itself. The hypothesis of the non-existence of perpetual motion as a basis for thermodynamics affords such an example of a fundamental hypothesis suggested by experience; the same thing holds for the principle of inertia of Galileo. In the same category, moreover, we find the fundamental hypotheses of the theory of relativity, which theory has led to an unexpected expansion and broadening of the field theory, and to the superseding of the foundations of classical mechanics.

The successes of the Maxwell-Lorentz theory have given great confidence in the validity of the electromagnetic equations for empty space and hence, in particular, to the statement that light travels "in space" with a certain constant speed $c$. Is this law of the invariability of light velocity in relation to any desired inertial system valid? If it were not, then one specific inertial system or more accurately, one specific state of motion (of a body of reference), would be distinguished from all others. In opposition to this idea, however, stand all the mechanical and electromagnetic-optical facts of our experience.

For these reasons it was necessary to raise to the degree of a principle, the validity of the law of constancy of light velocity for all inertial systems. From this, it follows that the special coordinates $X_1$, $X_2$, $X_3$, and the time $X_4$, must be transformed according to the "Lorentz-transformation" which is characterized by invariance of the expression

$$ds^2 = dx_1^2 + dx_2^2 + dx_3^2 - dx_4^2$$

(if the unit of time is chosen in such a manner that the speed of light $c = 1$).

By this procedure time lost its absolute character, and was included with the "special" coordinates as of algebraically (nearly) similar character. The absolute character of time and particularly of simultaneity were destroyed, and the four dimensional description became introduced as the only adequate one.

In order to account, also, for the equivalence of all inertial systems with regard to all the phenomena of nature, it is necessary to postulate invariance of all systems of physical equations which express general laws, with regard to the Lorentzian transformation. The elaboration of this requirement forms the content of the special theory of relativity.

This theory is compatible with the equations of Maxwell; but, it is incompatible with the basis of classical mechanics. It is true that the equations of motion of the material point can be modified (and with them the expressions for momentum and kinetic energy of the material point) in such a manner as to satisfy the theory; but, the concept of the force of interaction, and with it the concept of potential energy of a system, lose their basis, because these concepts rest upon the idea of absolute instantaneousness. The field, as determined by differential equations, takes the place of the force.

Since the foregoing theory allows interaction only by fields, it requires a field theory of gravitation. Indeed, it is not difficult to formulate such a theory in which, as in Newton's theory, the gravitational fields can be reduced to a scalar which is the solution of a partial differential equation. However, the experimental facts expressed in Newton's theory of gravitation lead in another direction, that of the general theory of relativity.

Classical mechanics contains one point which is unsatisfactory in that, in the fundamentals, the same mass constant is met twice over in two different roles, namely as "inertial mass" in the law of motion, and as "gravitational mass" in the law of gravitation. As a result of this, the acceleration of a body in a pure gravitational field is independent of its material; or, in a coordinate system of *uniform acceleration* (accelerated in relation to an "inertial system") the motions take place as they would in a homogeneous gravitational field (in relation to a "motionless" system of coordinates). If one assumes that the equivalence of these two cases is complete, then one attains an adaptation of our theoretical thinking to the fact that the gravitational and inertial masses are identical.

From this it follows that there is no longer any reason for favoring, as a fundamental principle, the "inertial systems"; and, we must admit as equivalent in their own right, also *non-linear* transformations of the coordinates $(x_1, x_2, x_3, x_4)$. If we make such a transformation of a system of coordinates of the special theory of relativity, then the metric

$$ds^2 = dx_1^2 + dx_2^2 + dx_3^2 - dx_4^2$$

goes over to a general (Riemannian) metric of Bane

$$ds^2 = g_{\mu\nu} dx_\mu dx_\nu \quad \text{(Summed over } \mu \text{ and } \nu\text{)}$$

where the $g_{\mu\nu}$, symmetrical in $\mu$ and $\nu$, are certain functions of $x_1 \cdots x_4$ which describe both the metric property, and the gravitational field in relation to the new system of coordinates.

The foregoing improvement in the interpretation of the mechanical basis must, however, be paid for in that—as becomes evident on closer scrutiny—the new coordinates could no longer be interpreted, as results of measurements by rigid bodies and clocks, as they could in the original system (an inertial system with vanishing gravitational field).

The passage to the general theory of relativity is realized by the assumption that such a representation of the field properties of space already mentioned, by functions $g_{\mu\nu}$ (that is to say by a Riemann metric), is also justified in the *general* case in which there is no system of coordinates in relation to which the metric takes the simple quasi-Euclidian form of the special theory of relativity.

Now the coordinates, by themselves, no longer express metric relations, but only the "neighborliness" of the things described, whose coordinates differ but little from one another. All transformations of the coordinates have to be admitted so long as these transformations are free from singularities. Only such equations as are covariant in relation to arbitrary transformations in this sense have meaning as expressions of general laws of nature (postulate of general covariancy).

The first aim of the general theory of relativity was a preliminary statement which, by giving up the requirement of constituting a closed thing in itself, could be connected in as simple a manner as possible with the "facts directly observed." Newton's gravitational theory gave an example, by restricting itself to the pure mechanics of gravitation. This preliminary statement may be characterized as follows:

(1) The concept of the material point and of its mass is retained. A law of motion is given for it, this law of motion being the translation of the law of inertia into the language of the general theory of relativity. This law is a system of total differential equations, the system characteristic of the geodetic line.

(2) In place of Newton's law of interaction by gravitation, we shall find the system of the simplest generally covariant differential equations which can be set up for the $g_{\mu\nu}$-tensor. It is formed by equating to zero the once contracted Riemannian curvature tensor ($R_{\mu\nu} = 0$).

This formulation permits the treatment of the problem of the planets. More accurately speaking, it allows the treatment of the problem of motion of material points of practically negligible mass in the gravitational field produced by a material point which itself is supposed to have no motion (central symmetry). It does not take into account the reaction of the "moved" material points on the gravitational field, nor does it consider how the central mass produces this gravitational field.

Analogy with classical mechanics shows that the following is a way to complete the theory. One sets up as field equation

$$R_{ik} - \tfrac{1}{2} g_{ik} R = - T_{ik}$$

where $R$ represents the scalar of Riemannian curvature, $T_{ik}$ the energy tensor of the matter in a phenomenological representation. The left side of the equation is chosen in such a manner that its divergence disappears identically. The resulting disappearance of the divergence of the right side produces the "equations of motion" of matter, in the form of partial differential equations for the case where $T_{ik}$ introduces, for the description of the matter, only *four* further functions independent of each other (for instance, density, pressure, and velocity components, where there is between the latter an identity, and between pressure and density an equation of condition).

By this formulation one reduces the whole mechanics of gravitation to the solution of a single system of covariant partial differential equations. The theory avoids all internal discrepancies which we have charged against the basis of classical mechanics. It is sufficient—as far as we know—for the representation of the observed facts of celestial mechanics. But, it is similar to a building, one wing of which is made of fine marble (left part of the equation), but the other wing of which is built of low grade wood (right side of equation). The phenomenological representation of matter is, in fact, only a crude substitute for a representation which would correspond to all known properties of matter.

There is no difficulty in connecting Maxwell's theory of the electromagnetic field with the theory of the gravitational field so long as one restricts himself to space, free of ponderable matter and free of electric density. All that is necessary is to put on the right hand side of the above equation for $T_{ik}$, the energy tensor of the electromagnetic field in empty space and to associate with the so modified system of equations the Maxwell field equation for empty space, written in general covariant form. Under these conditions there will exist, between all these equations, a sufficient number of the differential identities to guarantee their consistency. We may add that this necessary formal property of the total system of equations leaves arbitrary the choice of the sign of the member $T_{ik}$, a fact which was later shown to be important.

The desire to have, for the foundations of the theory, the greatest possible unity has resulted in several attempts to include the gravitational field and the electromagnetic field in one formal but homogeneous picture. Here we must mention particularly the five-dimensional theory of Kaluza and Klein. Having considered this possibility very carefully I feel that it is more desirable to accept the lack of internal uniformity of the original theory, because I do not consider that the totality of the hypothetical basis of the five-dimensional theory contains less of an arbitrary nature than does the original theory. The same statement may be made for the projective variety of the theory, which has been elaborated with great care, in particular, by v. Dantzig and by Pauli.

The foregoing considerations concern, exclusively, the theory of the field, free of matter. How are we to proceed from this point in order to obtain a complete theory of atomically constructed matter? In such a theory, singularities must certainly be excluded, since without such exclusion the differential equations do not completely determine the total field. Here, in the field theory of general relativity, we meet the same problem of a theoretical field-representation of matter as was met originally in connection with the pure Maxwell theory.

Here again the attempt to construct particles out of the field theory, leads apparently to singularities. Here also the endeavor has been made to overcome this defect by the introduction of new field variables and by elaborating and extending the system of field equations. Recently, however, I discovered, in collaboration with Dr. Rosen, that the above mentioned simplest combination of the field equations of gravitation and electricity produces centrally symmetrical solutions which can be represented as free of singularity (the well known centrally symmetrical solutions of Schwarzschild for the pure gravitational field, and those of Reissner for the electric field with consideration of its gravitational action). We shall refer to this shortly in the paragraph next but one. In this way it seems possible to get for matter and its interactions a pure field theory free of additional hypotheses, one moreover whose test by submission to facts of experience does not result in difficulties other than purely mathematical ones, which difficulties, however, are very serious.

## 5. QUANTUM THEORY AND THE FUNDAMENTALS OF PHYSICS

The theoretical physicists of our generation are expecting the erection of a new theoretical basis for physics which would make use of fundamental concepts greatly different from those of the field theory considered up to now. The reason is that it has been found necessary to use—for the mathematical representation of the so-called quantum phenomena—new sorts of methods of consideration.

While the failure of classical mechanics, as revealed by the theory of relativity, is connected with the finite speed of light (its avoidance of being ∞), it was discovered at the beginning of our century that there were other kinds of inconsistencies between deductions from mechanics and experimental facts, which inconsistencies are connected with the finite magnitude (the avoidance of being zero) of Planck's constant $h$. In particular, while molecular mechanics requires that both heat content and (monochromatic) radiation density of solid bodies should decrease in *proportion* to the decreasing absolute temperature, experience has shown that they decrease much more rapidly than the absolute temperature. For a theoretical explanation of this behavior it was necessary to assume that the energy of a mechanical system cannot assume any sort of value, but only certain discrete values whose mathematical expressions were always dependent upon Planck's constant $h$. Moreover, this conception was essential for the theory of the atom (Bohr's theory). For the transitions of these states into one another—with or without emission or absorption of radiation—no causal laws could be given, but only statistical ones; and, a similar conclusion holds for the radioactive decomposition of atoms, which decomposition was carefully investigated about the same time. For more than two decades physicists tried vainly to find a uniform interpretation of this "quantum character" of systems and phenomena.

Such an attempt was successful about ten years ago, through the agency of two entirely different theoretical methods of attack. We owe one of these to Heisenberg and Dirac, and the other to de Broglie and Schrödinger. The mathematical equivalence of the two methods was soon recognized by Schrödinger. I shall try here to sketch the line of thought of de Broglie and Schrödinger, which lies closer to the physicist's method of thinking, and shall accompany the description with certain general considerations.

The question is first: How can one assign a discrete succession of energy value $H_\sigma$ to a system specified in the sense of classical mechanics (the energy function is a given function of the coordinates $q_r$ and the corresponding momenta $p_r$)? Planck's constant $h$ relates the frequency $H_\sigma/h$ to the energy values $H_\sigma$. It is therefore sufficient to give to the system a succession of discrete *frequency* values. This reminds us of the fact that in acoustics, a series of discrete frequency values is coordinated to a linear partial differential equation (if boundary values are given) namely the sinusoidal periodic solutions. In corresponding manner, Schrödinger set himself the task of coordinating a partial differential equation for a scalar function $\psi$ to the given energy function $\varepsilon(q_r, p_r)$, where the $q_r$ and the time $t$ are independent variables. In this he succeeded (for a complex function $\psi$) in such a manner that the theoretical values of the energy $H_\sigma$, as required by the statistical theory, actually resulted in a satisfactory manner from the periodic solution of the equation.

To be sure, it did not happen to be possible to associate a definite movement, in the sense of mechanics of material points, with a definite solution $\psi(q_r, t)$ of the Schrödinger equation. This means that the $\psi$ function does not determine, at any rate *exactly*, the story of the $q_r$ as functions of the time $t$. According to Born, however, an interpretation of the physical meaning of the $\psi$ functions was shown to be possible in the following manner: $\psi\bar{\psi}$ (the square of the absolute value of the complex function $\psi$) is the probability density at the point under consideration in the configuration-space of the $q_r$, at the time $t$. It is therefore possible to characterize the content of the Schrödinger equation in a manner, easy to be understood, but not quite accurate, as follows: it determines how the probability density of a statistical ensemble of systems varies in the configuration-space with the time. Briefly: the Schrödinger equation determines the alteration of the function $\psi$ of the $q_r$ with the time.

It must be mentioned that the result of this theory contains—as limiting values—the result of the particle mechanics if the wave-length encountered during the solution of the Schrödinger problem is everywhere so small that the potential energy varies by a practically infinitely small amount for a change of one wavelength in the configuration-space. Under these conditions the following can in fact be shown: We choose a region $G_0$ in the configuration-space which, although large (in every dimension) in relation to the wave length, is small in relation to the practical dimensions of the configuration-space. Under these conditions it is possible to choose a function of $\psi$ for an initial time $t_0$ in such a manner that it vanishes outside of the region $G_0$, and behaves, according to the Schrödinger equation, in such a manner that it retains this property—approximately at least—also for a later time, but with the region $G_0$ having passed at that time $t$ into another region $G$. In this manner one can, with a certain degree of approximation, speak of the motion of the region $G$ as a whole, and one can approximate this motion by the motion of a point in the configuration-space. This motion then coincides with the motion which is required by the equations of classical mechanics.

Experiments on interference made with particle rays have given a brilliant proof that the wave character of phenomena of motion as assumed by the theory does, really, correspond to the facts. In addition to this, the theory succeeded, easily, in demonstrating the statistical laws of the transition of a system from one quantum condition to another under the action of external forces, which, from the standpoint of classical mechanics, appears as a miracle. The external forces were here represented by small additions of the potential energy as functions of the time. Now, while in classical mechanics, such additions can produce only correspondingly small alterations of the system, in the quantum mechanics they produce alterations of any magnitude however large, but with correspondingly small probability, a consequence in perfect harmony with experience. Even an understanding of the laws of radioactive decomposition, at least in their broad lines, was provided by the theory.

Probably never before has a theory been evolved which has given a key to the interpretation and calculation of such a heterogeneous group of phenomena of experience as has the quantum theory. In spite of this, however; I believe that the theory is apt to beguile us into error in our search for a uniform basis for physics, because, in my belief, it is an *incomplete* representation of real things, although it is the only one which can be built out of the fundamental concepts of force and material points (quantum corrections to classical mechanics). The incompleteness of the representation is the outcome of the statistical nature (incompleteness) of the laws. I will now justify this opinion.

I ask first: How far does the $\psi$ function describe a real condition of a mechanical system? Let us assume the $\psi_r$ to be the periodic solutions (put in the order of increasing energy values) of the Schrödinger equation. I shall leave open, for the time being, the question as to how far the individual $\psi_r$ are *complete* descriptions of physical conditions. A system is first in the condition $\psi_1$ of lowest energy $\varepsilon_1$. Then during a finite time a small disturbing force acts upon the system. At a later instant one obtains then from the Schrödinger equation a $\psi$ function of the form

$$\psi = \sum c_r \psi_r$$

where the $c_r$ are (complex) constants. If the $\psi_r$ are "normalized," then $|c_1|$ is nearly equal to 1, $|C_2|$ etc. is small compared with 1. One may now ask: Does $\psi$ describe a real condition of the system? If the answer is yes, then we can hardly do otherwise than ascribe* to this condition a definite energy $\varepsilon$, and, in particular, such an energy as exceeds $\varepsilon_1$ by a small amount (in any case $\varepsilon_1 < \varepsilon < \varepsilon_2$). Such an assumption is, however, at variance with the experiments on electron impact such

*Because, according to a well established consequence of the relativity theory, the energy of a complete system (at rest) is equal to its inertia (as a whole). This, however, must have a well defined value.

as have been made by J. Franck and G. Hertz, if, in addition to this, one accepts Millikan's demonstration of the discrete nature of electricity. As a matter of fact, these experiments lead to the conclusion that energy values of a state lying between the quantum values do not exist. From this it follows that our function $\psi$ does not in any way describe a homogeneous condition of the body, but represents rather a statistical description in which the $c_r$ represent probabilities of the individual energy values. It seems to be clear, therefore, that the Born statistical interpretation of the quantum theory is the only possible one. The $\psi$ function does not in any way describe a condition which could be that of a single system; it relates rather to many systems, to "an ensemble of systems" in the sense of statistical mechanics. If, except for certain special cases, the $\psi$ function furnishes only *statistical* data concerning measurable magnitudes, the reason lies not only in the fact that the *operation of measuring* introduces unknown elements, which can be grasped only statistically, but because of the very fact that the $\psi$ function does not, in any sense, describe the condition of *one* single system. The Schrödinger equation determines the time variations which are experienced by the ensemble of systems which may exist with or without external action on the single system.

Such an interpretation eliminates also the paradox recently demonstrated by myself and two collaborators, and which relates to the following problem.

Consider a mechanical system constituted of two partial systems $A$ and $B$ which have interaction with each other only during limited time. Let the $\psi$ function before their interaction be given. Then the Schrödinger equation will furnish the $\psi$ function after the interaction has taken place. Let us now determine the physical condition of the partial system $A$ as completely as possible by measurements. Then the quantum mechanics allows us to determine the function $\psi$ of the partial system $B$ from the measurements made, and from the $\psi$ function of the total system. This determination, however, gives a result which depends upon *which* of the determining magnitudes specifying the condition of $A$ has been measured (for instance coordinates *or* momenta).

Since there can be only *one* physical condition of $B$ after the interaction and which can reasonably not be considered as dependent on the particular measurement we perform on the system $A$ separated from $B$ it may be concluded that the $\psi$ function is *not* unambiguously coordinated with the physical condition. This coordination of several $\psi$ functions with the same physical condition of system $B$ shows again that the $\psi$ function cannot be interpreted as a (complete) description of a physical condition of a unit system. Here also the coordination of the $\psi$ function to an ensemble of systems eliminates every difficulty.*

The fact that quantum mechanics affords, in such a simple manner, statements concerning (apparently) discontinuous transitions from one total condition to another without actually giving a representation of the specific process, this fact is connected with another, namely the fact that the theory, in reality, does not operate with the single system, but with a totality of systems. The coefficients $c_r$ of our first example are really altered very little under the action of the external force. With this interpretation of quantum mechanics one can understand why this theory can easily account for the fact that weak disturbing forces are able to produce alterations of any magnitude in the physical condition of a system. Such disturbing forces produce, indeed, only correspondingly small alterations of the *statistical density* in the ensemble of systems, and hence only infinitely weak alterations of the $\psi$ functions, the mathematical description of which offers far less difficulty than would be involved in the mathematical representation of finite alterations experienced by part of the single systems. What happens to the single system remains, it is true, entirely unclarified by this mode of consideration; this enigmatic happening is entirely eliminated from the representation by the statistical manner of consideration.

But now I ask: Is there really any physicist who believes that we shall never get any inside view of these important alterations in the single systems, in their structure and their causal connections, and this regardless of the fact that these single happenings have been brought

---

*The operation of measuring $A$, for example, thus involves a transition to a narrower ensemble of systems. The latter (hence also its $\psi$ function) depends upon the point of view according to which this narrowing of the ensemble of systems is made.

so close to us, thanks to the marvelous inventions of the Wilson chamber and the Geiger counter? To believe this is logically possible without contradiction; but, it is so very contrary to my scientific instinct that I cannot forego the search for a more complete conception.

To these considerations we should add those of another kind which also voice their plea against the idea that the methods introduced by quantum mechanics are likely to give a useful basis for the whole of physics. In the Schrödinger equation, absolute time, and also the potential energy, play a decisive role, while these two concepts have been recognized by the theory of relativity as inadmissible in principle. If one wishes to escape from this difficulty he must found the theory upon field and field laws instead of upon forces of interaction. This leads us to transpose the statistical methods of quantum mechanics to fields, that is to systems of infinitely many degrees of freedom. Although the attempts so far made are restricted to linear equations, which, as we know from the results of the general theory of relativity, are insufficient, the complications met up to now by the very ingenious attempts are already terrifying. They certainly will rise sky high if one wishes to obey the requirements of the general theory of relativity, the justification of which in principle nobody doubts.

To be sure, it has been pointed out that the introduction of a space-time continuum may be considered as contrary to nature in view of the molecular structure of everything which happens on a small scale. It is maintained that perhaps the success of the Heisenberg method points to a purely algebraical method of description of nature, that is to the elimination of continuous functions from physics. Then, however, we must also give up, by principle, the space-time continuum. It is not unimaginable that human ingenuity will some day find methods which will make it possible to proceed along such a path. At the present time, however, such a program looks like an attempt to breathe in empty space.

There is no doubt that quantum mechanics has seized hold of a beautiful element of truth, and that it will be a test stone for any future theoretical basis, in that it must be deducible as a limiting case from that basis, just as electrostatics is deducible from the Maxwell equations of the electromagnetic field or as thermodynamics is deducible from classical mechanics. However, I do not believe that quantum mechanics will be the *starting point* in the search for this basis, just as, vice versa, one could not go from thermodynamics (resp. statistical mechanics) to the foundations of mechanics.

In view of this situation, it seems to be entirely justifiable seriously to consider the question as to whether the basis of field physics cannot by *any* means be put into harmony with the facts of the quantum theory. Is this not the only basis which, consistently with today's possibility of mathematical expression, can be adapted to the requirements of the general theory of relativity? The belief, prevailing among the physicists of today, that such an attempt would be hopeless, may have its root in the unjustifiable idea that such a theory should lead, as a first approximation, to the equations of classical mechanics for the motion of corpuscles, or at least to total differential equations. As a matter of fact up to now we have never succeeded in representing corpuscles theoretically by fields free of singularities, and we can, a priori, say nothing about the behavior of such entities. *One thing*, however, is certain: if a field theory results in a representation of corpuscles free of singularities, then the behavior of these corpuscles with time is determined solely by the differential equations of the field.

# 6. RELATIVITY THEORY AND CORPUSCLES

I shall now show that, according to the general theory of relativity, there exist singularity-free solutions of field equations which can be interpreted as representing corpuscles. I restrict myself here to neutral particles because, in another recent publication in collaboration with Dr. Rosen, I have treated this question in a detailed manner, and because the essentials of the problem can be completely shown by this case.

The gravitational field is entirely described by the tensor $g_{\mu\nu}$. In the three-index symbols $\Gamma_{\mu\nu}{}^{\sigma}$, there appear also the contravariants $g^{\mu\nu}$ which are defined as the minors of the $g_{\mu\nu}$ divided by the determinant

$g(= |g_{\alpha\beta}|)$. In order that the $R_{ik}$ shall be defined and finite, it is not sufficient that there shall be, for the environment of every part of the continuum, a system of coordinates in which the $g_{\mu\nu}$ and their first differential quotients are continuous and differentiable, but it is also necessary that the determinant $g$ shall nowhere vanish. This last restriction is, however, eliminated if one replaces the differential equations $R_{ik} = 0$ by $g^2 R_{ik} = 0$, the left hand sides of which are *whole* rational functions of the $g_{ik}$ and of their derivatives.

These equations have the centrally symmetrical solutions indicated by Schwarzschild

$$ds^2 = -\frac{1}{1 - 2m/r} dr^2 - r^2(d\theta^2 + \sin^2\theta\, d\varphi^2) + \left(1 - \frac{2m}{r}\right) dt^2$$

This solution has a singularity at $r = 2m$, since the co-efficient of $dr^2$ (i.e. $g_{11}$), becomes infinite on this hypersurface. If, however, we replace the variable $r$ by $\rho$ defined by the equation

$$\rho^2 = r - 2m$$

we obtain

$$ds^2 = -4(2m + \rho^2)d\rho^2 - (2m + \rho^2)^2(d\theta^2 + \sin^2\theta\, d\varphi^2)$$
$$+ \frac{\rho^2}{2m + \rho^2} dt^2$$

This solution behaves regularly for all values of $\rho$. The vanishing of the coefficient of $dt^2$ i.e. $(g_{44})$ for $\rho = 0$ results, it is true, in the consequence that the determinant $g$ vanishes for this value; but, with the methods of writing the field equations actually adopted, this does not constitute a singularity.

If $\rho$ extends from $-\infty$ to $+\infty$, then r runs from $+\infty$ to $r = 2m$ and then back to $+\infty$, while for such values of $r$ as correspond to $r < 2m$ there are no corresponding real values of $\rho$. Hence the Schwarzschild solution becomes a regular solution by representation of the physical space as consisting of two identical "shells" neighboring upon the hypersurface $\rho = 0$, that is $r = 2m$, while for this hypersurface the determinant $g$ vanishes. Let us call such a connection between the two (identical) shells a "bridge." Hence the existence of such a bridge between the two shells in the finite realm corresponds to the existence of a material neutral particle which is described in a manner free from singularities.

The solution of the problem of the motion of neutral particles evidently amounts to the discovery of such solutions of the gravitational equations (written free of denominators), as contain several bridges.

The conception sketched above corresponds, a priori, to the atomistic structure of matter insofar as the "bridge" is by its nature a discrete element. Moreover, we see that the mass constant $m$ of the neutral particles must necessarily be positive, since no solution free of singularities can correspond to the Schwarzschild solution for a negative value of $m$. Only the examination of the several-bridge-problem, can show whether or not this theoretical method furnishes an explanation of the empirically demonstrated equality of the masses of the particles found in nature, and whether it takes into account the facts which the quantum mechanics has so wonderfully comprehended.

In an analogous manner, it is possible to demonstrate that the combined equations of gravitation and electricity (with appropriate choice of the sign of the electrical member in the gravitational equations) produce a singularity-free bridge-representation of the electric corpuscle. The simplest solution of this kind is that for an electrical particle without gravitational mass.

So long as the important mathematical difficulties concerned with the solution of the several-bridge-problem are not overcome, nothing can be said concerning the usefulness of the theory from the physicist's point of view. However, it constitutes, as a matter of fact, the first attempt towards the consistent elaboration of a field theory which presents a possibility of explaining the properties of matter. In favor of this attempt one should also add that it is based on the simplest possible relativistic field equations known today.

# SUMMARY

Physics constitutes a logical system of thought which is in a state of evolution, and whose basis cannot be obtained through distillation by any inductive method from the experiences lived through, but which can only be attained by free invention. The justification (truth content) of the system rests in the proof of usefulness of the resulting theorems on the basis of sense experiences, where the relations of the latter to the former can only be comprehended intuitively. Evolution is going on in the direction of increasing simplicity of the logical basis. In order further to approach this goal, we must make up our mind to accept the fact that the logical basis departs more and more from the facts of experience, and that the path of our thought from the fundamental basis to these resulting theorems, which correlate with sense experiences, becomes continually harder and longer.

Our aim has been to sketch, as briefly as possible, the development of the fundamental concepts in their dependence upon the facts of experience and upon the strife towards the goal of internal perfection of the system. Today's state of affairs had to be illuminated by these considerations, as they appear to me. (It is unavoidable that historic schematic representation is of a personal color.)

I try to demonstrate how the concepts of bodily objects, space, subjective and objective time, are connected with one another and with the nature of the experience. In classical mechanics the concepts of space and time become independent. The concept of the bodily object is replaced in the foundations by the concept of the material point, by which means mechanics becomes fundamentally atomistic. Light and electricity produce insurmountable difficulties when one attempts to make mechanics the basis of all physics. We are thus led to the field theory of electricity, and, later on to the attempt to base physics entirely upon the concept of the field (after an attempted compromise with classical mechanics). This attempt leads to the theory of relativity (evolution of the notion of space and time into that of the continuum with metric structure).

I try to demonstrate, furthermore, why in my opinion the quantum theory does not seem likely to be able to produce a usable foundation for physics: one becomes involved in contradictions if one tries to consider the theoretical quantum description as a *complete* description of the individual physical system or happening.

On the other hand, up to the present time, the field theory is unable to give an explanation of the molecular structure of matter and of quantum phenomena. It is shown, however, that the conviction to the effect that the field theory is unable to give, by its methods, a solution of these problems rests upon prejudice.

# THE FUNDAMENTS OF
# THEORETICAL PHYSICS

From Albert Einstein: *Out of my Later Years,* Philosophical Library,
New York 1950.

Address before the Eighth American Scientific Congress, Washington,
May 15, 1490. First published in *Science,* vol. 91, May 1940.

Science is the attempt to make the chaotic diversity of our sense-expe-
rience correspond to a logically uniform system of thought. In this system single experiences must be correlated with the theoretic structure
in such a way that the resulting coordination is unique and convincing.

The sense-experiences are the given subject-matter. But the theory
that shall interpret them is man-made. It is the result of an extremely
laborious process of adaptation: hypothetical, never completely final,
always subject to question and doubt.

The scientific way of forming concepts differs from that which we
use in our daily life, not basically, but merely in the more precise definition of concepts and conclusions; more painstaking and systematic
choice of experimental material; and greater logical economy. By this
last we mean the effort to reduce all concepts and correlations to as
few as possible logically independent basic concepts and axioms.

What we call physics comprises that group of natural sciences
which base their concepts on measurements; and whose concepts and
propositions lend themselves to mathematical formulation. Its realm
is accordingly defined as that part of the sum total of our knowledge
which is capable of being expressed in mathematical terms. With the
progress of science, the realm of physics has so expanded that it seems
to be limited only by the limitations of the method itself.

The larger part of physical research is devoted to the development
of the various branches of physics, in each of which the object is the
theoretical understanding of more or less restricted fields of experience, and in each of which the laws and concepts remain as closely as

possible related to experience. It is this department of science, with its ever-growing specialization, which has revolutionized practical life in the last centuries, and given birth to the possibility that man may at last be freed from the burden of physical toil.

On the other hand, from the very beginning there has always been present the attempt to find a unifying theoretical basis for all these single sciences, consisting of a minimum of concepts and fundamental relationships, from which all the concepts and relationships of the single disciplines might be derived by logical process. This is what we mean by the search for a foundation of the whole of physics. The confident belief that this ultimate goal may be reached is the chief source of the passionate devotion which has always animated the researcher. It is in this sense that the following observations are devoted to the foundations of physics.

From what has been said it is clear that the word foundations in this connection does not mean something analogous in all respects to the foundations of a building. Logically considered, of course, the various single laws of physics rest upon this foundation. But whereas a building may be seriously damaged by a heavy storm or spring flood, yet its foundations remain intact, in science the logical foundation is always in greater peril from new experiences or new knowledge than are the branch disciplines with their closer experimental contacts. In the connection of the foundation with all the single parts lies its great significance, but likewise its greatest danger in face of any new factor. When we realize this, we are led to wonder why the so-called revolutionary epochs of the science of physics have not more often and more completely changed its foundation than has actually been the case.

The first attempt to lay a uniform theoretical foundation was the work of Newton. In his system everything is reduced to the following concepts: (1) Mass points with invariable mass; (2) action at a distance between any pair of mass points; (3) law of motion for the mass point. There was not, strictly speaking, any all-embracing foundation, because an explicit law was formulated only for the actions-at-a-distance of gravitation; while for other actions-at-a-distance nothing was established *a priori* except the law of equality of *actio* and *reactio.* Moreover, Newton himself fully realized that time and space were essential elements, as physically effective factors, of his system, if only by implication.

This Newtonian basis proved eminently fruitful and was regarded as final up to the end of the nineteenth century. It not only gave results for the movements of the heavenly bodies, down to the most minute details, but also furnished a theory of the mechanics of discrete and continuous masses, a simple explanation of the principle of the con-servation of energy and a complete and brilliant theory of heat. The explanation of the facts of electrodynamics within the Newtonian sys-tem was more forced; least convincing of all, from the very beginning, was the theory of light.

It is not surprising that Newton would not listen to a wave the-ory of light; for such a theory was most unsuited to his theoretical foundation. The assumption that space was filled with a medium con-sisting of material points that propagated light waves without exhibit-ing any other mechanical properties must have seemed to him quite artificial. The strongest empirical arguments for the wave nature of light, fixed speeds of propagation, interference, diffraction, polariza-tion, were either unknown or else not known in any well-ordered syn-thesis. He was justified in sticking to his corpuscular theory of light.

During the nineteenth century the dispute was settled in favor of the wave theory. Yet no serious doubt of the mechanical foundation of physics arose, in the first place because nobody knew where to find a foundation of another sort. Only slowly, under the irresistible pres-sure of facts, there developed a new foundation of physics, field-physics.

From Newton's time on, the theory of action-at-a-distance was constantly found artificial. Efforts were not lacking to explain gravi-tation by a kinetic theory, that is, on the basis of collision forces of hypothetical mass particles. But the attempts were superficial and bore no fruit. The strange part played by space (or the inertial system) within the mechanical foundation was also clearly recognized, and crit-icized with especial clarity by Ernst Mach.

The great change was brought about by Faraday, Maxwell and Hertz—as a matter of fact half-unconsciously and against their will. All three of them, throughout their lives, considered themselves adherents of the mechanical theory. Hertz had found the simplest form of the equations of the electromagnetic field, and declared that any theory leading to these equations was Maxwellian theory. Yet toward the end of his short life he wrote a paper in which he presented as the foundation of physics a mechanical theory freed from the force-concept.

For us, who took in Faraday's ideas so to speak with our mother's milk, it is hard to appreciate their greatness and audacity. Faraday must have grasped with unerring instinct the artificial nature of all attempts to refer electromagnetic phenomena to actions-at-a-distance between electric particles reacting on each other. How was each single iron filing among a lot scattered on a piece of paper to know of the single electric particles running round in a nearby conductor? All these electric particles together seemed to create in the surrounding space a condition which in turn produced a certain order in the filings. These spatial states, to-day called fields, if their geometrical structure and interdependent action were once rightly grasped, would, he was convinced, furnish the clue to the mysterious electromagnetic actions. He conceived these fields as states of mechanical stress in a space-filling medium, similar to the states of stress in an elastically distended body. For at that time this was the only way one could conceive of states that were apparently continuously distributed in space. The peculiar type of mechanical interpretation of these fields remained in the background—a sort of placation of the scientific conscience in view of the mechanical tradition of Faraday's time. With the help of these new field concepts Faraday succeeded in forming a qualitative concept of the whole complex of electromagnetic effects discovered by him and his predecessors. The precise formulation of the time-space laws of those fields was the work of Maxwell. Imagine his feelings when the differential equations he had formulated proved to him that electromagnetic fields spread in the form of polarized waves and with the speed of light! To few men in the world has such an experience been vouchsafed. At that thrilling moment he surely never guessed that the riddling nature of light, apparently so completely solved, would continue to baffle succeeding generations. Meantime, it took physicists some decades to grasp the full significance of Maxwell's discovery, so bold was the leap that his genius forced upon the conceptions of his fellow-workers. Only after Hertz had demonstrated experimentally the existence of Maxwell's electromagnetic waves, did resistance to the new theory break down.

But if the electromagnetic field could exist as a wave independent of the material source, then the electrostatic interaction could no longer be explained as action-at-a-distance. And what was true for electrical action could not be denied for gravitation. Everywhere Newton's actions-at-a-distance gave way to fields spreading with finite velocity.

Of Newton's foundation there now remained only the material mass points subject to the law of motion. But J. J. Thomson pointed out that an electrically charged body in motion must, according to Maxwell's theory, possess a magnetic field whose energy acted precisely as does an increase of kinetic energy to the body. If, then, a part of kinetic energy consists of field energy, might that not then be true of the whole of the kinetic energy? Perhaps the basic property of matter, its inertia, could be explained within the field theory? The question led to the problem of an interpretation of matter in terms of field theory, the solution of which would furnish an explanation of the atomic structure of matter. It was soon realized that Maxwell's theory could not accomplish such a program. Since then many scientists have zealously sought to complete the field theory by some generalization that should comprise a theory of matter; but so far such efforts have not been crowned with success. In order to construct a theory, it is not enough to have a clear conception of the goal. One must also have a formal point of view which will sufficiently restrict the unlimited variety of possibilities. So far this has not been found; accordingly the field theory has not succeeded in furnishing a foundation for the whole of physics.

For several decades most physicists clung to the conviction that a mechanical substructure would be found for Maxwell's theory. But the unsatisfactory results of their efforts led to gradual acceptance of the new field concepts as irreducible fundamentals—in other words, physicists resigned themselves to giving up the idea of a mechanical foundation.

Thus physicists held to a field-theory program. But it could not be called a foundation, since nobody could tell whether a consistent field theory could ever explain on the one hand gravitation, on the other hand the elementary components of matter. In this state of affairs it was necessary to think of material particles as mass points subject to Newton's laws of motion. This was the procedure of Lorentz in creating his electron theory and the theory of the electromagnetic phenomena of moving bodies.

Such was the point at which fundamental conceptions had arrived at the turn of the century. Immense progress was made in the theoretical penetration and understanding of whole groups of new phenomena; but the establishment of a unified foundation for physics seemed remote indeed. And this state of things has even been aggravated by subsequent developments. The development during the present century is characterized by two theoretical systems essentially independent of each other: the theory of relativity and the quantum theory. The two systems do not directly contradict each other; but they seem little adapted to fusion into one unified theory. We must briefly discuss the basic idea of these two systems.

The theory of relativity arose out of efforts to improve, with reference to logical economy, the foundation of physics as it existed at the turn of the century. The so-called special or restricted relativity theory is based on the fact that Maxwell's equations (and thus the law of propagation of light in empty space) are converted into equations of the same form, when they undergo Lorentz transformation. This formal property of the Maxwell equations is supplemented by our fairly secure empirical knowledge that the laws of physics are the same with respect to all inertial systems. This leads to the result that the Lorentz transformation—applied to space and time coordinates—must govern the transition from one inertial system to any other. The content of the restricted relativity theory can accordingly be summarized in one sentence: all natural laws must be so conditioned that they are covariant with respect to Lorentz transformations. From this it follows that the simultaneity of two distant events is not an invariant concept and that the dimensions of rigid bodies and the speed of clocks depend upon their state of motion. A further consequence was a modification of Newton's law of motion in cases where the speed of a given body was not small compared with the speed of light. There followed also the principle of the equivalence of mass and energy, with the laws of conservation of mass and energy becoming one and the same. Once it was shown that simultaneity was relative and depended on the frame of reference, every possibility of retaining actions-at-a-distance within the foundation of physics disappeared, since that concept presupposed the absolute character of simultaneity (it must be possible to state the location of the two interacting mass points "at the same time").

The general theory of relativity owes its origin to the attempt to explain a fact known since Galileo's and Newton's time but hitherto eluding all theoretical interpretation: the inertia and the weight of a body, in themselves two entirely distinct things, are measured by one and the same constant, the mass. From this correspondence follows that it is impossible to discover by experiment whether a given system of coordinates is accelerated, or whether its motion is straight and uniform and the observed effects are due to a gravitational field (this is the equivalence principle of the general relativity theory). It shatters the concepts of the inertial system, as soon as gravitation enters in. It may be remarked here that the inertial system is a weak point of the Galilean-Newtonian mechanics. For there is presupposed a mysterious property of physical space, conditioning the kind of coordination-systems for which the law of inertia and the Newtonian law of motion hold good.

These difficulties can be avoided by the following postulate: natural laws are to be formulated in such a way that their form is identical for coordinate systems of any kind of states of motion. To accomplish this is the task of the general theory of relativity. On the other hand, we deduce from the restricted theory the existence of a Riemannian metric within the time-space continuum, which, according to the equivalence principle, describes both the gravitational field and the metric properties of space. Assuming that the field equations of gravitation are of the second differential order, the field law is clearly determined.

Aside from this result, the theory frees field physics from the disability it suffered from, in common with the Newtonian mechanics, of ascribing to space those independent physical properties which heretofore had been concealed by the use of an inertial system. But it can not be claimed that those parts of the general relativity theory which can to-day be regarded as final have furnished physics with a complete and satisfactory foundation. In the first place, the total field appears in it to be composed of two logically unconnected parts, the gravitational and the electromagnetic. And in the second place, this theory, like the earlier field theories, has not up till now supplied an explanation of the atomistic structure of matter. This failure has probably some connection with the fact that so far it has contributed nothing to the understanding of quantum phenomena. To take in these phenomena, physicists have been driven to the adoption of entirely new methods, the basic characteristics of which we shall now discuss.

In the year nineteen hundred, in the course of a purely theoretic investigation, Max Planck made a very remarkable discovery: the law of radiation of bodies as a function of temperature could not be derived solely from the laws of Maxwellian electrodynamics. To arrive at results consistent with the relevant experiments, radiation of a given frequency had to be treated as though it consisted of energy atoms of the individual energy $h.v.$, where $h$ is Plank's universal constant. During the years following it was shown that light was everywhere produced and absorbed in such energy quanta. In particular Niels Bohr was able largely to understand the structure of the atom, on the assumption that atoms can have only discrete energy values, and that the discontinuous transitions between them are connected with the emission or absorption of such an energy quantum. This threw some light on the fact that in their gaseous state elements and their compounds radiate and absorb only light of certain sharply defined frequencies. All this was quite inexplicable within the frame of the hitherto existing theories. It was clear that at least in the field of atomistic phenomena the character of everything that happens is determined by discrete states and by apparently discontinuous transitions between them, Planck's constant $h$ playing a decisive role.

The next step was taken by De Broglie. He asked himself how the discrete states could be understood by the aid of the current concepts, and hit on a parallel with stationary waves, as for instance in the case of the proper frequencies of organ pipes and strings in acoustics. True, wave actions of the kind here required were unknown; but they could be constructed, and their mathematical laws formulated, employing Planck's constant $h$. De Broglie conceived an electron revolving about the atomic nucleus as being connected with such a hypothetical wave train, and made intelligible to some extent the discrete character of Bohr's "permitted" paths by the stationary character of the corresponding waves.

Now in mechanics the motion of material points is determined by the forces or fields of force acting upon them. Hence it was to be expected that those fields of force would also influence De Broglie's wave fields in an analogous way. Erwin Schrödinger showed how this influence was to be taken into account, re-interpreting by an ingenious method certain formulations of classical mechanics. He even succeeded in expanding the wave mechanical theory to a point where without the introduction of any additional hypotheses, it became applicable to any mechanical system consisting of an arbitrary number of mass points, that is to say possessing an arbitrary number of degrees of freedom. This was possible because a mechanical system consisting of $n$ mass points is mathematically equivalent to a considerable degree, to one single mass point moving in a space of $3\,n$ dimensions.

On the basis of this theory there was obtained a surprisingly good representation of an immense variety of facts which otherwise appeared entirely incomprehensible. But on one point, curiously enough, there was failure: it proved impossible to associate with these Schrödinger waves definite motions of the mass points—and that, after all, had been the original purpose of the whole construction.

The difficulty appeared insurmountable, until it was overcome by Born in a way as simple as it was unexpected. The De Broglie-Schrödinger wave fields were not to be interpreted as a mathematical description of how an event actually takes place in time and space, though, of course, they have reference to such an event. Rather they are a mathematical description of what we can actually know about the system. They serve only to make statistical statements and predictions of the results of all measurements which we can carry out upon the system.

Let me illustrate these general features of quantum mechanics by means of a simple example: we shall consider a mass point kept inside a restricted region $G$ by forces of finite strength. If the kinetic energy of the mass point is below a certain limit, then the mass point, according to classical mechanics, can never leave the region $G$. But according to quantum mechanics, the mass point, after a period not immediately predictable, is able to leave the region $G$, in an unpredictable direction, and escape into surrounding space. This case, according to Gamow, is a simplified model of radioactive disintegration.

The quantum theoretical treatment of this case is as follows: at the time $t_0$ we have a Schrödinger wave system entirely inside $G$. But from the time $t_0$ onwards, the waves leave the interior of $G$ in all directions, in such a way that the amplitude of the outgoing wave is small compared to the initial amplitude of the wave system inside $G$. The further these outside waves spread, the more the amplitude of the waves inside $G$ diminishes, and correspondingly the intensity of the later waves issuing from $G$. Only after infinite time has passed is the wave supply inside $G$ exhausted, while the outside wave has spread over an ever-increasing space.

But what has this wave process to do with the first object of our interest, the particle originally enclosed in *G*? To answer this question, we must imagine some arrangement which will permit us to carry out measurements on the particle. For instance, let us imagine somewhere in the surrounding space a screen so made that the particle sticks to it on coming into contact with it. Then from the intensity of the waves hitting the screen at some point, we draw conclusions as to the probability of the particle hitting the screen there at that time. As soon as the particle has hit any particular point of the screen, the whole wave field loses all its physical meaning; its only purpose was to make probability predictions as to the place and time of the particle hitting the screen (or, for instance, its momentum at the time when it hits the screen).

All other cases are analogous. The aim of the theory is to determine the probability of the results of measurement upon a system at a given time. On the other hand, it makes no attempt to give a mathematical representation of what is actually present or goes on in space and time. On this point the quantum theory of to-day differs fundamentally from all previous theories of physics, mechanistic as well as field theories. Instead of a model description of actual space-time events, it gives the probability distributions for possible measurements as functions of time.

It must be admitted that the new theoretical conception owes its origin not to any flight of fancy but to the compelling force of the facts of experience. All attempts to represent the particle and wave features displayed in the phenomena of light and matter, by direct course to a space-time model, have so far ended in failure. And Heisenberg has convincingly shown, from an empirical point of view, any decision as to a rigorously deterministic structure of nature is definitely ruled out, because of the atomistic structure of our experimental apparatus. Thus it is probably out of the question that any future knowledge can compel physics again to relinquish our present statistical theoretical foundation in favor of a deterministic one which would deal directly with physical reality. Logically the problem seems to offer two possibilities, between which we are in principle given a choice. In the end the choice will be made according to which kind of description yields the formulation of the simplest foundation, logically speaking. At the present, we are quite without any deterministic theory directly describing the events themselves and in consonance with the facts.

For the time being, we have to admit that we do not possess any general theoretical basis for physics, which can be regarded as its logical foundation. The field theory, so far, has failed in the molecular sphere. It is agreed on all hands that the only principle which could serve as the basis of quantum theory would be one that constituted a translation of the field theory into the scheme of quantum statistics. Whether this will actually come about in a satisfactory manner, nobody can venture to say.

Some physicists, among them myself, can not believe that we must abandon, actually and forever, the idea of direct representation of physical reality in space and time; or that we must accept the view that events in nature are analogous to a game of chance. It is open to every man to choose the direction of his striving; and also every man may draw comfort from Lessing's fine saying, that the search for truth is more precious than its possession.

# THE COMMON LANGUAGE
# OF SCIENCE

From Albert Einstein: *Out of my Later Years*, Philosophical Library, New York 1950.

Address broadcast to the meeting of the British Association for the Advancement of Science, Sep. 28, 1941.

First published in *Advancement of Science*, London, vol 2. no. 5.

The first step towards language was to link acoustically or otherwise commutable signs to sense-impressions. Most likely all sociable animals have arrived at this primitive kind of communication—at least to a certain degree. A higher development is reached when further signs are introduced and understood which establish relations between those other signs designating sense-impression. At this stage it is already possible to report somewhat complex series of impressions; we can say that language has come to existence. If language is to lead at all to understanding, there must be rules concerning the relations between the signs on the one hand and on the other hand there must be a stable correspondence between signs and impressions. In their childhood individuals connected by the same language grasp these rules and relations mainly by intuition. When man becomes conscious of the rules concerning the relations between signs the so-called grammar of language is established.

In an early stage the words may correspond directly to impressions. At a later stage this direct connection is lost insofar as some words convey relations to perceptions only if used in connection with other words (for instance such words as: "is," "or," "thing"). Then word-groups rather than single words refer to perceptions. When language becomes thus partially independent from the background of impressions a greater inner coherence is gained.

Only at this further development where frequent use is made of so-called abstract concepts, language becomes an instrument of reasoning in the true sense of the word. But it is also this development which turns language into a dangerous source of error and deception. Everything depends on the degree to which words and word-combinations correspond to the world of impression.

What is it that brings about such an intimate connection between language and thinking? Is there no thinking without the use of language, namely in concepts and concept-combinations for which words need not necessarily come to mind? Has not every one of us struggled for words although the connection between "things" was already clear?

We might be inclined to attribute to the act of thinking complete independence from language if the individual formed or were able to form his concepts without the verbal guidance of his environment. Yet most likely the mental shape of an individual, growing up under such conditions, would be very poor. Thus we may conclude that the mental development of the individual and his way of forming concepts depend to a high degree upon language. This makes us realize to what extent the same language means the same mentality. In this sense thinking and language are linked together.

What distinguishes the language of science from language as we ordinarily understand the word? How is it that scientific language is international? What science strives for is an utmost acuteness and clarity of concepts as regards their mutual relation and their correspondence to sensory data. As an illustration let us take the language of Euclidian geometry and Algebra. They manipulate with a small number of independently introduced concepts, respectively symbols, such as the integral number, the straight line, the point, as well as with signs which designate the fundamental operations, that is, the connections between those fundamental concepts. This is the basis for the construction, respectively definition, of all other statements and concepts. The connection between concepts and statements on the one hand and the sensory data on the other hand is established through acts of counting and measuring whose performance is sufficiently well determined.

The super-national character of scientific concepts and scientific language is due to the fact that they have been set up by the best

449

brains of all countries and all times. In solitude and yet in cooperative effort as regards the final effect they created the spiritual tools for the technical revolutions which have transformed the life of mankind in the last centuries. Their system of concepts have served as a guide in the bewildering chaos of perceptions so that we learned to grasp general truths from particular observations.

What hopes and fears does the scientific method imply for mankind? I do not think that this is the right way to put the question. Whatever this tool in the hand of man will produce depends entirely on the nature of the goals alive in this mankind. Once these goals exist, the scientific method furnishes means to realize them. Yet it cannot furnish the very goals. The scientific method itself would not have led anywhere, it would not even have been born without a passionate striving for clear understanding.

Perfections of means and confusion of goals seem—in my opinion—to characterize our age. If we desire sincerely and passionately the safety, the welfare and the free development of the talents of all men, we shall not be in want of the means to approach such a state. Even if only a small part of mankind strives for such goals, their superiority will prove itself in the long run.

# THE LAWS OF SCIENCE AND THE LAWS OF ETHICS

From Albert Einstein: *Out of my Later Years*, Philosophical Library, New York 1950.

First published as foreword to Philipp Frank, *Relativity—A Richer Truth*, Boston 1950.

Science searches for relations which are thought to exist independently of the searching individual. This includes the case where man himself is the subject. Or the subject of scientific statements may be concepts created by ourselves, as in mathematics. Such concepts are not necessarily supposed to correspond to any objects in the outside world. However, all scientific statements and laws have one characteristic in common: they are "true or false" (adequate or inadequate). Roughly speaking, our reaction to them is "yes" or "no."

The scientific way of thinking has a further characteristic. The concepts which it uses to build up its coherent systems are not expressing emotions. For the scientist, there is only "being," but no wishing, no valuing, no good, no evil; no goal. As long as we remain within the realm of science proper, we can never meet with a sentence of the type: "Thou shalt not lie." There is something like a Puritan's restraint in the scientist who seeks truth: he keeps away from everything voluntaristic or emotional. Incidentally, this trait is the result of a slow development, peculiar to modern Western thought.

From this it might seem as if logical thinking were irrelevant for ethics. Scientific statements of facts and relations, indeed, cannot produce ethical directives. However, ethical directives can be made rational and coherent by logical thinking and empirical knowledge. If we can agree on some fundamental ethical propositions, then other ethical propositions can be derived from them, provided that the original premises are stated with sufficient precision. Such ethical premises play a similar role in ethics, to that played by axioms in mathematics.

This is why we do not feel at all that it is meaningless to ask such questions as: "Why should we not lie?" We feel that such questions are meaningful because in all discussions of this kind some ethical premises are tacitly taken for granted. We then feel satisfied when we succeed in tracing back the ethical directive in question to these basic premises. In the case of lying this might perhaps be done in some way such as this: Lying destroys confidence in the statements of other people. Without such confidence, social cooperation is made impossible or at least difficult. Such cooperation, however, is essential to make human life possible and tolerable. This means that the rule "Thou shalt not lie" has been traced back to the demands: "Human life shall be preserved" and "Pain and sorrow shall be lessened as much as possible."

But what is the origin of such ethical axioms? Are they arbitrary? Are they based on mere authority? Do they stem from experiences of men and are they conditioned indirectly by such experiences?

For pure logic all axioms are arbitrary, including the axioms of ethics. But they are by no means arbitrary from a psychological and genetic point of view. They are derived from our inborn tendencies to avoid pain and annihilation, and from the accumulated emotional reaction of individuals to the behavior of their neighbors.

It is the privilege of man's moral genius, impersonated by inspired individuals, to advance ethical axioms which are so comprehensive and so well founded that men will accept them as grounded in the vast mass of their individual emotional experiences. Ethical axioms are found and tested not very differently from the axioms of science. Truth is what stands the test of experience.

# An Elementary Derivation of the Equivalence of Mass and Energy

From Albert Einstein: *Out of my Later Years*, Philosophical Library, New York 1950.

Originally published in *Technion Journal* 1946 [Yearbook of the American Society for the Advancement of the Hebrew Institute of Technology in Haifa.

The following derivation of the law of equivalence, which has not been published before, has two advantages. Although it makes use of the principle of special relativity, it does not presume the formal machinery of the theory but uses only three previously known laws:

1. The law of the conservation of momentum.

2. The expression for the pressure of radiation; that is, the momentum of a complex of radiation moving in a fixed direction.

3. The well known expression for the aberration of light (influence of the motion of the earth on the apparent location of the fixed stars—Bradley).

We now consider the following system. Let the body $B$ rest freely in space with respect to the system $K_0$. Two complexes of radiation $S$,

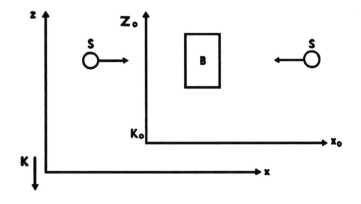

$S'$ each of energy $\frac{E}{2}$ move in the positive and negative $x_0$ direction respectively and are eventually absorbed by $B$. With this absorption

the energy of $B$ increases by $E$. The body $B$ stays at rest with respect to $K_0$ by reasons of symmetry.

Now we consider this same process with respect to the system $K$, which moves with respect to $K_0$ with the constant velocity $v$ in the negative $Z_0$ direction. With respect to $K$ the description of the process is as follows:

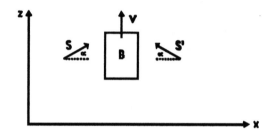

The body $B$ moves in the positive $Z$ direction with velocity $v$. The two complexes of radiation now have directions with respect to $K$ which make an angle $\alpha$ with the $x$ axis. The law of aberration states that in the first approximation $\alpha = \frac{c}{v}$, where $c$ is the velocity of light. From the consideration with respect to $K_0$ we know that the velocity $v$ of $B$ remains unchanged by the absorption of $S$ and $S'$.

Now we apply the law of conservation of momentum with respect to the $z$ direction to our system in the coordinate-frame $K$.

I. *Before the absorption* let $M$ be the mass of $B$; $Mv$ is then the expression of the momentum of $B$ (according to classical mechanics). Each of the complexes has the energy $\frac{E}{2}$ and hence, by a well known conclusion of Maxwell's theory, it has the momentum $\frac{E}{2c}$. Rigorously speaking this is the momentum of $S$ with respect to $K_0$. However, when $v$ is small with respect to $c$, the momentum with respect to $K$ is the same except for a quantity of second order of magnitude ($\frac{v^2}{c^2}$ compared to 1). The $z$-component of this momentum is $\frac{E}{2c}\sin\alpha$ or with sufficient accuracy (except for quantities of higher order of magnitude) $\frac{E}{2c}\alpha$ or $\frac{E}{2}\cdot\frac{v}{c^2}$ $S$ and $S'$ together therefore have a momentum $E\frac{v}{c^2}$ in the $z$ direction. The total momentum of the system before absorption is therefore

$$Mv + \frac{E}{c^2}\cdot v$$

II. *After the absorption* let $M'$ be the mass of $B$. We anticipate here the possibility that the mass increased with the absorption of the energy $E$ (this is necessary so that the final result of our consideration be consistent). The momentum of the system after absorption is then

$$M'v$$

We now assume the law of the conservation of momentum and apply it with respect to the $z$ direction. This gives the equation

$$Mv + \frac{E}{c^2}v = M'v$$

or

$$M' - M = \frac{E}{c^2}$$

This equation expresses the law of the equivalence of energy and mass. The energy increase $E$ is connected with the mass increase $\frac{E}{c^2}$. Since energy according to the usual definition leaves an additive constant free, we may so choose the latter that

$$E = Mc^2$$

# INDEX

## A

aberration, phenomenon of, 238
absence of matter, field equations for, 78–80
absolute space theory, 351
absolute velocity of a system, 36
abstraction, 405
acceleration
  body, constant of, 178
  co-ordinate, system of, 389
  force and, 2
  frames, bending light beams, 2
  freedom of, 38–39
  mass, resistance to, 393
  mean, 243–244
  uniform, 420
acoustical instrument, standing waves and,
    319–320, 329
action-at-a-distance theory, 438
addition
  tensors, 274
  of the velocities, theorem of (experiment of
    Fizeau), 159–161
analytic geometry, 386
antisymmetrical extension of a six-vector, 73
antisymmetrical tensors, 59–60
appearances, totality of physical, 349
atoms
  charge of, 306
  chemistry, 347
  crystals, 315–316
  discrete energy values, structure and, 443–444
  disintegrating, 395
  electron revolving around nucleus, 444
  as elementary quantum, 305
  energy level, 313–314
  Lorentz's theory of matter, 354–356
  mass, 304–305
  mechanics of a system, 412–413
  nucleus, 306, 444
  particles, defining, 283
  photons emitted by, 313–314
  size, calculating, 357–358
attraction, particle, 285–286
authority, suspicion of, 340
autobiographical notes (Einstein), 337–382
  books versus experiential learning, 343–344
  early quests for meaning, 339–341
  electro-magnetic foundation of physics, 360–362
  Euclidean plane geometry, 342–343
  field equations, finding for total field, 380–381
  general theory of relativity, 369–371, 374–375
  heat radiation investigations of Planck, 356–359
  Lorentz's theory of matter, 354–356
  mathematics education, 344–346
  Maxwell's theory, 353–354
  mechanics as basis of physics, 351–353
  motivation for writing, 339
  physics education, 346–351
  special theory of relativity, 364–365, 364–369
  statistical quantum theory, 371–372, 375–379

thinking, meaning of, 341–342
time of an event, 363-364
universal law of physical space, 371–372
universal principle, impossibility of, 362–363
axioms, truth of, 129–130

## B

bar magnet
  currents, 290
  field, 291
  induced currents, 295
behavior, probability and, 326
bending of light rays in gravitational field,
    43–45, 308
Bible, 339
billiard balls, 235–236
Black Holes, 126
Bohr, Niels, 314, 424, 443
Boltzmann, L., 106–107, 128, 358
books versus experiential learning, 343–344
Born, Max, 425
boundary conditions, 108–111
β-rays, 168
Brownian motion, law of the, 360–362

## C

calculation, result and, 115–116
calculus, 344
Cartesian co-ordinate system, 193, 194
  Gaussian curves, 196
  ideal rigid bodies, 267
  intervals, 268–269
  measurable distance between two points, 271
  space-lattice, members of, 276
  vectors, 274
cathode rays, 168
cause and effect, connecting, 237
centrifugal force, 189–190
chemical processes
  atomic hypothesis, 347
  elementary quantum, 302
chest, movement against gravitational field example,
    179–182
Christoffel, Elwin Bruno, 67–68
circle, 51
classical mechanics. See mechanics
classical physics, quantum physics versus, 331–332
clocks
  events, measurability of, 386
  gravitational fields and, 42–43
  ideal, 364
  intervals, rate of, 254
  kinematics, 47
  light, using as, 263–264
  in motion, behavior of, 157–158
  in motion, velocity of, 158
  moving, 14–16
  objective time, 409–410
  on a rotating body of reference, 1–2, 189–191
  simultaneity of, 147

clocks (*continued*)
  static gravitational field, 94–96
  synchronizing, 7, 9
  time, defining, 204
  velocity, 15–16
clouds, measuring height of, 133
coal mine, change and, 300
color, wave length listed by, 314–315
common language of science, 448–450
comprehensibility, 403
conductors, charged, 292–293
conservation
  of energy, 164, 392–393
  field equations of gravitation, deducing, 104
  in the general case, laws of
    gravitational field, theory of, 84–85
    mass and energy, principles of, 392–393
    thermal energy, 393
  of mass, 164, 392
  of momentum and energy, laws of, 366, 453
  of thermal energy, 393
constancy
  scalar of curvature, 121
  of velocities, law of, 142
constant limit, spatial infinity, 105–106
contact, permanent, 407
continua, mathematical treatment of, 197
continuity, co-variance of the equation of, 280
continuity-discontinuity quanta, 300–301
continuous medium
  mechanics of, 413
  motion, equations of, 278–280
continuum, Euclidean and non-Euclidean, 192–194
continuum, space-time
  character, note on, 64–65
  ether, role of, 244–245
  Euclidean geometry, 51, 198–199, 409
  four-dimensional, 111–112, 116, 254
  nature versus, 430
  not as Euclidean continuum, 200–202
contracovariant fundamental tensor, 63
contraction
  mixed tensor, 60–61
  tensors, 275
contravariant four-vectors, 56–57
contravariant tensors, 58
convection-currents, transformation of Maxwell-Hertz
    equation with, 26–31
conventions, 340–341
co-ordinates, system of, 132–134
  acceleration, 389
  arbitrarily moving, 387–388
  converting from stationary system, 9–14, 363
  equivalency, 370
  four to measure space and time, 53–55
  Galilean system, 137, 198, 397
  inertial, 335
  Lorentz transformation, 154–155
  Newton, 397
  rotating, 251–252
  tensors, defining by, 56, 274
  transformations, general theory of relativity, 421
cord, oscillation of, 318–319
corpuscles
  minimum of pressure/maximum of scalar of
    curvature, 121
  movement, 320
  Newton's theory, 308

relativity theory and, 431–433
Cosmological Considerations (cosmology), 105–107
  boundary conditions, 108–111
  calculation and result, 115–116
  spatially finite universe with uniform distribution
    of matter, 111–115
cosmological constant, 3, 126–127
covariant four-vectors, 57
covariant fundamental tensor, 62–63
covariant law for scalar field, 374
covariant partial differential equations, 422
covariant tensors, 58–59
crystals
  atoms, 315–316
  X rays diffraction through, 316, 317
curl of a contravariant vector, 73
current
  acting upon magnetic pole, 287–288
  associated with magnetic field, 290
  disconnected, spark and, 298–299
  induced, 295–299
  magnetic field, 287, 294
  Maxwell-Hertz equations, 330
curves
  Gaussian co-ordinates, 195–197
  tensor of curvature, 118
  variants for, 68

**D**
dark energy, 3
de Broglie, Louis, 320, 425, 444, 445
deflection, ray of light in gravitational field, 44–45,
    228–229
density
  energy-tensor, defining, 93–94
  mass, 256–257
  Newton's theory of mass, cosmological difficulties
    of, 256–257
  radiation, 359
Descartes, René, 408
deSitter, Akad. van Wetensch, 109, 111, 142
determinant of the fundamental tensor, 63
Dirac, Paul, 425
direction of travel, velocity and, 140
disintegrating atoms, 395
disk
  centrifugal force acting on, 189–190
  on globe, unbounded continuum, 258–261
distance
  between atoms in crystals, 316
  Euclidean geometry, 134
  force between two bodies, 287
  measurements, 132
  relativity of conception of, 151–152
  rigid body, 408
  two points on rigid body, 130–131
divergence
  of a contravariant vector, 72–73
  of a six-vector, 73–74
Doppler's principle for velocities, 23, 40

**E**
earlier and later events, 265
Eddington, Sir Arthur, 126
Einstein, Albert
  autobiographical notes, 337–382
  *The Evolution of Physics*, 283–336
  *The Meaning of Relativity*, 263–282
  *Out of My Later Years*, 383–456

*The Principle of Relativity,* 1–124
*Relativity—The Special and General Theory,*
     125–234
electrical current. See current
electricity. See also field theory
   charged conductors, 292–293
   currents associated with magnetic field, 290
   development of theory, 239
   discharge in a gas-containing tube, 312–313
   elementary quanta, 301–306
   elementary quanta of fluids, 302–303
   equilibrium, 355
   induced currents, 295–299
   and magnetism theory, 1, 338, 367
   mechanical interpretation difficulties, 287
   mechanics and, 414–415
   phenomenological physics, 413
   poles at rest, 293–294
   spark produced when current disconnected,
        298–299
electrodynamics, 187, 338
   electrodynamical part, 18–31
      convection-currents, transformation of Maxwell-
         Hertz equation with, 26–31
      light rays, transformation of the energy of, 23–26
      magnetic field in motion, 18–23
      Maxwell-Hertz equations, transformation, 18–23
      negative electrical masses, 168
      perfect reflectors, theory of the pressure of
         radiation exerted on, 23–26
   fundamental equations, 164–165
   kinematical part
      co-ordinates and times, converting from
         stationary system, 9–14
      length and time, relativity of, 7–9
      moving rigid bodies and moving clocks, equa-
         tions from, 14–16
      simultaneity, definition of, 5–7
      velocities, composition of, 16–18
   light rays, transformation of the energy of, 23–26
   Lorentz's theory, 161, 240
   magnetic field in motion, 18–23
   Maxwell-Hertz equations, transformation, 18–23
      perfect reflectors, theory of the pressure of
         radiation exerted on, 23–26
electromagnetic field, 235
   electric masses, introduction of, 350
   energy components of, 90–91
   equations for free space, 88–91
   ether as bearer of, 239–240, 245
   gravitational field and, 78–79, 422–423
   invention, 334–335
   special theory of relativity and, 367
   *in vacuo,* 243
electromagnetic foundation of physics, 360–362
electromagnetic phenomena, 144
electromagnetic waves, 440
electronic waves, diffraction of, 317, 322
electron, kinetic energy of, 29–30
electrons
   charges in different electric and magnetic external
      fields, 304–305
   influences on, 328
   Maxwell-Lorentz theory of, 119
   metal, extracting from, 307–308
   particle versus wave, 322–323
   photoelectric effect, 307–308
   probability waves, 330–331

showering in same direction, 304
standing wave, 320
wave-length of moving, 321–322
wave train, 444
electrostatics, 187, 294, 431
elementary quanta of matter and electricity,
     301–306, 324
ellipses of planetary orbits, 400
empty space
   equations of, 379–380
   Maxwell-Hertz equations, 32
   as seat of field, 416–417
energy
   conservation of, 164, 392–393
   electromagnetic field components, 90–91
   increasing, 164–165
   inert mass, increasing, 368–369
   kinetic and potential, division into, 353
   law of conservation of, 366, 453
   level, atoms, 313–314
   mass and, equivalence of, 392, 394, 453–455
   potential, 352–353, 430
equality of inertial and gravitational mass, 179–182
equations, general laws of nature, 52
equilibrium, 355
equivalence
   co-ordinate, system of, 370
   principle of, 389
ether, 5, 235
   as bearer of electromagnetic field, 239–240, 245
   mechanics of theory, 415
   relativity, theory of, 237–248
   space-time continuum, role in, 244–245
ethics, laws of science and, 451–452
Euclidean geometry, 337, 342
   autobiographical notes, 342–343
   continuum
      Minkowski, 198–199
      non-Euclidean and, 192–194
   curvature of space and, 400
   distances, 134
   flat model of universe, 125, 247–248
   four-dimensional space, 113-114, 172-173
   ideal rigid bodies, 267
   logical process, 129-130
   measurements by rules of, 50
   plane, infinite continuum of, 258
   postulates in *Elements,* 247
   simplicity of, 252
   solid bodies, 251
   space-time continuum, 51, 408-409
   straight lines, properties of, 268
Euler, Leonhard, 87–88, 347
events
   earlier and later, 265
   measurability of, 386
   simultaneity, 5–7, 366, 386
   time of, 363–364
*The Evolution of Physics* (Einstein and Infield),
     283–336
   field, relativity
      field as representation, 285–294
      two pillars of the field theory, 295–299
   quanta
      continuity-discontinuity, 300–301
      elementary quanta of matter and electricity,
         301–306, 324
      of light, 306–312

*The Evolution of Physics (continued)*
light spectra, 312–316
physics and reality, 333–335
probability waves, 323–333
waves of matter, 316–323
exact formulation, 203–205
expansion of universe, measurement of, 126
experience, 167–170, 385
books versus, 343–344
experimental confirmation, 225–232
light, deflection by gravitational field, 228–229
Mercury, motion of the perihelion of, 226–227
red, displacement of spectral lines towards, 230–232
extension of covariant tensor, 70–71

**F**
Faraday, Michael, 166, 295, 350, 415–416, 439
field of force. See gravitational field
field, relativity
forces, transition to, 353–354
as representation, 285–294
two pillars of the field theory, 295–299
field representation, 287
fields, 187, 415–416
field theory of gravitation, 235–236, 439
equations
finding for total field, 380–381
general form of, 82–84
gravitational field components, 77–78
limitations, 435
Lorentz, 373–374
Maxwell, 415–416
Newton, 372
physics and reality, 414–418
finite universe, 106, 212–215
FitzGerald, George Francis, 170
Fizeau, Armand
addition of the velocities, theorem of, 159–161
theory of the stationary luminiferous ether, 238
flat model of the universe, 125, 247–248
flatness of universe, 126
flat space, physics in. See relativity, special theory of
fluids
bodies, difference of two, 47–48
elementary quanta, 302–303
force
acceleration and, 2
expression for, 352–353
laws of, 411
lines of
of the gravitational field, 286
induction phenomena, 297
magnetic field, 288–289
metal plates, 303–304
on material point, 412
potential energy of system, 412–413
four–dimensional space
continuum, defining, 111–112, 116, 254
Euclidean geometry, 172–173
Minkowski, 223–224
time, 171–172
four-dimensional straight line movement in gravitational field, 78
Franck, J., 428
free space, Maxwell's electromagnetic field equations for, 88–91
frequency values, 425

frictionless adiabatic fluid, Euler's equations for, 87–88
friction, mass and energy, 393
function of the co-ordinates of the cord, 329
fundamental tensor ($\square_{uv}$) (insert correct symbols, please), 275
generally covariant equations, mathematical aids to formulation of, 62–66, 71–72
new tensors, formation of, 65–66

**G**
Galilean relativity. See The *Meaning of Relativity*
Galileo Gaililei
classical mechanics, 387
co-ordinates system, 137, 198, 397
uniform motion of translation, 138–139
mass, accelerated system of reference, 49
observable fact of experience, 48
references
general theory of relativity, 204
space free of gravitational fields, 50–51
uniform rectilinear motion, 185
transformation, 155
Lorentz transformation versus, 263
moving uniformly, 169
time, 172
velocities, addition of, 159
Gamow, George, 445
gas
molecules, Boltzmann's law of distribution, 106–107
particles
kinetic theory of, 357, 358, 414
method of statistics, 325–326
in tube, electricity discharge, 312–313
Gaussian co-ordinates, 195–197, 203–205
Gell-Mann, Murray, 283
general laws of nature, general co-variance for the equations expressing, 50–53
generally covariant equations, mathematical aids to formulation of, 55–77, 421
antisymmetrical extension of a six-vector, 73
contravariant and covariant four-vectors, 56–58
curl of a contravariant vector, 73
divergence
of a contravariant vector, 72–73
of a mixed tensor of the second rank, 74–75
of a six-vector, 73–74
fundamental tensor ($\square_{uv}$) (insert correct symbols, please), 62–66, 71–72
geodetic line, equation of, 66–68
multiplication of tensors, 60–62
particle, motion of, 66–68
Riemann-Christoffel tensor, 75–77
tensors
formation by differentiation, 68–71
of second and higher ranks, 58–60
general theory of relativity. See relativity, general theory of
geodetic line
equation of, 66–68
movement in gravitational field, 78
geometrical invariant, 271–272
geometrical propositions, physical meaning of, 129–131
geometry
bodies at rest, 47
experience and, 249–262

as intermediary between physical sciences and
math, 247–248
physical standpoint, 386
space and time, 386
globe, unbounded continuum, 258–261
gravitation
field equations, deducing from laws of
conservation, 104
mechanics and, 36–37
Newton's law of, 166, 190, 237, 391, 400, 411
gravitational field, 177–178, 237
absence of matter, field equations for, 78–80
acceleration, imparting, 49, 389
bending of light rays in, 43–45
centrifugal force, 189–191
clocks and, 42–43
conservation in the general case, laws of, 84–85
describing, 431–432
electricity, combined equations with, 433
electromagnetic field and, 78–79, 422–423
ether and, 245
field-components, expression for, 77–78
field equations of gravitation, general form
of, 82–84
Hamiltonian function, 80–82
law of, 372
light, deflection by, 228–229
lines of force, 286
material point, equations of motion of, 77–78
momentum and energy for matter, laws of, 85–86
momentum and energy, laws of, 80–82
Newton's theory, divergence from, 400
pure, 375
ray of light, transmitting curvilinearly, 185–186
role in structure of elementary particles of matter,
117–124
cosmological question, 122–124
defects in present (1919) view, 117–119
scalars, field equations freed of, 119–122
separate existence (Hamilton's Principle), 101
space and time absent, 187–188
gravitational lensing, 126
gravitational mass, 181, 368–369
gravitation, solution of problem of, 206–209
Grommer, J., 109

H
Hamilton's Principle
gravitational field, theory of, 80–82
invariants, theory of, conditioning properties of
field equations of gravities, 101–104
principle of variation and field–equations of
gravitation and matter, 99–100
separate existence of gravitational field, 101
Hawkings, Stephen, ix–xi
heat
phenomena, 393, 394, 413, 414
radiation investigations of Planck, 356–359
wires between charged plates, 304
Heisenberg, Werner, 425
Hertz, G., 428
Hertz, Heinrich, 239–240, 416, 439–440. See also
Maxwell-Hertz equations
heuristic value of theory of relativity, 162
Hilbert, D.
defects in theory, 117
variation, theory of relativity from, 99
homogeneity of space, 276

Hubble, Edwin, 3, 126
Huyghen's principle, 43–45, 97
hydrodynamics, 347, 413
hydrogen atom, 304–305

I
indivisible steps, change, 301–306, 324
induced currents, 295–299
induction, 297, 405
inertia, 351
of a body, dependence on energy content, 32–34
β-rays, 168
constant controlling, 399
co-ordinates, system of, 335
disk, centrifugal force acting on, 189–190
law of, 137, 388, 411
law of constancy of light velocity, 419
light and, 387
Mach's theory, 351–352, 367, 438
magnetic field and, 417
material point of mass, 111
radiation conveying between emitting and
absorbing bodies, 34
inertial mass, 181
classical mechanics problem, 420
constant of accelerated body, 178
energy increasing, 368–369
equivalence, principle of, 389
gravitational, equality with, 179–182
gravitation of energy, 37–40
Infield, Leopold. See The Evolution of Physics
infinity, spatial, 105–106
inner multiplications of tensors, 61–62
intervals
Cartesian co-ordinate system, 268–269
rate of, 254
rigid body, 267–268
two points on rigid body, 267
invariants, theory of, 101–104
isotropy of space, 276

J
Jacobi's theorem, 272
Jupiter, 45

K
Kaluza, Theodor, 423
kinematics
electrodynamics of moving bodies, 5–18
as laws regarding measuring bodies and
clocks, 47
length and time, relativity of, 7–9
moving rigid bodies and moving clocks, equations
from, 14–16
simultaneity, definition of, 5–7
special theory of relativity and, 398, 400
velocities, composition of, 16–18
kinetic energy, 353
of the body, 166
mass, material point of, 163–164
kinetic theory, gas particles, 357, 358, 414
Kirchhoff, Gustav, 356

L
Laser Interferometer Gravitational wave Observatory
(LIGO), 248
Laser Interfoerometer Space Antenna (LISA), 248
later events, 265

law of conservation of momentum and of energy for the gravitational field, 81–82
law of constancy of light velocity, 419
law of constancy of velocities, 142
law of inertia, 137, 388, 411
law of motion, 168–169, 352, 375, 411
law of pressure, 453
law of the constancy of the velocity of light, 144
law of the parallelogram of velocities, 17
law of the transmission of light in *vacuo*, 155–156
laws of conservation. See conservation
length
    distance, relativity of the conception of, 151–152
    of interval, 268
    measurements of, 132
    of moving rods, 8–9, 14
    and time, relativity of, 7–9
    wave listed by color, 314–315
LeVerrier, Urbain, 98
light
    aberration of, 453
    bending, 2
    clock, using as, 263–264
    compact fields from which can't escape, 126
    curving by action of gravitational fields, 400
    deflection by gravitational field, 228–229
    Ether, theory of, 238
    gravitation and propagation of
        bending of light rays in gravitational field, 43–45
        gravitation of energy, 37–40
        physical nature of gravitational field, hypothesis of, 35–37
        time and velocity of light in gravitational field, 40–43
    homogeneous, extracting electrons from metal, 307–308
    law of constancy of velocity, 419
    law of the constancy of the velocity of, 144
    in liquid, traveling with particular velocity, 160–161
    measuring height of clouds, 133
    motion of material points, 415
    particle properties of, 284
    phenomenological physics, 413–414
    quanta, 306–312
    radiation, 312
    ray
        curvature of, 97
        transformation of the energy of, 23–26
        velocity of, 8, 11–13
    as shower of photons, 309–310
    source of, inertia and, 387
    spectra, 312–316
    speed of, 1, 2, 5, 366–367, 386
    stars, lines of light from, 96
    transmitting rectilinearly, 185–186
    *in vacuo*, constancy of, 398
    wave-motion of, 347, 438
lightning strike, simultaneity of, 145–146, 148
LIGO (Laser Interferometer Gravitational wave Observatory), 248
lines of force
    of the gravitational field, 286
    induction phenomena, 297
    magnetic field, 288–289
    metal plates, 303–304
line, straight, 130

movement
    not subject to external forces, 77–78
    relative to two different points, 138
    properties of, 268
liquid, light traveling through, 160–161
LISA (Laser Interfoerometer Space Antenna), 248
longitudinal mass, 29
Lorentz contraction, 51
Lorentz, Hendrick A., 144, 170. *See also* Maxwell-Lorentz theory
    covariant law for scalar field, 374
    electrodynamic theory, 161, 240
    empty space as seat of field, 416–417
    field-theory of gravitation and, 373–374
    stationary charges, 1
    theory of matter, 354–356
    variation, theory of relativity from, 99
Lorentz transformation, 26–27, 153–156
    addition of velocities, 160
    conditions, 198–199
    demand, 387–388
    Galileo's versus, 263
    limiting velocity, 157–158
    simple derivation of, 218–222
    space-time variables, replacing, 203
    velocities, 153–154
Luminiferous Ether, 235, 238

**M**
Mach, Ernst, 47, 184, 348, 414
    inertia, 351–352, 367, 438
    mean acceleration, 243–244
magnetic field, 177
    asymmetries, 4
    current acting upon magnetic pole, 287–288
    electrical current, 287, 294
    inertia and, 417
    lines of force, 288–289
    masses, 415
    in motion, 18–23
    positive force, 288
magnetomotive forces, 21
maps, distances and, 300
marble slab rods example, 192–194
mass
    conservation of, 164, 392
    defined, 393–394
    densities, 256–257
    discontinuous nature of, 302
    electric, introduction of, 350
    and energy, equivalence of, 392–393, 394, 453–455
    hydrogen atom, 304–305
    inertia, 111
    inertial and gravitational, equality of, 179–182
    kinetic energy of a material point of, 163
    magnetic field, 415
    negative electrical, 168
    reciprocal action between, 256
material particle. See particles
material phenomena, 86–98
    free space, Maxwell's electromagnetic field equations for, 88–91
    frictionless adiabatic fluid, Euler's equations for, 87–88
    Newton's theory as a first approximation, 92–94
    rods and clocks, behavior in static gravitational field, 94–96

material point
  describing, 410–411
  equations of motion of, 77–78
  forces on, 412
mathematics
  antisymmetrical extension of a six-vector, 73
  contravariant and covariant four-vectors, 56–58
  curl of a contravariant vector, 73
  divergence
    of a contravariant vector, 72–73
    of a mixed tensor of the second rank, 74–75
    of a six-vector, 73–74
  education in autobiographical notes, 344–346
  fundamental tensor ($\square_{uv}$) (insert correct symbols, please), 62–66, 71–72
  geodetic line, equation of, 66–68
  multiplication of tensors, 60–62
  particle, motion of, 66–68
  physics versus, 246
  real things, measuring, 249–253
  Riemann-Christoffel tensor, 75–77
  tensors
    formation by differentiation, 68–71
    of second and higher ranks, 58–60
matter
  absence, field equations for, 78–80
  density defining energy-tensor, 93–94
  elementary particles, gravitational field and structure of, 119–122
  elementary quanta, 301–306
  Lorentz's theory, 354–356
  molecular structure of, 435
  uniform distribution in spatially finite universe, 111–115
  waves of, 316–323
Maxwell-Hertz equations
  for currents, 330
  electromagnetic waves, 440
  for empty space, 32
  mechanics as basis of physics, 347
  transformation, 18–23
Maxwell, James Clerk
  asymmetries in moving bodies, 4
  electric field theory, 415–416
  electricity and magnetism theory, 1, 338, 367
  electrodynamics, 187
  electromagnetic action at distance, 166, 350, 415–416, 439–440
  electromagnetic field theory
    autobiographical notes, 353–354
    as foundation of electron theory of Lorentz, 281, 441–442
    fundamental equations of electrodynamics, 164–165
    speed of light and, 5
  empty space, equations of, 379–380
  fields as fundamental variables, 353–354
  free space, electromagnetic field equations for, 88–91
Maxwell-Lorentz theory
  body moving uniformly, 169
  electron, theory of, 119
  ether and, 239
  experimental arguments in favor of, 167–168
Maxwell-Poynting expressions, 91
mean acceleration, 243–244
The Meaning of Relativity (Einstein), 263–282
measurable distance between two points, 271

measurement
  distance, 132
  of events, 386
  real things with mathematics, 249–253
measuring rods. See rods
mechanics
  as basis of physics, 351–353
  electricity and, 414–415
  force laws and, 411
  gravitation and, 36–37
  inertial mass problem, 420
  natural phenomena, insufficiency to describe, 139
  observable fact of experience, 48
  physics and reality, 406–414
  physics, inadequacy as basis of, 349–351
  potential energy as function of configuration, 412
  relativity principle, 139
  space and time, 135–136
  special theory of relativity, 46
  unsatisfactory aspects of, 183–184
Mercury
  ellipses of planetary orbits, 400
  mass density, 256
  perihelion motion, 226–227, 391
  rotation, 98
metal plates
  electrons, extracting, 307–308
  lines of force, 303–304
method of statistics, 325–326
  quantum physics, 327–328
metrical character (curvature), four-dimensional space-time continuum, 111–112, 116
Michelson, Albert, 169–170, 235
Mie, G., 117, 119
Milky Way
  distribution of stars, 256–257
  mean density, 256
Millikan, Robert, 428
Minkowski, Peter, 89
  four-dimensional space, 171–173, 223–224, 365
  objects to which motion cannot be applied, 242
  space-time continuum as Euclidean continuum, 198–199
mixed multiplications of tensors, 61–62
mixed tensors, 59
molecules
  elementary quanta of matter, 301–302
  matter, structure of, 435
momentum
  conservation of, 366, 453
  gravitational field, theory of, 80–82, 85–86
money, change and, 300–301
Morley, Edward, 170
  Ether, 235
  speed of light, 1
motion. See also kinematics; wave
  clocks and, 204
  coordinate systems in any state of, 443
  corpuscles, 320
  describing, need for second body and, 397
  distant masses, 48
  law of, 2, 168–169, 178, 263, 346, 352, 355, 375, 411
  magnetic field creating electricity, 294
  material particle, equations of, 277
  of material point, 77–78, 93–94, 201–202, 415
  neutral particle, 433
  Newton's law of, 2, 178, 263, 346, 355

Morley, Edward (*continued*)
  nonuniform, 176
  rectilinear and uniform, body in, 387
  retardation, 182
  rigid bodies and moving clocks, equations from, 14–16
  tensors, equations of, 278–279
  theory of relativity and, 385
  uniform acceleration, 420
  uniform rectilinear, 185
motion of uniform translation. *See* relativity, special
  theory of; special theory of relativity
movement
  chest against gravitational field, 179–182
  straight line, 138
moving bodies
  electrodynamical part
    convection-currents, transformation of
      Maxwell-Hertz equation with, 26–31
    light rays, transformation of the energy of, 23–26
    magnetic field in motion, 18–23
    negative electrical masses, 168
    perfect reflectors, theory of the pressure of
      radiation exerted on, 23–26
  kinematical part
    co-ordinates and times, converting from
      stationary system, 9–14
    length and time, relativity of, 7–9
    moving rigid bodies and moving clocks, equa-
      tions from, 14–16
    simultaneity, definition of, 5–7
    velocities, composition of, 16–18
moving frames, transformations between. See Lorentz
  transforms
multiplication of tensors, 60–62, 274

N
*n*-dimensional metrical spaces, 371–372
negative electrical fluid, 303
negative electrical masses, 168
Newton, Sir Isaac
  absolute space theory, 351
  co-ordinate system, 397
  corpuscular theory, 308
  cosmological difficulties of theories
    mass densities, 256–257
    universe as a whole, considerations, 210–211
  equation of motion of material point, 93–94
  field-law of gravitation, 372
  finite universe, 106
  as foundation of physics, 437–438
  gravitation, law of, 166, 190, 237, 391, 400, 411
  heat phenomena, 414
  immediate action at a distance, 366
  light, particle properties of, 284, 415
  material phenomena, 92–94
  motion, laws of, 2, 178, 263, 346, 355
  sound-transmission theory, 347
Nishijima, Kazuhiko, 283
nodes, standing wave, 319
non-symmetrical tensor, 380
nuclear physics, 306
nuclear transformation processes, 166
nucleus, 306, 444

O
object, primitive concept of, 334
observable fact of experience, 48
optical phenomena, 414–415
orbit, mass-point in reference to inertial system, 371

orientation, rigid body, 268
outer multiplication of tensors, 60
*Out of My Later Years* (Einstein), 383–456
  common language of science, 448–450
  defining theory of relativity, 396–400
  ethics, laws of science and, 451–452
  general theory of relativity, 388–395
  mass and energy, elementary derivation of equiva-
    lence, 453–455
  physics and reality
    corpuscles, relativity theory and, 431–433
    field concept, 414–418
    mechanics and, 406–414
    method of science, general consideration
      concerning, 401–406
    quantum theory and, 424–431
    scientific system, stratification of, 404–406
    theory of relativity, 418–424
  theoretical physics, fundaments of, 436–447
  theory of relativity, 385–395

P
pans on a gas range example, 183–184
parallel displacement vectors, 381
partial differential equations, 413, 417, 425
particles
  attraction, 285–286
  division, finite, 283–284
  electrical elementary, describing, 254–255
  motion of
    equations of, 277
    generally covariant equations, mathematical aids
      to formulation of, 66–68
    neutral, 433
  physics, beginning of, 334
  probability waves, 283, 330–331
  waves, appearance as, 284
  wave versus, 322–323
pendulum example, mass and energy, 392
perceptions, comparing experiences, 265–266
perfect reflectors, theory of the pressure of radiation
  exerted on, 23–26
perihelion motion, Mercury, 226–227, 391
*perpetuum mobile*, 364–365
philosophy, effect on scientific thought, 266, 401
photoelectric effect, 284, 307–308, 320
photographic plate, 311–312, 322
photons, 284, 308
  emission, 313–314
  light as shower of, 309–310
  probability waves, 330–331
  X rays, 315
physical nature of gravitational field, hypothesis of,
  35–37
physical space, universal law of, 371
physics
  classical mechanics and, 411–412
  education, 346–351
  mathematics versus, 246
  reality and
    appearances, totality of physical, 349
    corpuscles, relativity theory and, 431–433
    field concept, 414–418
    mechanics and, 406–414
    method of science, general consideration
      concerning, 401–406
    phenomenological, 413–414
    quanta, 333–335

quantum theory and, 424–431
  scientific system, stratification of, 404–406
  theory of relativity, 418–424
pinhole
  electrons and photons velocity through, 323–324
  light beam through, 311–312, 322
Planck, Max, 308–309, 424–425
  heat radiation investigations, 356–359
  radiation of bodies as a function of temperature, 443
Poincaré, H., 252
  changes of state and changes of position, 407
  experience, relation to concepts, 266
Poisson's equation, 105, 372
poles at rest, 293–294
Popper, Karl, 247
position, changes in, 266–267, 407
positive electrical fluid, 303
positive magnetic force, 288
potential energy, 353, 430
Poynting, 91
pressure
  law of, 453
  minimum, 121
primary concepts, 404
*The Principle of Relativity* (Einstein), 1–124
  Cosmological Considerations, 105–116
    boundary conditions, 108–111
    calculation and result, 115–116
    Newtonian theory, 105–107
    spatially finite universe with uniform distribution of matter, 111–115
  electrodynamics of moving bodies, 4–31
    electrodynamical part, 18–31
    kinematical part, 5–18
  generally covariant equations, mathematical aids to formulation of, 55–77
    antisymmetrical extension of a six-vector, 73
    contravariant and covariant four-vectors, 56–58
    curl of a contravariant vector, 73
    divergence of a contravariant vector, 72–73
    divergence of a mixed tensor of the second rank, 74–75
    divergence of a six-vector, 73–74
    fundamental tensor ($\square_{\mu\nu}$) (insert correct symbols, please), 62–66, 71–72
    geodetic line, equation of, 66–68
    multiplication of tensors, 60–62
    particle, motion of, 66–68
    Riemann-Christoffel tensor, 75–77
    tensors, formation by differentiation, 68–71
    tensors of second and higher ranks, 58–60
  gravitational field, theory of
    absence of matter, field equations for, 78–80
    conservation in the general case, laws of, 84–85
    field-components, expression for, 77–78
    field equations of gravitation, general form of, 82–84
    Hamiltonian function, 80–82
    material point, equations of motion of, 77–78
    momentum and energy for matter, laws of, 85–86
    momentum and energy, laws of, 80–82
  gravitation fields, role in structure of elementary particles of matter, 117–124
    cosmological question, 122–124
    defects in present (1919) view, 117–119
    scalars, field equations freed of, 119–122

Hamilton's Principle
  invariants, theory of, conditioning properties of field equations of gravities, 101–104
  principle of variation and field-equations of gravitation and matter, 99–100
  separate existence of gravitational field, 101
inertia of a body, dependence on energy content, 32–34
light, influence of gravitation on the propagation of, 35–45
  bending of light rays in gravitational field, 43–45
  gravitation of energy, 37–40
  physical nature of gravitational field, hypothesis of, 35–37
  time and velocity of light in gravitational field, 40–43
material phenomena, 86–98
  free space, Maxwell's electromagnetic field equations for, 88–91
  frictionless adiabatic fluid, Euler's equations for, 87–88
  Newton's theory as a first approximation, 92–94
  rods and clocks, behavior in static gravitational field, 94–96
postulate of relativity, fundamental considerations
  extension, need for, 47–50
  four co-ordinates to measurement in space and time, 53–55
  general laws of nature, general co-variance for the equations expressing, 50–53
  observations, 46–47
  space-time continuum, 50–53
principle of relativity, restricted sense, 138–140
principle of variation and field-equations of gravitation and matter, 99–100
principle-theories, 396–397
probability, 326
probability waves
  particles, defining, 283
  quanta, 323–333
propagation of light, apparent incompatibility with principle of relativity, 142–144
proposition, truth of, 344
Pythagorean theorem, 342

**Q**
quanta
  continuity-discontinuity, 300–301
  elementary quanta of matter and electricity, 301–306
  of light, 306–312
  light spectra, 312–316
  physics and reality, 333–335
  probability waves, 323–333
  waves of matter, 316–323
Quantum Field Theory, 236
Quantum Mechanics, 383–384
quantum physics, 327–328
quantum theory
  field theory, limitations of, 435
  particles, appearance as waves, 284
  physics and reality, 424–431

**R**
radiation
  acceleration, freedom of, 38–39
  of bodies as a function of temperature, 443
  density, 359

radiation (*continued*)
  diminishing energy, 34
  increasing energy, 164–165
  light spectra, 312
  photoelectric effect, 307–308
  pressure, law of, 453
  thermodynamics, 356, 362–363
  transparent bodies, refraction-indices of, 349–350
radioactive disintegration, 326–327
railroad embankment examples
  distance, relativity of the conception of, 151–152
  nonuniform motion, 176
  reference-body, choosing, 174–175
  relativity of simultaneity, 148–150
  retardation of motion, 182
  simultaneity and time, 145–147
  uniformly moving co-ordinate system, 138–140
  velocities, addition of, 141
ray of light, moving, 8, 11–13
real things, measuring, 249–253
rectilinear and uniform motion, body in, 387
red, displacement of spectral lines towards, 230–232
refraction-indices of transparent bodies, 349–350
Reissner, Hans, 423
relativity, general theory of
  autobiographical notes, 369–371, 374–375
  mechanics, 139
  *Out of My Later Years*, 388–395
relativity, special theory of, 364–369
  mechanics, 46
  physical interpretation of space and time in classical mechanics, 386–388
relativity, theory of
  defining, 396–400
  ether and, 237–248
  geometry and experience, 249–262
  physics and reality, 418–424
*Relativity—The Special and General Theory* (Einstein), 125–234
  addition of the velocities, theorem of (experiment of Fizeau), 159–161
  classical mechanics and, unsatisfactory aspects of, 183–184
  clocks and measuring–rods on a rotating body of reference, 189–191
  co-ordinate, system of, 132–134
  distance, relativity of conception of, 151–152
  equality of inertial and gravitational mass, 179–182
  Euclidean and non-Euclidean continuum, 192–194
  exact formulation, 203–205
  experience and, 167–170
  experimental confirmation, 225–232
    light, deflection by gravitational field, 228–229
    Mercury, motion of the perihelion of, 226–227
    red, displacement of spectral lines towards, 230–232
  Galilean system of co-ordinates, 137
  Gaussian co-ordinates, 195–197
  general results, 163–166
  geometrical propositions, physical meaning of, 129–131
  gravitational field, 177–178
  gravitation, solution of problem of, 206–209
  heuristic value of theory of relativity, 162
  inferences, 185–188
  Lorentz transformation, 153–156, 218–222

measuring-rods and clocks in motion, behavior of, 157–158
Minkowski's four-dimensional space, 171–173, 223–224
principle of relativity, restricted sense, 138–140
propagation of light, apparent incompatibility with principle of relativity, 142–144
simultaneity, relativity of, 148–150
space and time in classical mechanics, 135–136
space-time continuum as Euclidean continuum, 198–199
space-time continuum is not Euclidean continuum, 200–202
special and general principle, 174–176
structure of space, 233–234
theorem of addition of velocities in classical mechanics, 141
time, idea of in physics, 145–147
universe as a whole, considerations
  "finite" and "unbounded" universe, possibility of, 212–215
  Newton's theory, cosmological difficulties of, 210–211
  structure of space, 216–217
religion, experience with, 339–340
resonators, oscillation of all, 358–359
rest
  bodies at, 9
  geometry, 47
  poles at, 293–294
result, calculation and, 115–116
Riemann, Bernhard
  four-dimensional continuum of space-time, 254
  metric, 443
  *n*-dimensional metrical spaces, 371–372
  tensor of curvature, 118
Riemann-Christoffel tensor, 75–77
rigid bodies
  changes in position, 266–267
  distance, 408
  distance between two points, 130–131
  interval, 267–268
  moving, 14–16
  in nature, 409
  orientation, 268
rigid surfaces. See coordinates, system of
rods
  analytic geometry, 386
  ideal, 364
  kinematics, 47
  length of interval, 268
  marble slab example, 192–193
    temperature, 193–194
  in motion, behavior of, 157–158
  moving, length of, 8–9, 14
  objects above surface of earth, 132–133
  on a rotating body of reference, 1–2, 189–191
  static gravitational field, 94–96
Rosen, Robert, 423, 431
rotation
  co-ordinate, system of, 251–252
  Mercury, 98

S
scalar field, 367–368
  covariant law for, 374

scalar of curvature, 121
scalars, field equations freed of
  covariant law, 374
  gravitation fields, role in structure of elementary
    particles of matter, 119–122
Schrödinger, Erwin, 425–428, 444, 445
Schwarzschild, Karl, 126, 423, 432–433
scientific description, basis of, 132
scientific system, stratification of, 404–406
sense experiences, 401–403
simultaneity
  clocks, 147
  definition of, 5–7, 386–387
  events, 366
  railroad embankment examples, 145–147
  relativity of, 148–150, 410
six-vector, divergence of, 73–74
size, atoms, 357–358
skew-symmetry tensors, 275, 282
solenoid, magnetic field, 291–292, 295
solid bodies, Euclidean geometry, 251
space
  curvature, Euclidean geometry and, 399
  empty as seat of field, 416–417
  empty, equations of, 379–380
  structure in universe as a whole, 216–217
  structure of, 233–234
  in time in pre-relativity physics, 265–282
space and time
  absent gravitational fields, 187–188
  accelerated frames, bending lightbeams, 2
  in classical mechanics, 135–136
  geometrical behavior, 400
  in geometry, 386
  Newtonian basis, 437–438
  rigid bodies and, 409
space-time continuum
  character, note on, 64–65
  ether, role of, 244–245
  Euclidean geometry, 51, 198–199, 409
  four-dimensional, 111–112, 116, 254
  nature versus, 430
  not as Euclidean continuum, 200–202
  postulate of relativity, fundamental considerations,
    50–53
spark produced when current disconnected, 298–299
spatial infinity, 110–111
  constant limit, 105–106
spatially finite universe with uniform distribution of
  matter, 111–115
special and general principle, 174–176
special theory of relativity. See relativity, special
  theory of
spectral lines, displacement towards red, 230–232
spectroscope, 312–313
speed of light, 1, 5, 366–367, 386
sphere, lines in space model, 286
standing wave, 318–319, 320
stars
  Boltzmann's law of distribution for gas molecules,
    106–107
  distribution, 256–257
  lines of light from surface, 96
state, changes of, 407
statement of set of rules, 403
stationary charges, 1

stationary system, 9–14, 363
statistical quantum theory
  merits of, 378–379
  relativity, theory of, and, 375–376
  Riemann's $n$-dimensional metrical spaces, 371–372
  u-function (insert symbol please), 376–377
statistics, method of, 325–326
  quantum physics, 327–328
steam, pans on a gas range example, 183–184
stone, gravitational action of earth on, 177–178
straight line, 130
  movement
    not subject to external forces, 77–78
    relative to two different points, 138
  properties of, 268
subtraction of tensors, 274
sun
  radiation emitted by, 312
  rays traveling to (See gravitation)
  viewing light through spectroscope, 313
supernova explosions, 127
symmetry
  antisymmetrical extension of a six-vector, 73
  antisymmetrical tensors, 59–60
  tensors, 59, 275, 371–372, 380
synchronizing clocks, 7, 9

T
temperature
  heat phenomena, 393, 394
  radiation and, 358, 359–360, 443
  rods on marble slab example, 193–194
  wires between charged plates, 304
tensors, 272
  addition and subtraction, 274
  antisymmetrical, 59–60
  contraction, 275
  covariant fundamental, 62–63
  of curvature, 118
  defining by co-ordinates, 56, 274
  formation by differentiation, 68–71
  fundamental ($\Box_{uv}$) (insert correct symbols, please),
    62–66, 274
  motion, equations of, 278–279
  multiplication, 60–62, 274
  new, formation of, 65–66
  non-symmetrical, 380
  proof, 275–277
  Riemann-Christoffel, 75–77
  of second and higher ranks, 58–60
  symmetrical, 59, 275, 371–372, 380
  theorem, 275
  transformation, 273–274
  vectors, 274
  virial theorem, 278
test body, 285–286
theorem of addition of velocities in classical
  mechanics, 141
theory of the stationary luminiferous ether, 238
thermal energy, conservation of, 393
thermodynamics, 356, 360, 362–363
  mechanical interpretation, 414
  perpetuum mobile, 364–365
  as principle-theory, 397
thinking, meaning of, 341–342
Thomson, J.J., 440

time. *See also* simultaneity; space-time continuum
absolute, 335, 430
of an event, 363–364
clocks, 204
constant flow of, 263–264
force between two bodies, 287
four-dimensional space, 171–172
Galilei transformation, 172
gravitational fields and, 42–43
idea of in physics, 6, 145–147
length and, relativity of, 7–9
of light in gravitational field, 40–43
motion of a material body, 202
objective, introduction of, 409–410
railroad embankment examples, 145–147
simultaneity of events, 5–7, 386–387
space in pre-relativity physics, 265–282
speed of light, 5, 366–367, 386
stationary system, converting from, 9–14, 363
subjective feeling of, 334
time and space
absent gravitational fields, 187–188
accelerated frames, bending lightbeams, 2
in classical mechanics, 135–136
geometrical behavior, 400
in geometry, 386
Newtonian basis, 437–438
rigid bodies and, 409
tract
defined, 253
light paths, 254
train travel, change and, 300
trajectory, space and time in classical mechanics, 135–136
transformation
general theory of relativity, 421, 443
tensors, 273–274
translation, uniform motion, 138–139
transparent bodies, refraction-indices of, 349–350
transverse mass, 29
tube, oscillation of, 318

U
u-function (insert symbol please) statistical quantum theory, 376–377
ultraviolet photon, 315
unbounded universe, possibility of, 212–215
uniform acceleration, 420
uniform rectilinear and nonrotary motion, 175
uniform rectilinear motion, 185
universal law of physical space, 371–372
universal principle, impossibility of, 362–363
universe
"finite" and "unbounded" possibilities of, 212–215, 255–262
finite nature of, 106
flatness theory, 125, 247–248
Newton's theory, cosmological difficulties of, 210–211
structure of space, 216–217

V
variation, Hamilton's Principle of, 99–100
vectors, 272
Cartesian co-ordinate system, 274
parallel displacement, 381
velocity
absolute of a system, 36
addition of, 159–161
β-rays, 168
clocks, 15–16, 158
composition of, 16–18
direction of travel, 140
electrical current, 287
electrons and photons through pinholes, 323–324
kinetic energy
of the body, 166
of a material point of mass, 163–164
law of constancy of, 142
law of constancy of light, 419
of light in gravitational field, 40–43
limiting, 157–158
Lorentz transformation, 153–154
propagation of light, 142–143
ray of light, moving, 8, 11–13
simultaneity and time, 149–150
in tensor equation, 277
theorem of the addition of, 141
violin string, oscillation of, 318–319
virial theorem of tensors, 278
volume scalar, 64

W
wave
of matter, 316–323
particle versus, 322–323
wave theory, 310
length listed by color, 314–315
of light, 347, 438
standing wave, 318–319
sunlight viewed through spectroscope, 313
weight of a system, 368
conservation of energy, 393
constant controlling, 399
Weyl, Hermann, 117
Wheeler, John Archibald, 2, 247
Wilkinson Microwave Anisotropy Probe (WMAP) satellite, 126, 248
wires
broken, 298
heated between charged plates, 304
surface bound by, induction and, 297
wondering nature, 341–342

X
X rays
diffraction through crystal, 316, 317
photons comprising, 315
wave lengths, 322